Solitons in Optical Fibers

Solitons in Optical Fibers
Fundamentals and Applications

Linn F. Mollenauer and James P. Gordon

AMSTERDAM • BOSTON • HEIDELBERG • LONDON
NEW YORK • OXFORD • PARIS • SAN DIEGO
SAN FRANCISCO • SINGAPORE • SYDNEY • TOKYO
Academic Press is an imprint of Elsevier

Elsevier Academic Press
30 Corporate Drive, Suite 400, Burlington, MA 01803, USA
525 B Street, Suite 1900, San Diego, California 92101-4495, USA
84 Theobald's Road, London WC1X 8RR, UK

This book is printed on acid-free paper. 

Library of Congress Cataloging-in-Publication Data

Mollenauer, L. F. (Linn Frederick), 1937-
Solitons in optical fibers : fundamentals and applications / Linn F. Mollenauer, James P. Gordon.
p. cm.
Includes bibliographical references and index.
ISBN-13: 978-0-12-504190-4 (hardcover : alk. paper)
ISBN-10: 0-12-504190-X (hardcover : alk. paper) 1. Optical fibers. 2. Solitons. I. Gordon, James P. II. Title.
TA1800.M65 2006
621.36'92–dc22

2005031663

British Library Cataloguing-in-Publication Data
A catalogue record for this book is available from the British Library

ISBN 13: 978-0-12-504190-4
ISBN 10: 0-12-504190-X

For all information on all Elsevier Academic Press publications
visit our Web site at www.books.elsevier.com

Printed in the United States of America
06 07 08 09 10 11 9 8 7 6 5 4 3 2 1

About the Authors

Linn F. Mollenauer received a Bachelor of Engineering Physics degree from Cornell University in 1959 and a PhD in physics from Stanford University in 1965. He joined the technical staff of Bell Laboratories, Holmdel, in 1972, where, for many years, he focused on the experimental study of solitons and other nonlinear effects in optical fiber pulse propagation and their application to ultra-long-distance transmission. Retired from Bell Labs at the end of 2003, he is now a visiting scientist in the Applied and Engineering Physics Dept. at Cornell and a part-time professor at the University of Arizona at Tucson. Dr. Mollenauer is a fellow of the Optical Society of America, the American Physical Society, the IEEE, the American Association for the Advancement of Science, and in 1993, he was elected to the National Academy of Engineering. He was an IEEE/LEOS Distinguished Traveling Lecturer for 1990–1991, and a Morris Loeb Lecturer at Harvard in 1996. Other honors include the R. W. Wood prize of the Optical Society (1982), the Bell Laboratories Distinguished Technical Staff Award, the Ballantine Medal of the Franklin Institute (1986), a Rank Prize in opto-electronics (1991), and the Charles Hard Townes Award of the OSA (1997). Finally, in 2001, he received the IEEE/LEOS Quantum Electronics Award and was elected a Bell Labs Fellow.

James P. Gordon was educated at MIT (BS physics, 1949) and Columbia University (PhD physics, 1955). He joined the Electronics Research division of Bell Telephone Laboratories in 1955. From 1958–1980 he was head of the Quantum Electronics Research Department. In the following years he returned to a research position. In the academic year 1962–1963 he was a visiting professor at the University of California at San Diego. He retired from Bell Labs in 1996, remaining in a consultant status. His research interests have spanned a number of areas connected with quantum electronics and optics. His PhD research project under Prof. Charles H. Townes resulted in the creation and understanding of the first pair of (ammonia) masers, which initiated the field of quantum electronics. Some highlights of his later work are an extension of Shannon's information capacity formula to include the effects of field quantization, a result important to the understanding of the limits of optical communications, and a theory of the diffusion of neutral atoms in an optical tweezers trap, which for a time was required reading for those in the field. In recent years he has been involved, with his present coauthor, in a broad study of solitons in optical fibers and of their possible use for communications. Dr. Gordon is a member of the National Academies of Sciences and Engineering. He is a fellow of the OSA and IEEE. He has received the Charles Hard Townes award, the Born award, and the Frederick Ives medal from the OSA.

Contents

Preface

With the recent commercial success of at least two ultra-long-haul, all-optical, dense wavelength division multiplexing (WDM) fiber optic transmission systems (Lucent's *LambdaXtreme* and one by Marconi), solitons have at long last found their proper place in fiber optic transmission. Hence the principal motivation for writing this book: to provide engineers and applied physicists with a satisfying understanding of the underlying physical principles and a thorough description of the practice and capabilities of this new technology. At the same time, we have attempted to write a book that could serve as the text for a college or graduate-level course on fiber optic transmission, with emphasis on the fundamental propagation equation and its consequences, but as well including a discussion of amplifier spontaneous emission noise and other limiting factors, such as "polarization-mode dispersion" and other fiber defects.

Another principal goal in the writing of this book has been to achieve genuine accessibility. Unfortunately, most other books about solitons have been written by and for mathematical specialists, and thus tend to be at a level of abstraction that is off-putting for the applied physics and engineering fraternities. At the other extreme are the numerous book chapters and engineering review articles that discuss various nonlinear phenomena in fiber optic transmission. All too often, however, these tend to be very narrowly focused on whatever list of nonlinear "defects" or "penalties" the author deems important at the moment, and all within the context of a certain limited range of transmission modes. In this book, we have striven to achieve the middle ground. That is, while the nonlinear Schrödinger (NLS) equation is central to most of what we do in this book, right from the beginning we attempt to dissect that equation and show the action and physical consequences of each term in it. Thus, while never shying away from whatever mathematical treatment is needed for a genuine understanding of the physics, we have also attempted to illustrate all major results with real-world examples containing realistic parameters. We have also illustrated as much as possible with graphs and experimental data. Thus, we hope that we have succeeded in bringing the seemingly exotic world of soliton phenomena down to earth, and to demonstrate that it is really not that hard to understand after all.

The outline of this book is straightforward. Chapter 1 introduces the NLS equation and the ordinary soliton, while Chapter 2 treats the more complex but also more practically useful "dispersion-managed solitons." In Chapter 3 we take a temporary break from soliton propagation in order to have a thorough discussion of optical amplifiers and amplifier spontaneous emission noise, the most fundamentally limiting factor in ultra-long-haul transmission. Chapter 4 discusses the soliton interactions in a very general and fundamental way; featured are those important to wavelength division multiplexing, as well as soliton interactions within a given channel and with noise. Chapters 5 and 6 discuss WDM with ordinary and dispersion-managed solitons, respectively, both chapters including important experimental results along with theory. Chapter 7 contains a relatively brief but easy to understand discussion of polarization-mode dispersion, and Chapter 8 discusses some issues of experimental techniques and hardware important both to solitons and to long-haul transmission in general. Finally, as a kind of epilogue, we have included a brief history of solitons, from the brilliant observations of John Scott Russell in the early 19th century to the present. In order not to disrupt the main text, however, the historical material has been deferred to an appendix.

This book represents the distillate of more than two decades (the period from approximately 1980–2003) of our collaborative work at Bell Laboratories in Holmdel, NJ, with one of us (Mollenauer) being the experimentalist, and the other (Gordon), the theorist. That work was driven by the belief that because ultra-long-haul transmission is inescapably nonlinear, the only proper way to design systems is to start with a fundamental understanding of pulse propagation in (nonlinear) fibers and to make full use of the only pulses that are natural to that environment, namely, solitons. This concept is in contrast to the approach most often taken, which is to embrace legacy, essentially short-haul technology, and to try somehow to stretch it to fit the new requirements. Although it took a while to become fully evident, the record now shows that the former approach has indeed produced the better results.

We would also like to acknowledge that from the very beginning in our work, we were much indebted to others, including former Bell Labs colleagues Rogers Stolen, Akira Hasegawa, Stephen Evangelides, and Pavel Mamyshev; former postdocs Fedor Mitschke and Kevin Smith; and the late Professor Hermann Haus of MIT. Nick Doran, both in his days at British Telecom and later as a professor at the University of Austin, and Professor Curtis Menyuk and associates at the University of Maryland, Baltimore, also generated many important ideas, especially with regard to dispersion-managed solitons. More recently, the extensive numerical simulations of Nadja Mamysheva and of Chongjin Xie have been invaluable, and it has been highly stimulating to discuss ideas and to work with Chris Xu, Xing Wei, and Xiang Liu. The experimental work has had the skillful inputs of Jay Cloonan, Mike Neubelt, Andrew Grant, and Inuk Kang. Finally, but by no means least, one of us (Mollenauer) has had the most patient help with computer programming and administration from Jürgen Gripp, Andrew Grant, and Laura Luo, and colleague Ildar Gabitov of the Department of Applied Math at the University of Arizona, Tucson.

Linn Mollenauer and James Gordon
Monmouth County, New Jersey
August, 2005

Chapter 1

The Nonlinear Schrödinger Equation and Ordinary Solitons

1.1. Introduction

The basic equation governing the propagation of pulses in optical fibers is known as the nonlinear Schrödinger equation, or NLS equation for short. It is of fundamental importance to almost everything we shall discuss in this book. In this chapter we shall derive that equation and examine some of its most immediate consequences. Before beginning a detailed study, however, it is useful to have a first look at this innocent-looking equation and its most significant properties. The NLS equation is

$$-i\frac{\partial u}{\partial z} = \frac{1}{2}\frac{\partial^2 u}{\partial t^2} + |u|^2 u. \tag{1.1}$$

As its name suggests, Eq. (1.1) is similar to the well-known Schrödinger equation of quantum mechanics. Here, of course, it has nothing to do with quantum mechanics. Rather, it is just Maxwell's equations, adapted to field propagation in single-mode optical fiber. However, the analogy to quantum mechanics may be instructive to some, as the nonlinear term (the second on the right) is analogous to a negative potential energy, which allows the possibility of self-trapped pulse solutions. These are the solitons, which are the central concern of this book.

In a single-mode fiber, there is only one possible spatial behavior in the transverse dimensions x and y, so that we need to deal only with appropriate averages of the field quantities over those dimensions. Thus, the NLS equation involves only distance along the propagation direction, z, and time, t. The complex quantity $u(z, t)$ is proportional to the light field, with the "central optical frequency" (a frequency arbitrarily chosen somewhere near the mean frequency of the pulse) and the mean time of flight to location z removed; the latter is done so that the pulse always remains in view. The absolute magnitude of u represents the amplitude envelope of the pulse. The phase of u, both in its average change with z and in its variation across the pulse, is equally important in determining the way a pulse propagates.

When polarization is an issue, two functions $u(z, t)$ are required, one for each of the two possible (orthogonal) polarization states. Mild birefringence of the (nominally cylindrical) fibers tends to make the path-average behavior of these two functions nearly identical, however, so usually just one u can be taken as representative, to tell us all we really want to know. (A detailed look at polarization and its effects is taken up in Chapter 7.)

The solutions $u(z, t)$ of Eq. (1.1) directly yield the temporal form of the pulse as seen by an observer at the location z; thus, in the context of telecommunications, time and distance are typically on the scales of picoseconds and kilometers, respectively. Of course, u can be easily transformed to yield the complementary picture where the observer takes a snapshot of the pulse (i.e., observes it as a function of z) at various different times. In the latter picture, one has such familiar terms as group velocity and group velocity dispersion. In the former picture, one has (transit) time delay and time delay dispersion. In this book, as in almost all work on pulse propagation in fibers, we shall adhere to this former view.

It should be emphasized that t in Eq. (1.1) represents retarded time, i.e., ordinary time, but with the transit time delay of a pulse at the central frequency subtracted off. Also, in the frequency spectrum of a pulse satisfying Eq. (1.1), the frequency is the actual frequency minus the central frequency. This makes it much easier to see changes in the pulse shape and spectrum, and deviations from the expected arrival time of a pulse, especially after it has traveled a great distance down the fiber.

The first term on the right of Eq. (1.1), the one involving the second derivative with respect to time, describes the effects of chromatic dispersion. (This is more evident in the Fourier transform of the equation, where the second derivative with respect to time becomes more simply minus the square of the frequency.) It is important to note that this linear term, when acting by itself, does nothing to change the frequency spectrum of the pulse. It serves only to broaden (or narrow) the pulse in time.

The second term on the right of Eq. (1.1) is the nonlinear term. Note that it is just the pulse's intensity envelope times u itself. It is based on the fact that the index of refraction, as will be detailed shortly, is dependent on the light intensity. It is important to note that this term, when acting by itself, does nothing to change the pulse shape in time. It serves only to broaden (or narrow) the pulse in the frequency domain.

We will study many modifications of the NLS equation pertinent to light field propagation in glass fibers. The foremost is the power loss (or gain) in the fiber. Loss is primarily due to absorption and scattering. Gain can be incorporated in the fiber, as we shall detail later. When loss or gain is taken into account, Eq. (1.1) becomes

$$-i\frac{\partial u}{\partial z} = \frac{1}{2}\frac{\partial^2 u}{\partial t^2} + |u|^2 u - i\frac{1}{2}\alpha\, u, \tag{1.1a}$$

where a positive or negative value of α implies, respectively, gain or loss. The factor of one-half in the α term causes α itself to represent power (or energy) loss or gain per unit length.

Equation (1.1) is applicable when the loss/gain term can be neglected (as can happen, even in very long fibers, through the application of loss-canceling optical gain). It is a much studied equation, as it is one of the few that support solitons. Solitons are a class of solitary wave pulses that can pass through one another with no scattering. Zakarov and Shabat [1, 2] most elegantly showed that the general solution of Eq. (1.1) consists of solitons accompanied by smaller dispersive fields called "radiation." They used a method called the "inverse scattering transform," which is a kind of nonlinear Fourier transform. (There will be more on the inverse scattering transform and its consequences in Chapter 4.)

Equation (1.1) has the special solution

$$u(z, t) = \text{sech}(t)\exp(iz/2), \tag{1.2}$$

known as the fundamental soliton. It is a particular pulse of unit amplitude whose mean frequency is the central frequency. [For the benefit of those unfamiliar with it, the hyperbolic secant (sech) function, defined as $\text{sech}(t) = 2/[\exp(t)+\exp(-t)]$, has a shape similar to the more familiar Gaussian function, but it has a narrower peak and broader wings.] Note that since the phase term in Eq. (1.2) has no dependence on t, the soliton is completely nondispersive. That is, its shape does not change with z either in the temporal domain, as shown explicitly here, or in the frequency domain. It happens that the Fourier transform of a sech function is also a sech function. In particular, the Fourier transform of Eq. (1.2) is proportional to $\text{sech}(\pi\omega/2)$. This invariance with propagation occurs, for the soliton, since the dispersive and nonlinear terms of the NLS equation cancel each other's effects, leaving only a phase shift of the whole pulse. It may at first seem somewhat mysterious that the dispersive term, which affects the pulse only in the time domain, can cancel the nonlinear term, which affects the pulse only in the frequency domain. As we shall soon see, however, the explanation of this seeming paradox is quite straightforward, since over very short distances both terms modify only the phase of the pulse.

At this point the reader may well ask, "Where are the fundamental physical constants, such as the speed of light or the nonlinear index coefficient, and where are the parameters particular to the problem, such as the pulse width and power, and the fiber's dispersion parameter and core area?" The answer is that they have been incorporated into definitions of special "soliton units" used to measure distance, time, and power, in such a way that the NLS equation has the simple appearance of Eq. (1.1). This scheme of special units provides yet another simplification, this time to facilitate solution through numerical simulation, and by often enabling interpretation of the results in terms of a range of pulse parameters, instead of for just one particular set. The scaling of the soliton units with pulse and fiber parameters is of fundamental importance, both for understanding and for their engineering consequences. Thus we shall examine that behavior thoroughly in the following sections.

The theme of this book is the study of solitons in communications systems. Thus the pulses we are concerned with are always "narrow band" in the sense that their spectral width is much less than their mean frequency. As such, we note that the NLS equation, either as displayed in Eq. (1.1), or with minor modification, tends to embrace all dispersive and nonlinear phenomena that may be of interest here. For example, the basic NLS equation supposes that a quadratic function adequately describes the dispersion relation $k(\omega)$ over the bandwidth of concern. For shorter pulses, this may not be a completely adequate description. Usually, however, departures from quadratic dispersion are small and can be treated as perturbations of the basic equation. Another example is the Raman effect, which can transfer energy from higher frequencies to lower frequencies within a pulse or between pulses of different frequencies. The NLS equation supposes that the nonlinearity of the refractive index is instantaneous. The Raman effect can be heuristically regarded as resulting from a small delay in the establishment of the nonlinear index. For pulses longer than a few picoseconds in typical glass fibers, these effects are small and can often be accounted for in simple ways, but they cannot be neglected. Other yet smaller effects are present, but can usually be ignored for our purposes.

1.2. Fiber Dispersion and Nonlinearity

1.2.1. Dispersion Relations and Related Velocities

If "weak" monochromatic light at some angular frequency ω enters a fiber, the wavelength λ_{fiber} of the resulting lightwave in the fiber is determined mainly by the refractive index of the fiber and to a lesser extent by its guiding properties.

The phase ϕ of the lightwave has the form $\phi(z, t) = kz - \omega t$, where the wavenumber k is equal to $2\pi/\lambda_{fiber}$, and z is the distance along the fiber. Central to the problem of lightwave propagation in the fiber is the dependence of the wavenumber k on the frequency. This is the dispersion relation $k = k(\omega)$. For plane waves in vacuum it is simply $k = \omega/c$. For plane waves in an isotropic transparent medium it is $k = n\omega/c$, where n is the refractive index at frequency ω. For a single-mode transmission line consisting of fiber core and cladding, we can also use the same form $k = n\omega/c$ with the caveat that n has now an effective value intermediate between the values for core and cladding, depending on the transverse mode shape.

An observer moving with velocity $v = dz/dt$ will observe the phase ϕ to change with time according to

$$\frac{d\phi(z, t)}{dt} = kv - \omega. \tag{1.3}$$

If we require the phase to be constant in Eq. (1.3), the needed velocity is $v = \omega/k$. This is the velocity with which any point of constant phase on the wave travels down the fiber. It is called the phase velocity v_p, and its value is just c/n. If we add a second wave at a slightly different frequency, the combined wave will be modulated, with greater amplitude where the two frequency components are in phase and add, and lesser where they are out of phase and subtract. To follow the modulation envelope, our observer must travel at a velocity such that the rate of change of the phase difference between the two waves is zero. From Eq. (1.3) we thus require $k_1 v - \omega_1 = k_2 v - \omega_2$, where the indices 1 and 2 refer to the two waves. The required velocity is $v = (\omega_1 - \omega_2)/(k_1 - k_2)$. The group velocity $v_g(\omega)$ is this velocity in the limit of a small frequency difference. Thus, we have

$$v_g(\omega) = \frac{d\omega}{dk}. \tag{1.4}$$

Phase and group velocities usually differ, hence one will generally observe the phase moving with respect to the envelope of a modulated wave. Finally, consider a pulse having a continuous band of frequency components. Such a pulse is most highly peaked if all of its frequency components have the same phase at some common time and place. The pulse envelope will tend to move with its average group velocity, but if the group velocity varies with frequency, i.e., there is some group velocity dispersion, then the pulse will spread as it propagates, becoming chirped as its various frequency components separate in time. We shall see this behavior in more detail in the following sections.

1.2.2. Inverse Group Velocity and Retarded Time

In dealing with lightwave propagation in fibers, it is natural to observe a pulse as a function of time at various locations along the fiber. To record the progress of a pulse, we therefore plot power versus t for a succession of values of z. This has two consequences for how we choose to represent time and velocity. First, the inverse of Eq. (1.4),

$$v_g^{-1}(\omega) = \frac{dk}{d\omega}, \tag{1.5}$$

or inverse group velocity (rate of change of time with respect to distance), is clearly more immediately useful than is the group velocity itself. Second, to avoid having to deal with awkwardly large times as z becomes large, the time window must be moved as z is varied. However, we can easily manage this movement by invoking the retarded time mentioned in the introduction (Section 1.1). Once again, a pulse traveling with the group velocity v_g will appear to be stationary in a retarded time frame t' such that $t' = t - v_g^{-1}z$. This is a standard trick used to simplify the analysis.

1.2.3. The Dispersion Parameter

As applied to optical fibers, the term "group velocity dispersion" almost always refers to variation with frequency or wavelength of the inverse group velocity defined by Eq. (1.5). (More accurately, it is sometimes called "time delay dispersion.") Thus, the term can refer either to the quantity $d^2k/d\omega^2$ (often referred to in the literature as "β_2"), or, as is more frequently encountered, especially in engineering papers, the wavelength derivative of v_g^{-1} (almost always called "D"). Thus, D is related to $d^2k/d\omega^2$ or the refractive index by

$$D = \frac{d}{d\lambda}\left(v_g^{-1}\right) = -\frac{2\pi c}{\lambda^2}\frac{d^2k}{d\omega^2} = -\frac{\lambda}{c}\frac{d^2n}{d\lambda^2}. \tag{1.6}$$

(Incidentally, here, and from now on throughout this book, the symbol λ, without a subscript, will always refer to the *vacuum* wavelength.) D is usually expressed in units of picoseconds of delay, per nanometer of wavelength change, per kilometer of fiber length, while $d^2k/d\omega^2$ ("β_2") is expressed in the related units of picoseconds squared per kilometer. By a fortunate accident, for wavelengths in the middle of the 1500-nm band, the numerical values of these two quantities, when expressed in those units, respectively, are roughly the same, save for algebraic sign. The conversion factor, at $\lambda = 1555$ nm, is $(-1.27)x$ ps/nm-km $\cong x$ ps^{-2}. Thus, for example, a dispersion $D = 16$ ps/nm-km (typical of standard single-mode fiber) becomes $\beta_2 \cong -20$ ps^2/km.

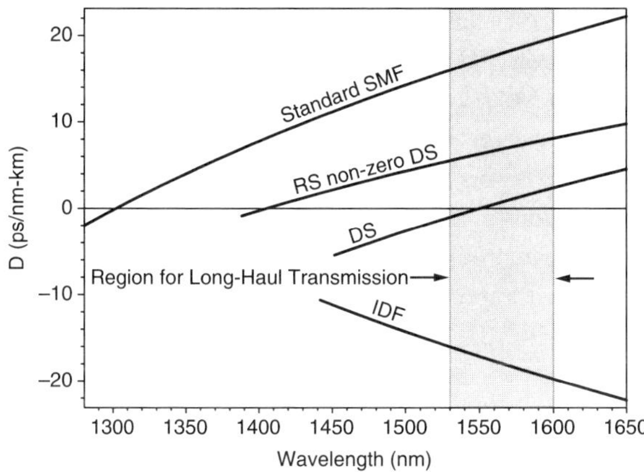

FIGURE 1.1 Dispersion parameters for several transmission fiber types as a function of wavelength. For interpretation of the abbreviations, see text.

Figure 1.1 shows D as a function of wavelength for several typical single-mode transmission fibers, viz., "standard single-mode fiber" or simply "standard SMF," "dispersion-shifted (DS) fiber," "reduced-slope, non-zero dispersion-shifted fiber" (RS non-zero DS), and "inverse dispersion fiber" (IDF). Another fiber type of considerable technological importance is the so-called "dispersion-compensating fiber," or DCF, whose $D \approx -100$ ps/nm-km is beyond the scale of Fig. 1.1. The dispersion of standard SMF reflects almost entirely the material properties of the silica-glass core from which it is made. For the other types, however, various amounts of "modal dispersion" have been added to the material dispersion to create the respective dispersion curves. [The required modal dispersion is created through adjustment of the fiber's "index profile," i.e., the (often complex) way the index of refraction varies outward through the fiber's core and into the cladding.] Note that the zero dispersion wavelength of the standard single-mode fiber occurs at about 1300 nm. In ordinary fibers, the addition of modal contributions can only push that zero dispersion point to longer wavelengths. Thus, solitons, which require $D > 0$ (anomalous dispersion), can exist in standard SMF only for wavelengths greater than about 1300 nm, and only for even greater wavelengths in the others. (The very recent invention of "photonic-crystal fibers" [3–6] has lifted this barrier, however, since at least some of them possess $D > 0$ for wavelengths considerably shorter than 1300 nm [7, 8].)

For modern silica-glass transmission fibers, the loss factor tends to be dominated by Rayleigh scattering, which scales as λ^{-4}, until the onset of fundamental infrared absorption in the 1600-nm region reverses the rapid fall of loss with increasing wavelength. The loss rate minimum thus tends to occur at roughly 1570 nm for most transmission fibers. That minimum value is about 0.19 dB/km for standard SMF, rises to about 0.21 dB/km for the dispersion-shifted fibers, and tends to become much larger (~0.7 dB/km) for the very strongly negative D fibers such as DCFs. This rise in loss results from increased microscopic irregularities (hence increased Rayleigh scattering) that accompany the more complex index profiles. There is a similar monotonic trend in the fiber's effective core area. Thus, A_{eff}, in units of micrometers squared, ranges from a high of 80 μm^2 for standard SMF, through values in the neighborhood of 50 μm^2 for the dispersion-shifted fibers, to a low of about 15 μm^2 for DCFs.

1.2.4. *Fiber Nonlinearity*

The induced polarization in a nonlinear dielectric takes the form

$$\mathbf{P} = \epsilon_0 \left[\chi^{(1)} \cdot \mathbf{E} + \chi^{(2)} \cdot \mathbf{EE} + \chi^{(3)} \cdot \mathbf{EEE} + \cdots \right],$$

where $\mathbf{P}$ and $\mathbf{E}$ are the polarization and electric field vectors, respectively, and the susceptibilities $\chi^{(n)}$ are nth rank tensors. Since the glass of optical fibers is isotropic, one has simply $\chi^{(1)} = n^2 - 1$, where n is the index of refraction, while $\chi^{(2)} = 0$. The effects of $\chi^{(3)}$ of interest here are nonlinear refraction and four-wave mixing. Raman scattering becomes important for shorter pulses than we consider here; third harmonic generation is negligibly small.

In silica-glass fibers, because of their isotropy, and because of the relatively small value of $\chi^{(3)}$, the index can be written, with great accuracy, as

$$n\left(\omega, |\mathbf{E}|^2\right) = n(\omega) + n_2|\mathbf{E}|^2; \tag{1.7}$$

here n_2 is related to $\chi^{(3)}$ by

$$n_2 = \frac{3}{8n}\chi^{(3)}_{xxxx}, \tag{1.8}$$

where $\chi^{(3)}_{xxxx}$ is a scalar component of $\chi^{(3)}$, appropriate to whatever polarization state the light may have at the moment.

Even the highest quality transmission fibers are mildly birefringent, however, so that the polarization state of the light tends to change significantly on a scale of no more than a few meters. In the meantime, the nonlinear effects of interest in long-distance transmission tend to require many kilometers of path for their development. Thus, in general, in optical fibers we are usually only interested in n_2 as suitably averaged over all possible polarization states. In silica-glass fibers, if we write the nonlinear index as $n_2 I$, where I is the intensity in watts per centimeter squared, then the latest measurements of n_2 yield such a polarization-averaged value of about 2.6×10^{-16} cm^2/W.

1.3. The NLS Equation: Derivation and Fundamental Consequences

1.3.1. Derivation

We now consider lightwave propagation in a fiber that has both group velocity dispersion and index nonlinearity. Let the lightwave in the line be represented by a scalar function $U(z, t)$ proportional to the complex field amplitude, such that the power P in the line is given by

$$P = P_c |U|^2. \tag{1.9}$$

The proportionality constant P_c can be considered as a power unit. For frequencies near some central frequency ω_0, the generic dispersion relation $k = n\omega/c$ can be expanded to the approximate form

$$k = k_0 + k'(\omega - \omega_0) + \frac{1}{2}k''(\omega - \omega_0)^2 + k_{NL}P. \tag{1.10}$$

Here the primes indicate partial differentiation with respect to ω, and k_{NL} is given by

$$k_{NL} = \frac{2\pi\, n_2}{\lambda\, A_{eff}}. \tag{1.11}$$

Equation (1.11) is derived from the relation $k = 2\pi n/\lambda$ applied to the second term on the right of Eq. (1.7). The quantity A_{eff} is the effective fiber core area, i.e., that area which yields the proper average intensity when the pulse power is divided by it. Note that Eq. (1.10) has the form of a Taylor series expansion of $k(\omega, P)$ in the neighborhood of $(\omega_0, 0)$. As such, it adequately describes the propagation of monochromatic waves $U = u_0 \exp(ikz - i\omega t)$ in the line so long as the frequency does not stray too far from ω_0. It leads directly to the nonlinear Schrödinger equation. For solitons in fibers, the last two terms of Eq. (1.10) are of comparable importance, k_{NL} is positive, k'' is negative, and succeeding higher order terms [e.g., $k'''(\omega - \omega_0)^3$, $k_2'(\omega - \omega_0)P$, etc.] can be neglected or adequately treated as perturbations. For the moment, we assume that the line has no loss or gain; i.e., that the constants k_0, k', k'', and k_{NL} are all real.

The expression for the inverse group velocity, namely

$$v_g^{-1} = \frac{\partial k}{\partial \omega} = k' + k''(\omega - \omega_0), \tag{1.12}$$

identifies k' as the inverse group velocity at frequency ω_0, and k'' as its frequency dispersion constant. (Thus, k'' is just the quantity we have already discussed in Section 1.2.3, but this time expressed as a partial derivative, on account of the assumed additional dependence of k on power.)

The term $k_{NL}P$ in Eq. (1.10) represents the primary nonlinear effect, self-phase modulation, resulting from the intensity dependence of the refractive index of the fiber. Note from Eq. (1.12) that the dependence of the group velocity on power is among the higher order terms not included in Eq. (1.10).

To derive the NLS equation, it is only necessary to consider the propagation of a monochromatic field. [If higher terms are included in Eq. (1.10) one must consider two or more frequencies.] Thus, let $U = u_0 \exp(-i\omega t + ikz)$. If we now remove the central frequency and the corresponding wavenumber from U by defining

$$u(z,t) = Ue^{i(\omega_0 t - k_0 z)} \tag{1.13a}$$

so that when U is written out explicitly, $u(z,t)$ becomes

$$u(z,t) = u_0 e^{i[(k-k_0)z-(\omega-\omega_0)t]}, \tag{1.13b}$$

then the wave equation for u that is necessary and sufficient to reproduce exactly our initial dispersion relation Eq. (1.10) is

$$-i\frac{\partial u}{\partial z} = (k-k_0)u = ik'\frac{\partial u}{\partial t} - \frac{1}{2}k''\frac{\partial^2 u}{\partial t^2} + k_{NL}P_c|u|^2 u. \tag{1.14}$$

This can be shown by inserting Eq. (1.13b) in Eq. (1.14).

The standard form of the propagation equation is generated from Eq. (1.14) by transforming to the retarded time frame (this eliminates the k' term) and by choosing unit values of time and distance such that $k'' = -1$ and $k_{NL} = 1$ when measured in those units and the power unit already mentioned. The appropriate new variables are

$$t' = (t - k'z)/t_c, \tag{1.15a}$$

$$z' = z/z_c, \tag{1.15b}$$

where the unit values t_c, z_c, and P_c satisfy the relations

$$t_c^2/z_c = -k'' = \lambda^2 D/(2\pi c), \tag{1.16a}$$

$$z_c P_c = 1/k_{NL}. \tag{1.16b}$$

In terms of these new variables, the resulting equation (after we drop the primes on z and t) is clearly just the NLS equation, Eq. (1.1).

There is an important arbitrariness left in the definitions of the three unit values z_c, t_c, and P_c, since there are only two relations [Eqs. (1.16a) and (1.16b)] that they must satisfy. One unit value may be chosen freely, and thus different real-world fields can be represented by the same solution of Eq. (1.1), and vice versa.

In particular, if one solution of Eq. (1.1) is $u(t, z)$, then different scalings of the same real-world field give other solutions of the form $Au(At, A^2z)$, where A is the ratio of the values of t_c. This scaling transformation of the solutions of Eq. (1.1) can be verified by direct substitution.

For the accurate study of sub-picosecond pulses with correspondingly wide bandwidths, it is necessary to modify the dispersion relation, Eq. (1.10) and thus the propagation equation. Higher order terms such as third-order dispersion [proportional to $(\omega - \omega_0)^3$] and power dependence of the group velocity [proportional to $P(\omega - \omega_0)$] become appreciable and need to be added. Furthermore, the Raman effect, which is reproduced in first order by a delay of a few femtoseconds in the nonlinear response of the fiber, begins to have noticeable consequences.

The reader may well note that this derivation did not begin with Maxwell's equations (as might have been anticipated), nor has it depended in any way on specific reference to lightwaves or optical fiber! (The words "lightwave" and "fiber" in the opening sentences were not intrinsic to the derivation.) Thus, this derivation can be taken to apply to *any* transmission line whose dispersive and nonlinear properties can be represented by Eq. (1.10). The transmission line need not even involve light—it could be, for example, a coaxial cable using a nonlinear dielectric or a string made of suitably nonlinear elastic. In this way, it is much more powerful, and affords greater insight, than direct transformation of Maxwell's equations. For the purposes of this book, of course, we shall specialize it to Maxwell's equations and optical fibers. That specialization is the subject of the next section on soliton units.

1.3.2. *Soliton Units*

Equation (1.1) is often referred to as a dimensionless form of the nonlinear Schrödinger equation. As already indicated in the introduction, however, it is more useful to think of it as having specific dimensions, with z, for example, being a distance always measured in units of z_c rather than in meters or kilometers or any other standard unit. Thus, $z = 2$ means a distance of $2 \times z_c$. Similarly, t_c and P_c become the units of time and power, respectively. Since we are primarily interested in solitons, it is convenient to tie these three units (which so far are very general in meaning) to the specific requirements of solitons. The canonical single soliton solution of Eq. (1.1), already displayed in Eq. (1.2), has a full width at half maximum power (FWHM) of $\Delta t = 2\cosh^{-1}(\sqrt{2}) \approx 1.763$. In order for this form to represent some soliton whose FWHM is τ (in picoseconds, for example), we need simply to take

$$t_c = \frac{\tau}{1.763}. \qquad (1.17)$$

The unit distance, z_c, is a characteristic length for effects of the dispersive term, and is given by

$$z_c = \frac{1}{(1.763)^2} \frac{2\pi c}{\lambda^2} \frac{\tau^2}{D}, \tag{1.18}$$

where c and λ are the light velocity and wavelength in vacuum, respectively, and where D is the dispersion constant, as already described in Section 1.2.3. When D is expressed in the usual units of picoseconds per nanometer per kilometer, τ in picoseconds, and for λ =1557 nm, Eq. (1.18) becomes

$$z_c \approx 0.25\tau^2/D, \tag{1.18a}$$

where z_c is in kilometers. Note that for the pulse widths ($\tau \sim$ 15–50 ps) and dispersion parameters ($D \sim$ 0.3–1 ps/nm-km) most desirable for long-distance soliton transmission, z_c is hundreds of kilometers.

The unit of power, P_c, is just the soliton peak power, and is given by the formula

$$P_c = \frac{1}{k_{NL} z_c} = \frac{A_{eff}}{2\pi n_2} \frac{\lambda}{z_c} = \left(\frac{1.763}{2\pi}\right)^2 \frac{A_{eff}\lambda^3}{n_2 c} \frac{D}{\tau^2}, \tag{1.19}$$

where n_2, the nonlinear coefficient, has the polarization-averaged value already cited (see Section 1.2.4). Thus, for $A_{eff} \sim 50\,\mu\text{m}^2$, and λ =1557 nm, one has $P_c \approx 476/z_c$, where P_c is in milliwatts and z_c is in kilometers. Note that for z_c of hundreds of kilometers, the soliton power is just a few milliwatts.

The soliton pulse energy, another very useful quantity, is given by

$$W_{sol} = P_c \int_{-\infty}^{\infty} \text{sech}^2(t/t_c)\, dt = 2P_c t_c = \frac{2}{1.763} P_c \tau = (1.346) P_c \tau, \tag{1.20}$$

or, substituting the expression for P_c in Eq. (1.20), we get

$$W_{sol} = \frac{1.763}{(2\pi)^2} \frac{A_{eff}\lambda^3}{n_2 c} \frac{D}{\tau}. \tag{1.20a}$$

Thus, again for $A_{eff} \sim 50\,\mu\text{m}^2$, and λ = 1557 nm, one has $W_{sol} \approx 2000(D/\tau)$, where W_{sol} is in femtojoules and D and τ are in the usual units (ps/nm-km and ps, respectively). Note that for D = 0.5 ps/nm-km and τ = 20 ps, W_{sol} is about 50 fJ. While this is a small energy, it comprises about 4×10^5 photons. As we will show later, this is what allows us to treat quantum effects in a simple way.

1.3.3. *Pulse Motion in the Retarded Time Frame*

Another important transformation of the solutions of Eq. (1.1) is that produced by a carrier frequency shift. Because the inverse group velocity dispersion constant has the value -1 in the soliton unit system, a frequency shift produces an inverse group velocity shift of equal magnitude. Thus, for the same solution as that given by Eq. (1.2), one finds yet other solutions, frequency shifted by Ω (in units of t_c^{-1}), of the form

$$u(t + \Omega z, z)e^{-i(\Omega t + \Omega^2 z/2)}. \tag{1.21}$$

This transformation also applies to *any* solution of Eq. (1.1).

1.3.4. A Useful Property of Fourier Transforms

Shortly, we shall have need of the following simple relation between the Fourier transforms of $u(t)$ and those of its time derivatives: Let $u(t)$ and $\tilde{u}(\omega)$ be Fourier transforms of each other, i.e.,

$$u(t) = \frac{1}{\sqrt{2\pi}} \int_{-\infty}^{+\infty} \tilde{u}(\omega)e^{-i\omega t} d\omega \quad \text{and} \quad \tilde{u}(\omega) = \frac{1}{\sqrt{2\pi}} \int_{-\infty}^{+\infty} u(t)e^{i\omega t} dt. \tag{1.22}$$

We are using the tilde symbol to imply a function of frequency. Successive time differentiations of the first expression in Eq. (1.22) show that $\partial u(t)/\partial t$ and $-i\omega\tilde{u}(\omega)$ are also Fourier transforms of each other, as are $\partial^2 u(t)/\partial t^2$ and $-\omega^2\tilde{u}(\omega)$, and so on.

1.3.5. Action of the Dispersive Term in the NLS Equation

To obtain the action of the dispersive term alone, we temporarily turn off the nonlinear term, so that Eq. (1.1) becomes

$$\frac{\partial u}{\partial z} = \frac{i}{2}\frac{\partial^2 u}{\partial t^2}. \tag{1.23}$$

The problem is most easily solved in the frequency domain. The Fourier transform of this last equation yields

$$\frac{\partial \tilde{u}}{\partial z} = -\frac{i}{2}\omega^2\tilde{u}, \tag{1.24}$$

and its solution is

$$\tilde{u}(z, \omega) = \tilde{u}(0, \omega)e^{-i\omega^2 z/2}. \tag{1.25}$$

From the form of this general solution, it should be clear that the dispersive term merely rearranges the phase relations among existing frequency components; it adds no new ones. To find how the dispersion affects a pulse, we must transform back to the time domain. An example that has an instructive analytic solution is the Gaussian pulse. Taking $u(0,t) = e^{-t^2/2}$, we have $\tilde{u}(0,\omega) = e^{-\omega^2/2}$, and upon turning the crank we get

$$u(z,t) = \frac{1}{\sqrt{1+iz}} \exp\left[\frac{-t^2}{2(1+z^2)}(1-iz)\right]. \tag{1.26}$$

In the near field ($z^2 \ll 1$) the pulse gets some chirp (change of frequency across the pulse), but the pulse shape does not change (see Fig. 1.2). In the far field ($z^2 \gg 1$) the field approaches

$$u(z \gg 1, t) \approx \frac{1}{\sqrt{1+iz}} \exp\left(-\frac{t^2}{2z^2} + i\frac{t^2}{2z}\right), \tag{1.26a}$$

which shows how dispersion fans the field out into a spectrum of its various frequency components. The frequency is chirped; the local frequency, (-1) times the time derivative of the phase of u, is $\omega = -t/z$. Thus in the far field, the lines of constant frequency ω fan out according to $t(z) = -\omega z$. Also, there is a general rule that is demonstrated here: In the far field the intensity distribution $|u(z,t)|^2$ is proportional to the initial spectral distribution $|\tilde{u}(0,\omega)|^2$ evaluated at $\omega = t/z$.

As depicted in Fig. 1.2, note that the intensity envelope $|u|^2 \propto \exp[-t^2/(1+z^2)]$. Thus the pulse width grows as

$$\tau = \tau_0\sqrt{1+z^2}, \tag{1.27}$$

where τ_0 is the initial, minimum pulse width. We see that the change in τ is only to second order in z at the origin. This may also be seen directly from the differential equation, where we note that if $u(t)$ has a constant phase, then $\partial u/\partial z$ is everywhere in quadrature with u. This initially slow scaling of pulse width with z is important to creation of the soliton.

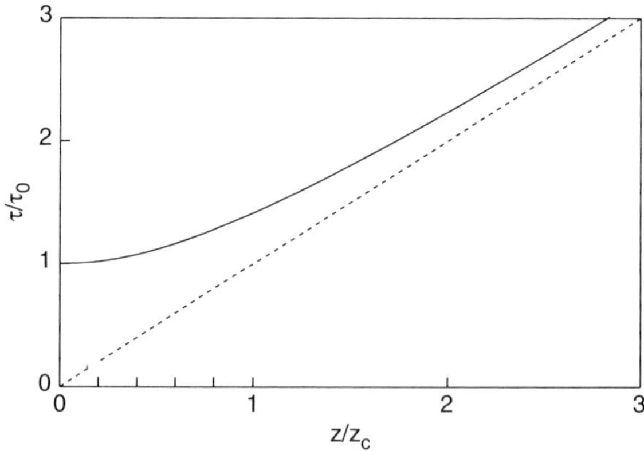

FIGURE 1.2 Dispersive broadening of a Gaussian pulse with distance.

1.3.6. *Action of the Nonlinear Term in the NLS Equation*

To observe the action of the nonlinear term of Eq. (1.1) alone, we turn off the dispersive term, so the equation becomes simply

$$\frac{\partial u}{\partial z} = i|u|^2 u. \tag{1.28}$$

The problem is most naturally solved in the time domain, where the general solution is

$$u(z, t) = u(0, t)e^{i|u|^2 z}. \tag{1.29}$$

From the form of this general solution, it should be clear that the nonlinear term modifies the phase shift across the pulse, $\phi(t)$, but not the intensity envelope. Thus, it only adds new frequency components. This is the phenomenon of "self-phase modulation," or SPM for short.

To get the spectral spreading, we must transform back to the frequency domain. Once again, for example, let $u(0, t) = e^{-t^2/2}$. In that case, one has

$$\tilde{u}(z, \omega) = \frac{1}{\sqrt{2\pi}} \int_{-\infty}^{\infty} u(0, t)e^{i|u|^2 z} e^{i\omega t} dt = \frac{1}{\sqrt{2\pi}} \int_{-\infty}^{\infty} e^{-t^2/2} e^{ize^{-t^2}} e^{i\omega t} dt. \tag{1.30}$$

For $z \gg 1$, this integral produces a multipeaked spectrum, where the number of peaks and the overall spectral width increase directly with z (see Fig. 1.3). The numbers at each spectrum indicate the peak nonlinear phase shift. However, for $z \ll 1$, the integral is approximately

$$\frac{1}{\sqrt{2\pi}} \int_{-\infty}^{\infty} e^{-t^2/2} \left(1 + ize^{-t^2}\right) e^{i\omega t} dt = \tilde{u}(0, \omega) + \frac{iz}{\sqrt{3}} \tilde{u}(0, \omega)^3. \tag{1.31}$$

Note that once again, the new component is in quadrature with the original pulse, so the increase in net spectral width scales only as z^2. Thus, the initial increase, here in bandwidth, is also only to second order in z. *This behavior is equally important to the creation of the soliton as was the increase in pulse width from the dispersive term.*

1.4. The Soliton

1.4.1. *Origin of the Soliton*

We are now finally in a position to discuss the origin of the soliton. In the introduction (Section 1.1), we stated that the soliton is that pulse for which the nonlinear

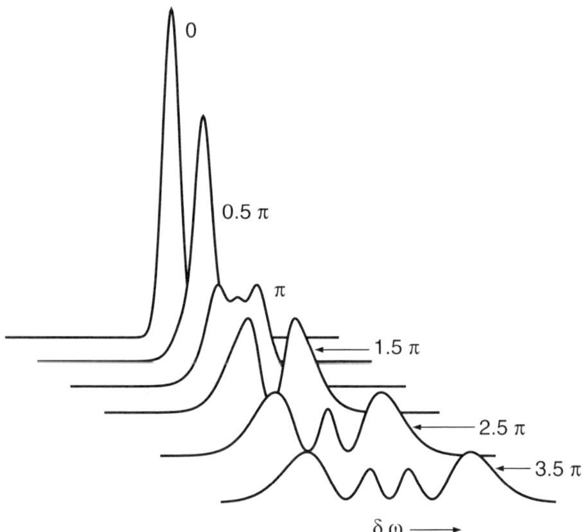

FIGURE 1.3 Spectral broadening of a Gaussian pulse at zero dispersion. The number at each spectrum indicates the corresponding peak nonlinear phase shift.

and dispersive terms of the NLS equation cancel each other's effects. At first, it may seem mysterious that the tendencies to spectral and temporal broadening can cancel one another. As we have just taken pains to show, however, whenever one starts from an unchirped pulse, such as sech(t), there is no broadening of either kind to first order in z (to be thought of as dz). Instead, the first-order effects of both terms are just complementary phase shifts $d\phi(t)$. We have already seen how the nonlinear term generates $d\phi(t) = |u(t)|^2 dz$. For the dispersive effect, first, we recognize that if $f(z, t)$ is real, then the general equation

$$\frac{\partial u}{\partial z} = if(z, t)u \tag{1.32}$$

simply generates the phase change $d\phi(t) = f(0, t)dz$ in the distance dz. We then write the reduced NLS equation in the form

$$\frac{\partial u}{\partial z} = \left(\frac{i}{2u}\frac{\partial^2 u}{\partial t^2}\right)u. \tag{1.33}$$

Thus, the dispersive term generates

$$d\phi = \left(\frac{1}{2u}\frac{\partial^2 u}{\partial t^2}\right)dz. \tag{1.34}$$

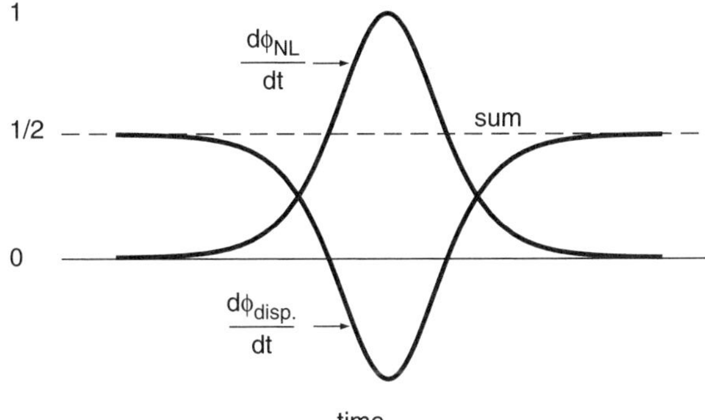

FIGURE 1.4 Dispersive and nonlinear phase shifts of a soliton pulse and their sum.

For $u = \text{sech}(t)$, these terms are, respectively,

$$d\phi_{NL} = \text{sech}^2(t)\, dz \quad \text{and} \quad d\phi_{disp.} = \left[\frac{1}{2} - \text{sech}^2(t)\right] dz. \tag{1.35}$$

Note that these differentials sum to a constant (see Fig. 1.4), which, when integrated, simply yields a phase shift of $z/2$ common to the entire pulse. In this way, we arrive at the simplest form for the soliton, already displayed in Eq. (1.2).

It should also be noted that a common phase shift does nothing to change the temporal or spectral shapes of a pulse. Thus, as already advertised, the soliton remains completely nondispersive in both the temporal and frequency domains. Nevertheless, the associated wavenumber shift of $z/(2z_c)$, or simply $z/2$ in soliton units, is important in understanding interaction of the soliton with perturbing nonsoliton field components.

1.4.2. Path-average Solitons

For reasons of economy, the loss-canceling optical fiber amplifiers of a long fiber transmission line are usually spaced apart by a distance, which we shall call the amplifier span, or L_{amp}, of several tens of kilometers. This spacing results in a rather large periodic variation in the signal intensity, as illustrated in Fig. 1.5. In addition, the dispersion parameter D may vary significantly within each amplifier span (again, see Fig. 1.5).

Clearly, in that case, the differential phase shifts of the dispersive and nonlinear terms [see Eq. (1.35)] do not cancel in every element dz of the fiber. Nevertheless, if the condition

$$z_c \gg L_{amp} \tag{1.36}$$

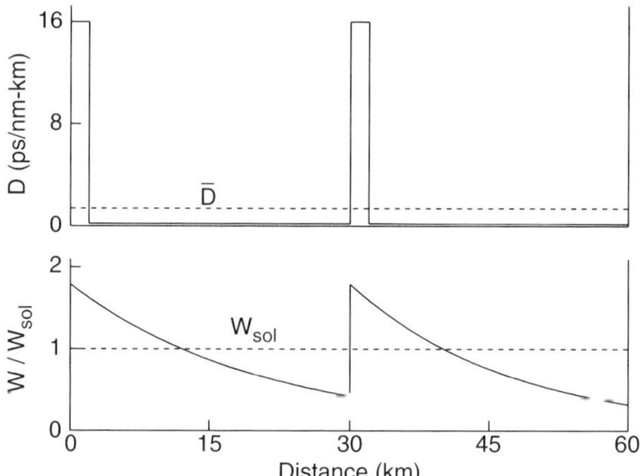

FIGURE 1.5 Sample of transmission line used for numerical test of the path-average soliton concept. Here the desired $\bar{D}$ is obtained by combining short lengths of high D fiber with dispersion-shifted fiber (for which $D \simeq 0$), so there are large variations in D, periodic with the amplifier spacing, as well as in the pulse energy.

is satisfied, and if, furthermore, the path-average values, $\bar{I}$ and $\bar{D}$ of the intensity and dispersion, respectively, are the same for every amplifier span, then one can still have, at least in a practical sense, perfectly good solitons. The reason is that when the inequality of Eq. (1.36) is well satisfied, as already shown, neither the temporal nor spectral shapes of the soliton are significantly affected within each span. Thus, all that matters is that, over each amplifier span, the path-average dispersive and nonlinear phase shifts cancel (sum to a standard constant). Figure 1.6 illustrates how very well this concept of "path-average" solitons [9–13] can work.[1] Through numerical simulation, it shows $\tau = 50$ ps solitons before and after traversal of 15,000 km of transmission line whose spans are those of Fig. 1.5. For this case, $z_c/L_{amp} \approx 16$. The difference between the output and input solitons, which can only be seen on the logarithmic plot, appears in the form of very low-intensity tails on the pulse. This defect, usually known as "dispersive wave radiation," represents the nonsoliton component of the pulse. When it is as small as shown here, it is usually of no practical import. Even when z_c/L_{amp} is as small as three or four, the path-average soliton concept still works fairly well.

[1] Although the meaning of the name tends to be obscure, the "guiding center solitons" of Ref. [13] are essentially path-average solitons. We prefer the latter name, however, as it is more accurately descriptive, is better known, and avoids confusion with the more apt use of the word "guiding" in connection with jitter reducing filters (see Chapter 3).

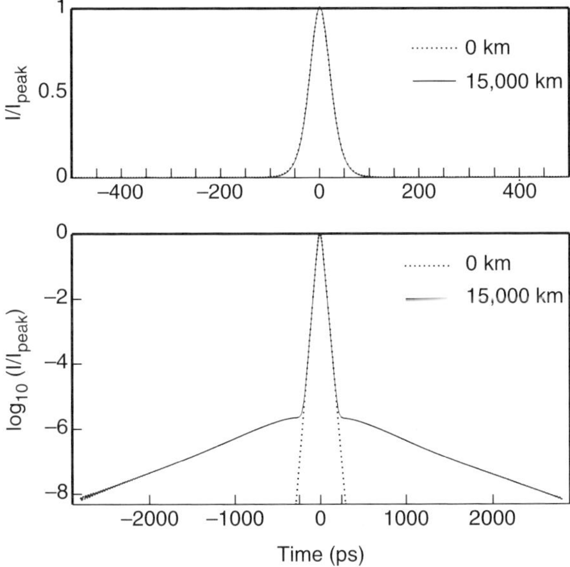

FIGURE 1.6 Solitons, for which $z_c \simeq 440$ km, at input and after traversing 15,000 km of the transmission line of Fig. 1.5. Note that the defects (dispersive tails on the pulse) are extremely small. The defects have also been computed analytically.

There is another, complementary, insightful way of understanding the behavior of path-average solitons, wherein the periodic fluctuations of the amplifier spans are seen as a perturbation to provide phase matching between the solitons on the one hand, and the linear, or dispersive, waves on the other. As we have just seen in the foregoing discussion, the dispersion relation for solitons is just $k_{sol} = 1/2$ (in soliton units), while that for the linear waves is $k_{lin} = -(1/2)\omega^2$. Clearly, the amplifier spans provide $k_{pert} = 2\pi z_c / L_{amp}$. The phase-matching condition is

$$k_{pert} = k_{sol} - k_{lin}. \tag{1.37}$$

If, as illustrated in Fig. 1.7, k_{pert} is so large that the phase matching occurs only where the spectral density of the soliton is small, then the path-average solitons work well. On the other hand, if k_{pert} approaches 1/2, then the phase matching will be to a region of high spectral density, where a large fraction of the soliton's energy will drain away into dispersive (linear) waves, and the path-average soliton will not work well. In this regard it is important to note that the inverse group velocity of the linear waves is proportional to $k_{pert} - 1/2$. Hence when $k_{pert} \gg 1/2$, the linear waves created by the perturbation quickly leave the vicinity of the soliton, reducing their interaction with the soliton. The opposite occurs as k_{pert} approaches 1/2.

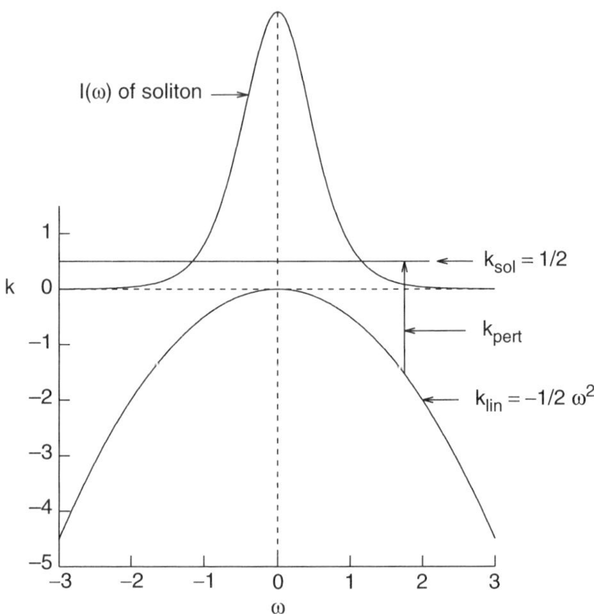

FIGURE 1.7 Dispersion relations ($k(\omega)$) for the soliton and for linear waves, and the spectral density of the soliton. A perturbation of wave vector k_{pert} phase matches linear waves of frequency ω to the corresponding region of the soliton's spectrum.

It is interesting that, historically, the concept of path-average solitons, and the associated resonance condition for disaster ($k_{pert} = 1/2$), were first encountered [9] in these terms of phase matching.

1.4.3. Soliton Transmission in Dispersion-tapered Fiber

With ever-increasing bit rate, eventually the soliton pulse width, and hence z_c, become so short that it is no longer possible to satisfy the inequality of Eq. (1.36) in a satisfactory way. Nevertheless, in principle at least, there is still a way to have perfect soliton transmission with lumped amplifiers, and that is to taper the fiber's dispersion parameter $D(z)$ to the same exponential decay curve as that of the intensity itself. That is, $D(z)$ should be given by

$$D(z) = \frac{\alpha L_{amp} \bar{D}}{1 - \exp(-\alpha L_{amp})} e^{-\alpha z} = D_0 e^{-\alpha z} \tag{1.38}$$

so that Eq. (1.1a) becomes

$$-i\frac{\partial u}{\partial z} = \frac{D_0}{2\bar{D}}e^{-\alpha z}\frac{\partial^2 u}{\partial t^2} + |u|^2 u - i(\alpha/2)u. \tag{1.39}$$

Clearly, the phase shifts generated by the dispersive and nonlinear terms of Eq. (1.39) will cancel in each and every segment dz of the amplifier span, so the solitons will be without perturbation. It is also easily understood that an N-step approximation to the ideal dispersion taper of Eq. (1.38) can be tremendously helpful, even when N is as small as two or three. Later, in the chapter on WDM with ordinary solitons, we shall show how that tapered dispersion can help to avoid excessive growth of four-wave mixing components (see Chapter 5).

1.4.4. More General Forms for the Soliton

Any single real-world soliton can be expressed in the simple form of Eq. (1.2) by the appropriate choice of scale and central frequency. Alternatively, application to Eq. (1.2) of the scale and frequency transformations discussed in Sections 1.3.1 and 1.3.3, respectively, yields the more general form

$$u = A\,\mathrm{sech}[A(t - t_0 + \Omega z)]\exp[-i\Omega t + i\frac{1}{2}\left(A^2 - \Omega^2\right)z + i\phi_0]. \tag{1.40}$$

The form given by Eq. (1.40) is necessary to the consideration of perturbation or multisoliton problems, such as soliton–soliton collisions in WDM, for example. The four parameters (A, Ω, t_0, ϕ_0) completely describe any single soliton. The first two give the amplitude and frequency of the soliton, while the second two describe the initial conditions at $z = 0$.

1.4.5. Numerical Solution of the NLS Equation: The Split-step Fourier Method

The NLS equation is generally difficult to solve analytically. Numerical solution, however, can be remarkably efficient, when it is based on the "split-step Fourier" method (Fig. 1.8). The method is based on the fact that the effects of the dispersive term are most naturally dealt with in the frequency domain, while those of the

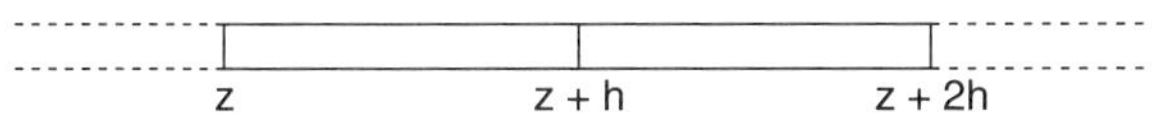

FIGURE 1.8 Scheme of the split-step method.

nonlinear term are best handled in the time domain. Thus, each increment h in z is treated in two consecutive steps, as follows:

$$\text{step 1:} \qquad u(z,t) \to \tilde{u}(z,\omega); \qquad \tilde{u}_{new}(z,\omega) = \tilde{u}(z,\omega)e^{-i(\omega^2/2)h};$$

and

$$\text{step 2:} \qquad \tilde{u}_{new}(z,\omega) \to u_{new}(z,t); \qquad u(z+h,t) = u_{new}(z,t)e^{i|u|^2 h}.$$

That is, in step 1, $u(z,t)$ is Fourier transformed to $\tilde{u}(z,\omega)$, and then, to reflect the dispersive effects of the element h, an intermediate state, $\tilde{u}_{new}(z,\omega)$, is computed from $\tilde{u}(z,\omega)$ according to the *analytic* solution of Eq. (1.23). In step 2, $\tilde{u}_{new}(z,\omega)$ is Fourier transformed back to make an intermediate state, $u_{new}(z,t)$. Then, from $u_{new}(z,t)$, $u(z+h,t)$ is computed according to the *analytic* solution of Eq. (1.27), so that it now reflects the nonlinear effect of the element h as well. To increase the accuracy of this scheme, in essence because one does not know whether the dispersive change or the nonlinear change should come first, it is best to do a half step of nonlinear change at the beginning of the sequence and then to do another half step of nonlinear change at the end. Based on the ideas just discussed with respect to path-average solitons, one can easily see that reasonable accuracy can often be obtained with relatively large step sizes. Finally, note that fiber loss, amplifier gain, filter response functions, and other frequency-dependent factors are obviously most easily applied in the frequency domain.

Chapter 2

Dispersion-managed Solitons

2.1. Introduction

For those steeped in the lore of ordinary solitons, dispersion management, with its strong periodic variation in pulse properties, is strongly counterintuitive. Nevertheless, in 1995, in a seminal paper, Suzuki *et al.* [14] reported experimental demonstration of essentially error-free 20-Gbit/s soliton transmission over 9000 km through strong reduction of the normally limiting Gordon–Haus jitter. The trick was to use a periodic dispersion "map" that allowed the soliton pulse energy to remain fixed in the face of greatly reduced $\bar{D}$. A rash of papers soon followed, pursuing the enhanced energy [15–18] and general pulse dynamics [19–24] of these "dispersion-managed solitons." Thus this early, almost exclusively numerical work was largely focused on the behavior of isolated pulses, as were some following real-world experiments [25–30].

Papers on "dense WDM" (wavelength division multiplexing involving many, closely spaced channels) with dispersion-managed solitons tended to come somewhat later [31–35]. Nevertheless, as we shall soon see, dispersion management is of tremendous importance to dense WDM. Dispersion-managed solitons in particular offer unique advantages for the formation of ultra-long-haul, dense WDM, all-optical networks. Before we can even begin to discuss these advantages, however, we must first have the quick overview of dispersion management and dispersion-managed solitons provided by the following section.

2.1.1. Dispersion Management and Dispersion-managed Solitons

With dispersion management, the transmission line consists of segments of fiber whose individual dispersion parameters (D_{local}) are of alternating algebraic sign (Fig. 2.1). Furthermore, this arrangement, or "dispersion map," is ideally periodic (although, in practice, it need not be exactly so). For each map period, the accumulated dispersions of the two segments nearly cancel, so that the path-average dispersion parameter of the map, $\bar{D}$, is usually much smaller than either D^{+}_{local} or $|D^{-}_{local}|$. (Typically, $\bar{D}$ is no greater than a few tenths of a picosecond per nanometer-kilometer.) To support solitons, $\bar{D}$ is also positive (anomalous dispersion).

It is instructive first to consider pulse behavior at very low intensities, when only the dispersive term of the NLS equation is important. In response to the relatively large, alternating D_{local} values, the pulse width tends to undergo a significantly large fractional change, periodic with the map. This pulse "breathing" is accompanied and promoted by a similarly periodic variation in the chirp parameter, with the chirp passing through zero at or near the center of each fiber segment. (Note that the sign of the chirp at the end of each segment is that required for pulse compression in the following segment.) But on a distance scale typically many times greater than the map period, there is also a gradual net broadening of the

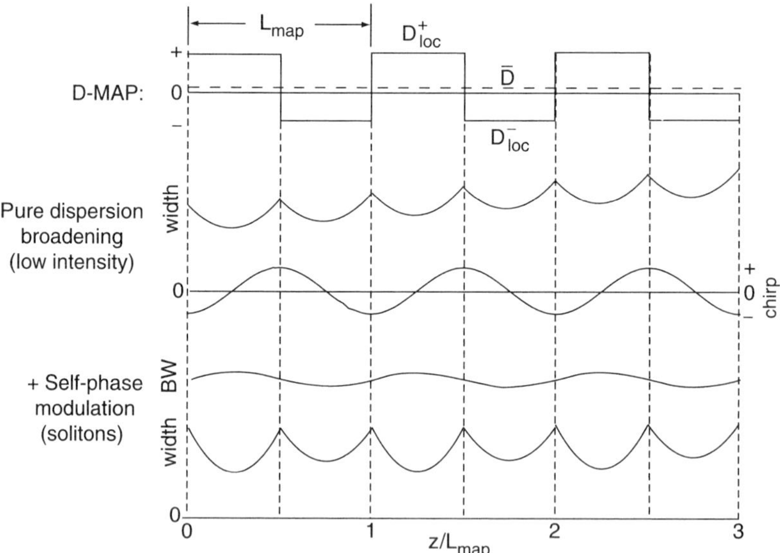

FIGURE 2.1 Dispersion-managed solitons in a nutshell. Top: Scheme of a prototypical dispersion map. Middle section: Pulse behavior (pulse width above and chirp below) in the linear limit. Bottom section: Pulse behavior at the power level for solitons. Note that the plots of pulse bandwidth (above) and pulse width (below) refer to the same zero.

pulses, in response to the effect of $\bar{D}$. This behavior is illustrated in the upper half of Fig. 2.1.

Now, to obtain dispersion-managed solitons, we merely need increase the pulse intensity until self-phase modulation (see Section 1.3.6) produces a phase shift across the pulse that just cancels out the net phase shift produced by the dispersive term within each map period. This periodic cancellation of phase shifts in turn eliminates the net pulse broadening from $\bar{D}$, so that the pulse behavior now becomes truly periodic (the bottom curve of Fig. 2.1). The cancellation of phase shifts is similar to that obtaining with ordinary solitons (Fig. 1.4), but with the important difference that here, in general, the variation of phase across the pulse is quite large. The variation becomes zero, of course, only at the points of zero chirp.

While the pulse field-envelope shape function of the ordinary soliton is sech(t), that of the dispersion-managed soliton is essentially Gaussian. This is because of the fact that in each segment, the large dispersive term (which scales with D_{local}) dominates the much smaller nonlinear term (which scales with $\bar{D}$), and because the solution to the NLS equation, in the case of pure dispersion (as already shown in Section 1.3.5), is a Gaussian.

The bottom half of Fig. 2.1 also shows the variation of pulse bandwidth (BW) within the map. The fractional change, though small relative to the breathing in pulse width, is nevertheless of great importance. That is, it tends to significantly *increase* the amount of SPM required to generate the nonlinear phase shift needed to cancel that from the dispersive term. That increase in SPM comes about as follows: First, note that for a Gaussian pulse undergoing purely dispersive broadening [Eq. (1.26)], the coefficient of t^2 in the phase term scales as $z/(1+z^2)$, or, in the near field, simply as z, thus as z/z_c in ordinary units. Since $1/z_c$ [see Eq. (1.18)] scales as D/τ^2, or equivalently, as $D \times (BW)^2$, the net phase shift from the dispersive term scales approximately as $\int_0^{L_{map}} D(z)(BW)^2\,dz$, and not just as $\int_0^{L_{map}} D(z)\,dz$. Thus, the fact that the average $(BW)^2$ is larger in the D^+ segment than in the D^- segment can bring about a large increase (typically, by a factor of several times) in the *net* dispersive phase shift created by the two segments. The resultant increase in required SPM, which we shall soon calculate more accurately, means a corresponding increase in soliton pulse energy. This is the origin of the "energy enhancement factor" that excited so much interest early on [15–18], and that, as we shall soon see, has very important practical consequences.

It should be noted that the lengths of the individual segments of the dispersion map need not be equal or nearly equal, as might be inferred from Fig. 2.1. Rather, the only requirement is that the sum $D^+L^+ + D^-L^- = \bar{D}L_{map}$, where D^+, L^+ and D^-, L^- refer to the anomalous and normal dispersive segments of the map, respectively. Thus, for example, it is common to have an 80- to 100-km span of non-zero dispersion-shifted fiber ($D^+ \sim$ 4–8 ps/nm-km) compensated by a several-kilometer-long coil of DCF ($D^- \sim -100$ ps/nm-km. Save for the very different lengths of the two segments, however, the pulse behavior in that case is just like that shown in Fig. 2.1.

Finally, it will be undoubtedly noted that we have not yet discussed the effects of fiber loss, nor have we discussed schemes for the placement of amplifiers with respect to the dispersion map. Although it is common engineering practice to place amplifiers (or points for injection of pump power for Raman gain) at the end of each map period, that need not necessarily be the arrangement. We shall discuss these matters more fully in a later section.

2.1.2. *Why Dispersion Management?*

Dispersion management was invented to meet certain needs of dense WDM [14, 36–38]. In the first place, four-wave mixing between adjacent channels (a potentially most harmful effect, since it tends to cause severe amplitude and timing jitter) is efficiently repressed by the large phase mismatch provided by the large $|D_{local}|$ values of the individual fiber segments. For example, consider a system with the parameters $D_{local} \approx 6$ ps/nm-km and $\Delta f = 50$ GHz (typical for a 10-Gbit/s per channel system). Equation (4.19) then yields $\Delta k = 0.76$/km. Because of this large phase mismatch, the E field of the four-wave mixing product spirals rapidly in tight circles in the complex plane, and hence cannot grow to significant size. [Note that, in this example, the circle is completed (and hence nearly closes on itself, especially in Raman-amplified systems) once every $2\pi/0.76 = 8.2$ km.]

Second, fibers having the rather small D values needed for ordinary solitons, constant over the wide wavelength bands required for dense WDM, simply do not exist, and probably never will. For dispersion management, however, it is possible to use *combinations* of fiber for which the path-average dispersion is nearly constant. That is, as can be seen from the curves of Fig. 1.1, the dispersion parameters of most fibers can be fairly represented over a typical WDM band by the linear approximation $D(\lambda) = D(\lambda_0) + S \times (\lambda - \lambda_0)$, where λ_0 is a wavelength in the middle of the band and where $S = \partial D/\partial\lambda$ there. Clearly, if the ratio S/D (or D/S) is the same for the D^+ and D^- fibers of the map, the (nearly zero) net dispersion will tend to be constant. Several examples of fiber pairs permitting this extra degree of freedom in map design are shown in Table 2.1.

Figure 2.2 shows the variation in path-average dispersion parameter typically obtained from such fiber combinations. Note that the variation in $\bar{D}$ for a 50-nm-wide band (sufficient for 125 channels at 50 GHz/channel), centered about the peak in the curve, is only about $\pm 27\%$ of the median value for that band. This residual variation results primarily from fourth-order dispersion.

TABLE 2.1 Parameters of Some Fiber Types Suitable for Use in Dispersion-managed WDM Transmission

Fiber Type[a]	D (ps/nm-km) (at 1555 nm)	S(ps/nm^2) ($\partial D/\partial\lambda$)	D/S (nm)	A_{eff} (μm^2)
Standard	+16.7	+0.056	298	80
Low-slope DS	+6.6	+0.045	147	50
IDF	−17.7	−0.057	310	35
High-slope DCF	−105	−0.35	300	20
Ultra-slope DCF	−115	−0.78	147	18

[a]Note that the combination of standard fiber with either IDF or high-slope DCF on the one hand, or of low-slope DS with ultra-slope DCF on the other, should allow for nearly constant $\bar{D}$.

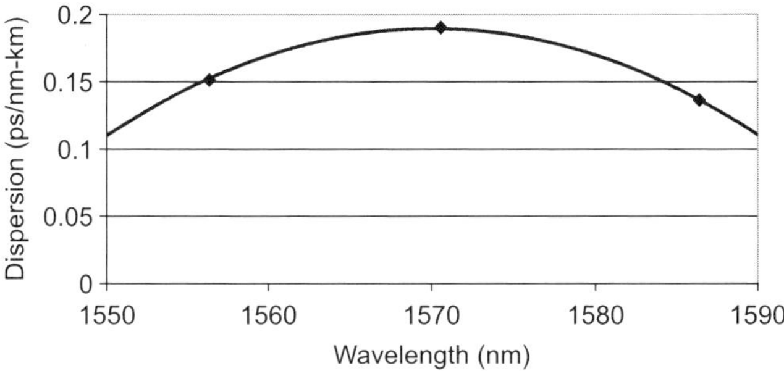

FIGURE 2.2 Measured path-average dispersion parameter ($\bar{D}$) for a 100-km-span low-slope DS fiber, compensated with an ≈5.6-km-long coil of matching DCF (see Table 2.1), as a function of wavelength. The data make an excellent fit to a shallow parabola whose peak is at the center of the intended WDM band.

2.1.3. *Why Dispersion-managed Solitons?*

It will be noted that, thus far, the cited needs of dense WDM to be met by dispersion management do not necessarily require dispersion-managed *solitons*. But when the goal is to provide the backbone of an all-optical network, then the periodicity of the solitons' behavior, unique to them, becomes vital. The many important issues surrounding the periodic nature of soliton transmission can be best discussed in terms of Fig. 2.3. Although the choice is arbitrary, it is convenient to let the map periods begin and end at the unchirped pulse positions in each coil of DCF, as in Fig. 2.3. Note that the accumulated linear dispersion values shown there are the discrete values obtaining at the end of each period, so that they correspond to the product of $\bar{D}$ and the particular transmission distance. Note further that the pre-compensation ("pre-comp") coil is really an integral part of the first map period,

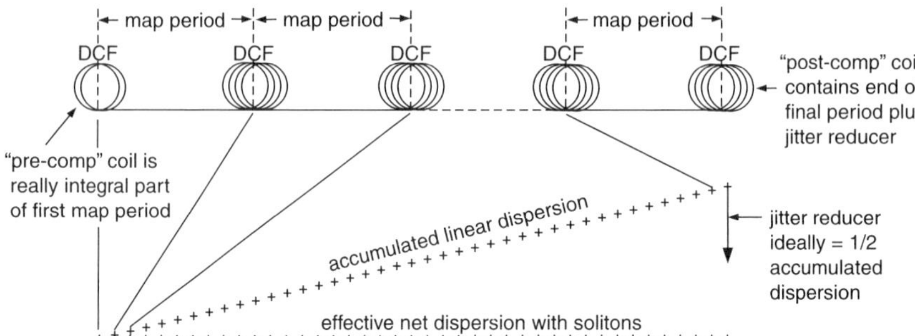

FIGURE 2.3 Dispersion compensation in a DMS system. In this view, the map periods begin and end at the unchirped pulse positions in each coil of DCF; the accumulated linear dispersion values shown here are those discrete values obtained at the end of each such period. Note that the pre-comp and post-comp coils are really just integral parts of the map periods (see text).

and that the post-compensation ("post-comp") coil, save for an additional "jitter reducer," is likewise an integral part of the final map period. The jitter reducer ideally represents a dispersion equal to $-1/2$ of the accumulated linear dispersion, but the exact value is not at all critical, and thus, in practice, it can be set to the best value for the longest distance to be encountered in the system. Finally, note that the "effective net dispersion" for solitons is always zero at the end of each map period. This scheme of dispersion compensation has the following important consequences:

1. The pulse parameters (temporal width, bandwidth, energy, chirp, etc.) are identical at the end of each map period, and the pulses are always well resolved from each other in time. This periodic behavior in turn means that:

   (a) The data can be read instantly anywhere, or at least at the end of any map period.

   (b) Standard (unchirped, minimum width) soliton pulses can be injected anywhere (i.e., at the beginning of any map period).

2. The pre- and post-comp dispersion values are independent of distance.

These properties are exactly as required for the creation of an all-optical network, and for efficient, inexpensive system monitoring. Their compatibility with the use of standard parts (pre- and post-comp coils) are also very important for the reduction of system cost and for the ease of system assembly. Once again, these properties are uniquely supplied by dispersion-managed solitons.

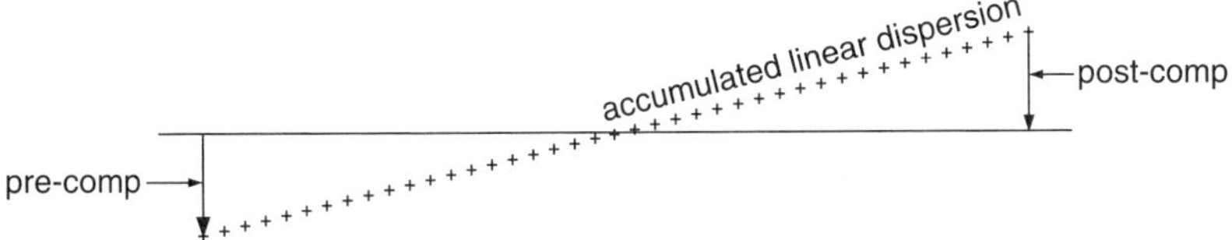

FIGURE 2.4 Dispersion compensation in a non-DMS system (see text).

It is instructive to look at the dispersion compensation scheme most often used in non-DMS systems. There it is common to use much greater pre-comp dispersion, so that the accumulated linear dispersion tends to pass through zero somewhere near the halfway point of the net transmission distance (see Fig. 2.4). This scheme represents an attempt to reduce cross-phase modulation (XPM) from interchannel collisions by greatly broadening the pulses and thus making their peak intensities lower over at least most of the path. Unfortunately, however, that action simultaneously greatly increases nonlinear penalties from certain intrachannel effects, such as adjacent-pulse interaction and intrachannel four-wave mixing. [The four-wave mixing tends to produce ghost pulses in the positions of "zeros" (bit slots where there are no pulses) by transferring energy from adjacent ones [39].] Furthermore, in strong contrast to solitons, most of the accumulated linear dispersion is not compensated by self-phase modulation. In consequence, one has the following facts:

1. Over much of the path, the pulses are strongly overlapped, so that the data are not immediately readable.
2. The pre- and post-compensation dispersion values must be carefully tuned for each distance. (The total of pre- and post-dispersion compensation required is roughly proportional to the total distance.)
3. Even for a fixed distance, dispersion tends to make it impossible to properly compensate all wavelengths of a wide WDM band with just one set of pre- and post-compensation coils.

These facts argue strongly against the creation of an all-optical network and efficient, inexpensive system monitoring!

2.1.4. A Shortcut for Computing DMS Behavior

Thus far, the discussion of dispersion-managed solitons has been largely qualitative. For real system design, however, we must compute exact pulse behavior, often for many different possible dispersion maps, amplifier span gain profiles, and initial pulse parameters. To do all of this computation by exact numerical solution of the NLS equation is tedious and time consuming. One can create an efficient shortcut, however, by taking advantage of the fact that, as already stated, in a DMS system, the pulse shape is Gaussian to a very good approximation. That is, by applying that assumed pulse shape, or "Ansatz," to the NLS equation, one can create an equivalent set of ordinary differential equations, or ODEs, that are much easier and faster to solve. Although several other ODE (largely variational [40–42]) approaches have been used by others, the nonvariational ODE method [43] we describe here is especially efficient and easy to understand.

We write the (Gaussian) signal pulse in the following general form:

$$u(t) = \sqrt{W}(\eta/\pi)^{1/4} \exp\left[-\frac{1}{2}(\eta + i\beta)t^2\right], \tag{2.1}$$

where $1/\sqrt{\eta}$ is a measure of the pulse width, and β is the chirp parameter. Let η_0 refer to the unchirped pulse, i.e., $\eta = \eta_0$ when $\beta = 0$. Clearly, if we know the complex number $\eta + i\beta$, and the pulse energy W, we then know all of the pulse properties. In particular, for the pulse width in time, we have

$$1/\sqrt{\eta} = \tau/\sqrt{4 \ln 2} = \tau/1.6651\ldots, \tag{2.2}$$

where τ is the intensity FWHM. The phase and frequency shifts across the pulse are, respectively,

$$\phi(t) = -\frac{1}{2}\beta t^2 \tag{2.3a}$$

and

$$\delta\omega(t) = -\beta t. \tag{2.3b}$$

Finally, the spectrum of the pulse [the Fourier transform, $\tilde{u}(\omega)$, of Eq. (2.1)] yields the spectral intensity

$$|\tilde{u}|^2 \propto \exp\left[-\eta\omega^2/\left(\eta^2 + \beta^2\right)\right], \tag{2.4}$$

which has an FWHM of

$$\Delta f = (1.6651\ldots/2\pi)\sqrt{(\eta^2 + \beta^2)/\eta}. \tag{2.5}$$

Thus, the time–bandwidth product is $\tau\Delta f = 0.441\sqrt{1 + (\beta/\eta)^2}$.

By applying the above Ansatz to the NLS equation, we get the following complex ODE (really a pair of coupled ODEs) for η and β:

$$\frac{dq}{dz} = i[1 - Kq^2(\Re(1/q))^{3/2}], \tag{2.6}$$

where

$$q(z, K) = \eta_0/(\eta + i\beta). \tag{2.7}$$

The distance z is always measured in units of the characteristic dispersion length, which, for a Gaussian pulse, is

$$z_c = \frac{1}{4\ln 2}\frac{2\pi c}{\lambda^2}\frac{\tau_0^2}{D}. \tag{2.8}$$

Note that, save for the leading numerical coefficient (which makes it about 1.12× longer), this "Gaussian" dispersion length is the same as the "sech" dispersion length of Eq. (1.18). The nonlinear coefficient K is calculated as

$$K = \frac{(2\pi)^2}{\sqrt{2}\,4\ln 2}\frac{n_2 c}{\lambda^3 A_{eff}}\frac{\tau_0^2}{D}P = \frac{P}{P_c}, \tag{2.9}$$

where A_{eff} is the fiber core area, τ_0 refers to the unchirped pulse, P is the peak pulse power, and P_c is the peak power of an ordinary soliton of pulse width τ_0 in fiber of the (local) dispersion parameter D. Note that $K = 1$ corresponds to ordinary solitons (although the Gaussian pulse shape is not quite right in that case), and that dispersion-managed solitons tend to correspond to $|K| \ll 1$. Note also that z_c, K, and P_c are negative when D is negative. Although this convention and, in particular, the concept of a negative dispersion length and negative unit power, may seem strange at first, it is self-consistent, and avoids a certain awkwardness that would occur without it.

If we let $z = 0$ correspond to the unchirped pulse ($\eta = \eta_0, \beta = 0$), then $q(0, K) = 1$. The solution to Eq. (2.6) then has the general form

$$q(z, K) = 1 + iz + Kf(z, K). \tag{2.10}$$

Note that the linear solution $q = 1 + iz$ is the well-known solution for a Gaussian pulse subject to pure dispersion, as already discussed in Section 1.3.5 [see Eq. (1.26)]. For $K \neq 0$, a numerical solution is usually required to obtain the quantity

$$q(z, K) - 1 - iz = Kf(z, K), \tag{2.11}$$

the so-called nonlinear residue.

Note that loss and the effects of Raman gain can be included by writing

$$K(z) = K_0 I(z), \tag{2.12}$$

where $I(z)$ is a normalized intensity or energy profile.

Writing the real and imaginary parts of q as x and iy, respectively, the pulse width, bandwidth, and chirp parameter can be obtained from them as follows: First, the pulse width in time is

$$\frac{1}{\sqrt{\eta}} = \frac{1}{\sqrt{\eta_0}}\sqrt{\frac{x^2+y^2}{x}} \qquad \left(\tau = \tau_0\sqrt{\frac{x^2+y^2}{x}}\right), \tag{2.13}$$

while the bandwidth is

$$\delta\omega = \sqrt{\frac{\eta^2+\beta^2}{\eta}} = \sqrt{\frac{\eta_0}{x}} \qquad \left(\Delta f = \frac{1.665}{2\pi}\sqrt{\frac{\eta_0}{x}}\right) \tag{2.14}$$

and the chirp parameter is

$$\beta = -\frac{\eta_0 y}{x^2+y^2}. \tag{2.15}$$

Finally, for $\lambda = 1550\,\text{nm}$ and $n_2 = 2.6 \times 10^{-16}\,\text{cm}^2/\text{W}$, the pulse energy is

$$W\,(\text{fJ}) = 2548.2\frac{50}{A_{eff}\,(\mu\text{m}^2)}\frac{D\,(\text{ps/nm-km})}{\tau_0\,(\text{ps})}K. \tag{2.16}$$

An efficient computer program based on the Maple mathematics package, to produce ODE solutions as outlined herein, is reproduced in Appendix A. From input data consisting of details of the dispersion map, Raman pumping conditions, and input pulse parameters, the program first calculates the signal energy profile $I(z)$ and, from that, $K(z)$. It then uses the Maple program "dsolve" to obtain solutions of Eq. (2.6), and, finally, it graphs the various pulse parameters as functions of z. The program is very efficient, so that on a reasonably fast personal computer, one can obtain a full picture of pulse behavior in a given map in just a matter of minutes. Thus, the program has proved to be a very useful engineering tool, enabling the exploration of system performance over a wide range of map designs and pulse parameters in a relatively short time. It has also engendered a deeper understanding of the fundamentals of dispersion-managed soliton transmission. Much of the pulse behavior presented in this chapter was computed with this program.

2.1.5. *Pulse Behavior in Lossless Fiber*

Before going on to study pulse behavior in"real" dispersion maps, where loss and gain tend to complicate matters, it is instructive to survey behavior in lossless fiber. Accordingly, Figs. 2.5–2.7 plot the most important pulse parameters (pulse width, bandwidth, and chirp) as functions of distance (normalized to $|z_c|$), with

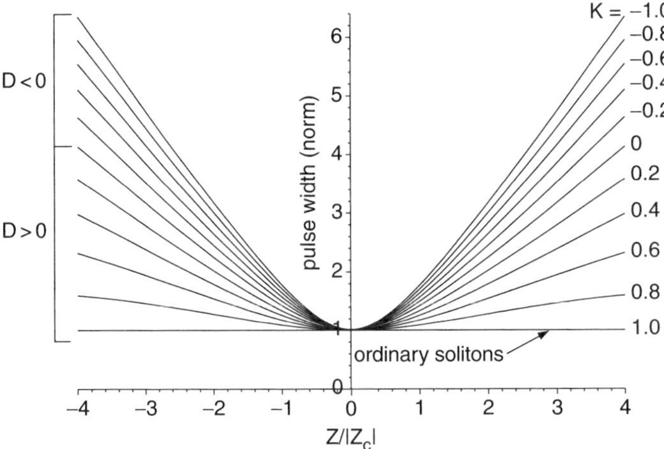

FIGURE 2.5 Pulse width (normalized to that of the unchirped pulse) in lossless fiber as a function of distance for various values of the nonlinear parameter K. Note that the pulse broadening in time, driven by the dispersive term, undergoes a monotonic reduction as K ranges from -1 to 1.

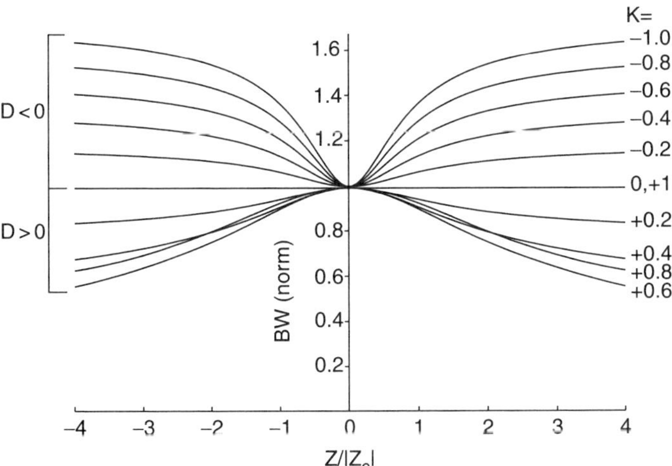

FIGURE 2.6 Pulse bandwidth (normalized to that of the unchirped pulse) in lossless fiber as a function of distance, for various values of the nonlinear parameter K. The change in bandwidth is, of course, driven by the nonlinear term. Indeed, note that for $D < 0$, the rate of growth in bandwidth with z increases monotonically with increasing $|K|$. For $D > 0$, however, the bandwidth declines, and the rate of decline increases with increasing K only until $K \sim 0.6$, when it begins to diminish, reaching 0 at $K = 1$. Finally, for $K > 1$, the bandwidth once again begins to grow with z.

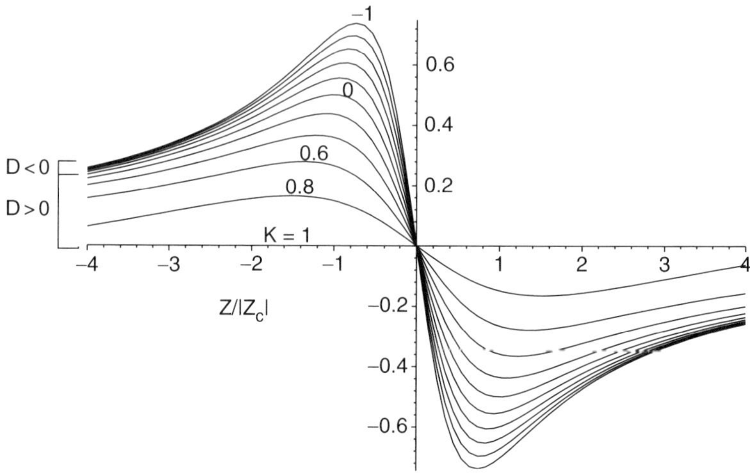

FIGURE 2.7 Chirp parameter (normalized to η_0) in lossless fiber as a function of distance, for various values of the nonlinear parameter K. Note that beginning with the unchirped pulse, the chirp at first increases almost linearly with distance (as the frequency components of the pulse just begin to separate), then peaks and declines as the separation becomes complete (when the range of frequencies is spread out over ever greater time).

the nonlinear coefficient K as the parameter. Note that for all three parameters, the dispersive and nonlinear effects always enhance each other for $D < 0$, but tend to diminish each other for $D > 0$. And of course, for the ordinary soliton ($K = 1$), the dispersive and nonlinear effects disappear altogether.

It should be noted that these plots were computed by the ODE method of the previous section, and hence are based on the Gaussian Ansatz. Thus, while they are accurate, at least for $|K| \ll 1$, they represent a certain degree of approximation for large $|K|$. Nevertheless, they give a rather complete and insightful picture, over the entire range $-1 \leq K \leq 1$, of the interaction of the dispersive and nonlinear terms of the NLS equation. Therefore, it is recommended to study them thoroughly and with careful attention to detail.

2.1.6. *Adjacent-pulse Interaction*

There is one more important general aspect of pulse behavior in dispersion maps to be considered before we can go on to a detailed examination of specific maps, and that is the potential for an interaction between adjacent pulses in a data stream, mitigated by excessive pulse breathing. That is, when adjacent pulses overlap, each can induce a frequency shift on the other through cross-phase modulation (see Chapter 4, Section 4.1.1). The interaction can perhaps best be understood

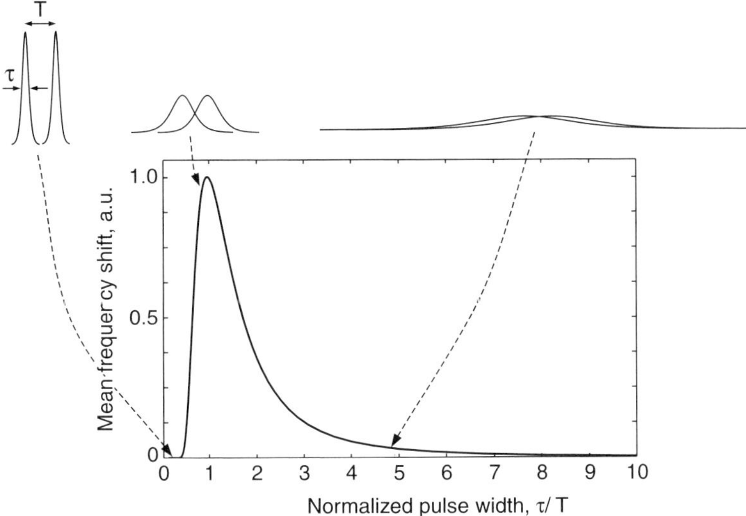

FIGURE 2.8 Adjacent-pulse interaction (frequency shifts induced by cross-phase modulation) as a function of pulse width normalized to the bit period T.

from the middle inset of Fig. 2.8. Consider, for example, the action of the rising intensity envelope of the later of the two overlapping pulses on the earlier one. The corresponding rising index change leads to the growth of a time-dependent phase shift of the earlier pulse. Since the induced phase shift increases with time across the pulse, it corresponds to a negative frequency shift. In like manner, the intensity envelope of the earlier pulse induces a positive frequency shift on the later pulse. The fiber's dispersion then converts these frequency shifts into time shifts. Since the effect tends to grow slowly, only the path-average dispersion matters in the long run. Since, for dispersion-managed solitons, one always has $\bar{D} > 0$, the net effect is that of an effective attraction between the two pulses in time.

Figure 2.8 shows how the interaction varies with the pulse width τ, as normalized to the bit period T. The calculation for the graph assumes constant pulse widths, and hence does not average the interaction properly over the range of pulse widths afforded by the pulse breathing. Thus, in particular, it tends to greatly overestimate the effect in the limit of $\tau/T \leq 0.5$. Also, it does not take into account interference between the two pulses. It should be noted that constructive interference in the region of overlap (the usual case, i.e., where the adjacent pulses are nominally in-phase) enhances the attraction, while destructive interference in the region of overlap produces a repulsion. [Later, in Chapter 4, we shall see how these interference effects are strongly operative in the case of ordinary solitons; in particular, see Eq. (4.58) and the discussion immediately following it.] When dispersion-managed solitons overlap only slightly, however, they tend to be highly chirped, so that the relative phases of the overlapping parts vary rapidly in (retarded) time. Thus, in that case, interference effects tend to be washed out. Despite all of these caveats, the graph of Fig. 2.8 shows the general trend rather well, viz., that while the interaction is negligible or at least very small for $\tau/T \leq 0.5$, it grows to a maximum in the neighborhood of $\tau/T \sim 1$, and then declines again for $\tau/T > 1$.

In any event, to avoid the error-inducing time displacements that may result from the adjacent-pulse interaction, it is wise to restrict the maximum pulse width in breathing to $\tau_{max} \leq T/2$. The breathing is so restricted in all of the examples to follow in this chapter.

Incidentally, it should be noted that the other extreme, where $\tau/T \gg 1$, is often called the "quasi-linear" regime, from the somewhat misguided idea that the resultant low peak intensities of the pulses will significantly reduce unwanted nonlinear effects. While that may be true for isolated pulses, the simultaneous overlap and interference of many data pulses from a real data stream will result in high intensities and serious nonlinear effects.

2.2. Pulse Behavior in Maps Having Gain and Loss

We are now in a position to study the behavior of pulses in maps having gain and loss and other "real-world" properties. In particular, in the following sections we shall study behavior in a few practical maps as a function of those parameters that can be changed once the span lengths, fiber types, and amplification scheme of a map have been fixed, viz., the unchirped pulse width τ_0 and the path-average dispersion $\bar{D}$.

2.2.1. *A Prototypical Real-world Map*

The dispersion map shown in Fig. 2.9 [44] corresponds, with one exception, to that used in an actual, commercial, ultra-long-haul dense WDM transmission system. The mid-span, backward Raman pumping, used here to produce a nearly perfect bilateral symmetry in the signal intensity profile, is not used commercially for reasons of economy. Nevertheless, it is a very good map to display all of the significant features. As will be shown later, the changes that occur when the commercial pumping scheme is used are not particularly significant. In the following sections, we shall study the pulse behavior in this prototypical map thoroughly.

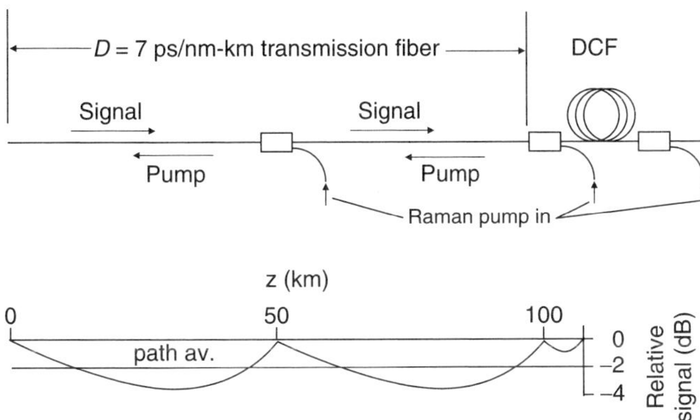

FIGURE 2.9 Top: The prototypical dispersion map and pumping scheme for Raman gain. The little rectangular boxes represent WDM couplers for the efficient injection of Raman pump light at about 100-nm shorter wavelength than that of the signals. Bottom: The signal energy profile resulting from the Raman gain. Note its nearly perfect bilateral symmetry about the middle of the main span.

2.2.2. *Pulse Behavior in the Prototypical Map for Optimum Parameters*

Figures 2.10–2.12 show pulse behavior (pulse width, bandwidth, and chirp) as functions of distance, for fixed, "optimum" values of the unchirped pulse width and of the path-average dispersion ($\tau_0 = 30$ ps and $\bar{D} = 0.15$ ps/nm-km, respectively), in the prototypical map of Fig. 2.9.

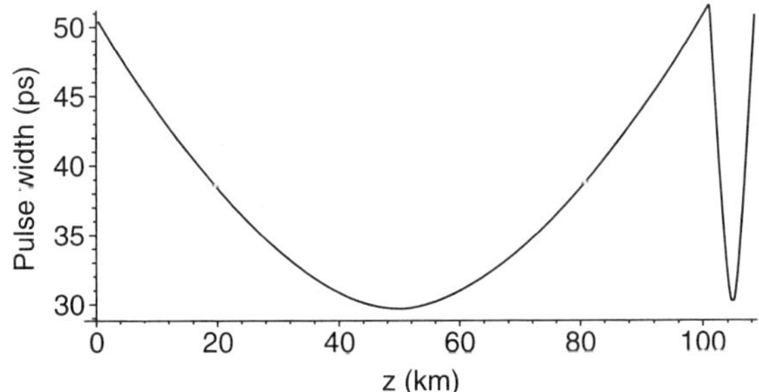

FIGURE 2.10 Pulse width τ vs. z in the map of Fig. 2.9, for $\tau_0 = 30$ ps and $\bar{D} = 0.15$ ps/nm-km. Note that the maximum pulse width, τ_{max}, does not significantly exceed 50 ps, or half the bit period for a transmission rate of 10 Gbit/s.

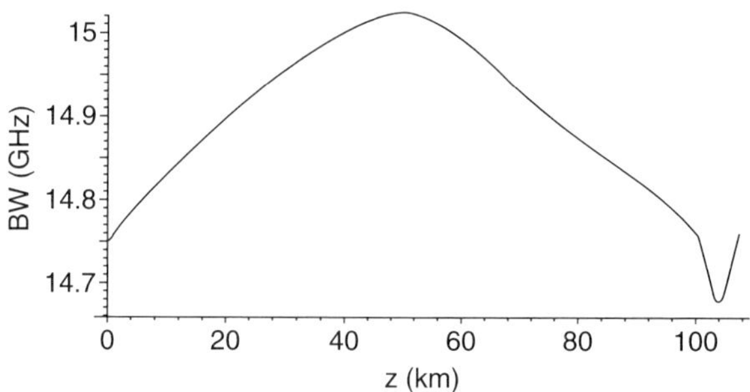

FIGURE 2.11 Bandwidth vs. z in the map of Fig. 2.9, for $\tau_0 = 30$ ps and $\bar{D} = 0.15$ ps/nm-km. Note the much smaller fractional breathing of the BW versus that of τ shown in Fig. 2.10. Note also that just as was seen in Fig. 2.6, the BW is at a maximum at the unchirped pulse position in the D^+ section, while it is at a minimum in the corresponding position in the D^- section.

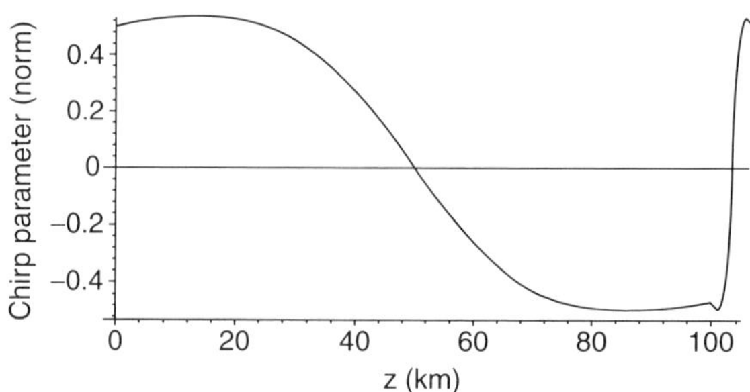

FIGURE 2.12 Chirp vs. z in the map of Fig. 2.9, for $\tau_0 = 30$ ps and $\bar{D} = 0.15$ ps/nm-km. Note that the positions of zero chirp correspond exactly to the maxima and minima of the pulse width and pulse bandwidth as seen in the two previous figures.

2.2.3. *Pulse Behavior as a Function of Map Strength*

Many important things can be learned from a study of the pulse behavior in a map as the unchirped pulse width is varied. The study is conveniently carried out in terms of the map strength parameter, defined as

$$S_{map} = (L^+/z_c^+ + |L^-/z_c^-|)/2, \tag{2.17}$$

where z_c^+ and z_c^- are the characteristic dispersion lengths [Eq. (2.8)] for the unchirped pulses in the D^+ and D^- segments of the map, respectively. Note that when the two terms of Eq. (2.17) are nearly equal (the usual case), S_{map} is essentially just the length of the transmission fiber, as measured in dispersion units. Note further that S_{map} scales like $1/\tau^2$ [or, equivalently, as the $(BW)^2$] of the unchirped pulses. Serious dispersion management usually involves $S_{map} > 1$. On the other hand, for $S_{map} \ll 1$, where there is no significant pulse broadening, note that one just has path-average solitons.

Figure 2.13 shows the degree of pulse breathing in time as a function of S_{map}. Since the pulse breathing is strongly dominated by the dispersive term of the NLS equation, the behavior here is expected to be very similar to that shown in Fig. 1.2. Indeed, note the very close fit to the function $\sqrt{1 + (S_{map}/2)^2}$, until the ever-rising nonlinear contribution becomes significant for the larger values of S_{map}.

Figure 2.14 shows the degree of pulse breathing in bandwidth as a function of S_{map}. As discussed earlier, the degree of breathing here is much smaller (by

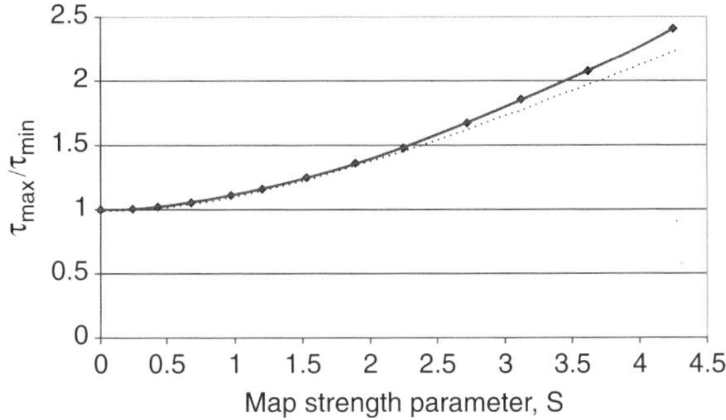

FIGURE 2.13 Degree of pulse breathing as a function of S_{map} for dispersion-managed solitons in the map of Fig. 2.9. $\bar{D}$ is fixed at 0.15 ps/nm-km. From right to left, the specific data points correspond to unchirped pulse widths (τ_0) of 24, 26, 28, 30, 33, 36, 40, 45, 50, 60, 75, and 100 ps, respectively. The faint dotted line is the function $\sqrt{1 + (S/2)^2}$.

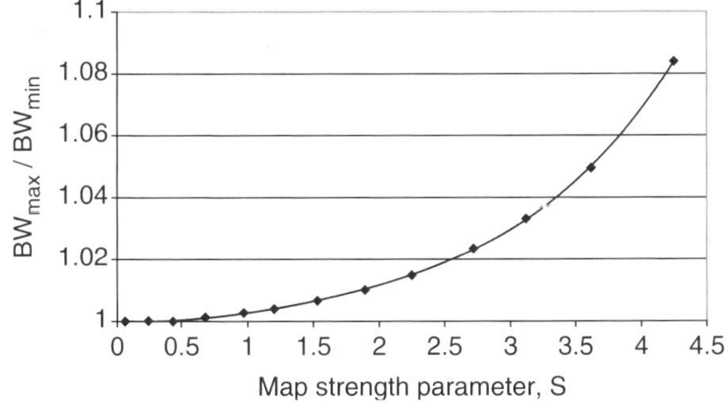

FIGURE 2.14 Degree of pulse breathing in bandwidth as a function of S_{map} for dispersion-managed solitons in the map of Fig. 2.9. Again, $\bar{D}$ is fixed at 0.15 ps/nm-km, and the specific data points have the same identity as in Fig. 2.13.

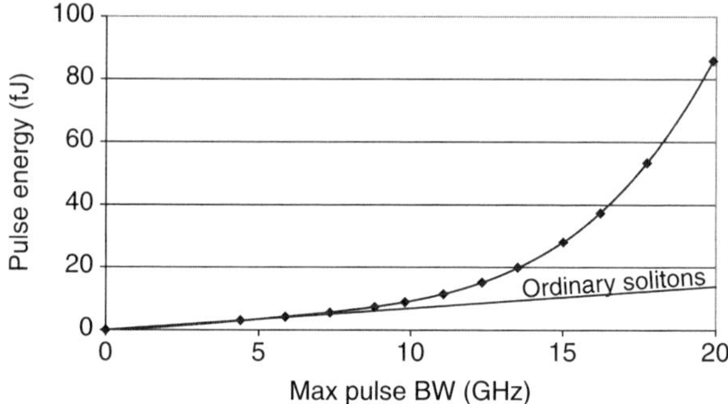

FIGURE 2.15 Path-average pulse energy as a function of pulse bandwidth for dispersion-managed solitons in the map of Fig. 2.9, with $\bar{D}$ once again fixed at 0.15 ps/nm-km. (The specific data points have the same identity as in Fig. 2.13.) The straight line shows the energy of ordinary solitons in a map of constant $D = 0.15$ ps/nm-km.

factors of ~20× or more) than that of Fig. 2.13, due to the much smaller size of the nonlinear term in relation to that provided by the local dispersion. Also, note the different shape of the curve, compared to that of Fig. 2.13. Here the data make approximate fit to a simple quadratic.

Figure 2.15 shows the path-average pulse energy for dispersion-managed solitons in the map of Fig. 2.9. Here the data are plotted as a function of the (maximum) pulse bandwidth, in order to facilitate comparison with the energy of ordinary solitons having the same path-average dispersion. ($\bar{D} = 0.15$ ps/nm-km.) The ever-increasing energy of the dispersion-managed solitons with respect to that of the ordinary solitons as the pulse bandwidth increases is the energy enhancement effect mentioned in Section 2.1.1. Figure 2.16 shows the energy enhancement factor that can be inferred from Fig. 2.15, plotted as a function of S_{map}. The good fit to the function $1 + 0.22S^2$ until $S > 3$ is in essential accord with that found in earlier fitting to numerical work [15–18].

Thus far, the value of $\bar{D}$ has been kept at a fixed value (0.15 ps/nm-km). It is useful, however, to see how the energy of a dispersion-managed soliton behaves as $\bar{D}$ changes. Figure 2.17 shows the path-average energy as a function of $\bar{D}$, with the unchirped pulse width as a parameter. Note how a particular pulse energy can be obtained from a much lower $\bar{D}$ with dispersion-managed solitons than with ordinary solitons of the same pulse width. From the figure it can be seen, for example, that ~30-fJ pulses of 30-ps width require $\bar{D}$ of only ~0.15 ps/nm-km when they are dispersion-managed solitons, while the corresponding ordinary solitons require $D \sim 0.5$ ps/nm-km for the same energy. This is another consequence of the energy enhancement effect. It is of great importance, since the reduction in $\bar{D}$ means a corresponding reduction in timing jitter from frequency deviations.

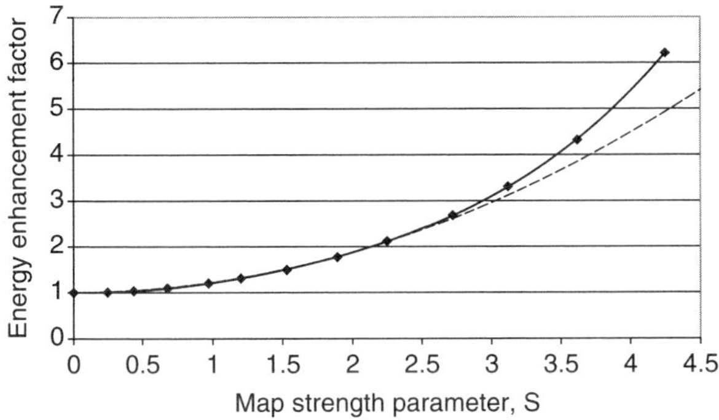

FIGURE 2.16 Energy enhancement factor for dispersion-managed solitons in the map of Fig. 2.9, with $\bar{D}$ fixed at 0.15 ps/nm-km, plotted as a function of S_{map}. (The specific data points have the same identity as in Fig. 2.13.) The dashed line represents the function $1 + 0.22S^2$.

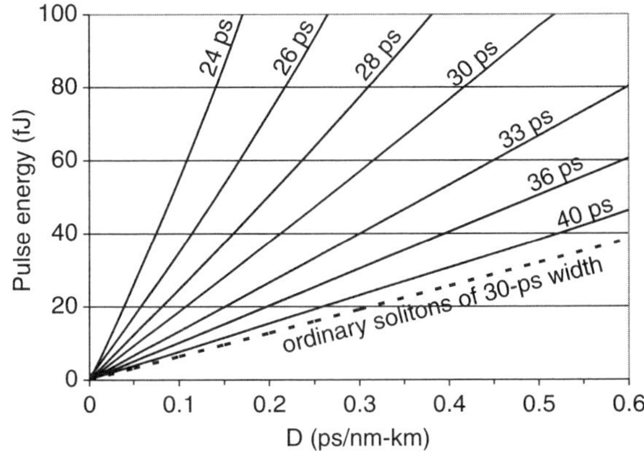

FIGURE 2.17 Path-average energy of dispersion-managed solitons in the map of Fig. 2.9, as a function of $\bar{D}$, with the unchirped pulse width as a parameter. The dashed line shows the corresponding energy of ordinary solitons of 30-ps pulse width in a map of constant $D = \bar{D}$.

2.2.4. *Pulse Behavior in a Map with Asymmetric Intensity Profile*

If we move the mid-span WDM pump coupler in Fig. 2.9 to the far left end of the 100-km span and turn it around such that the pump light for Raman gain travels with the entering signal, then the resulting 50% forward/50% backward Raman pumping leads to the signal intensity profile shown in Fig. 2.18. Note that the bilateral symmetry of the intensity profile shown in Fig. 2.9 has been destroyed, since the average intensity in the left half of the 100-km span is now

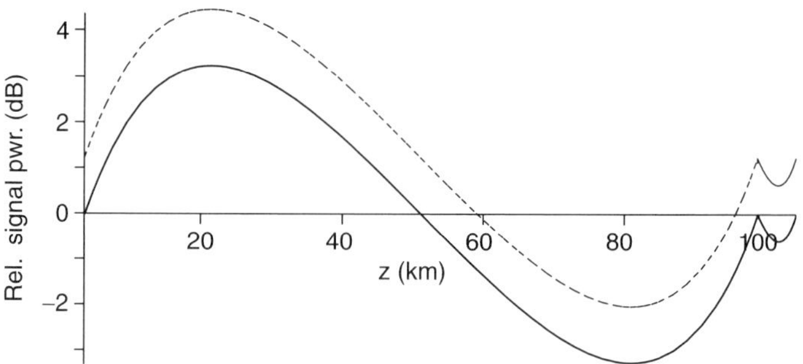

FIGURE 2.18 Signal intensity profile resulting when a 50% forward/50% backward Raman pumping scheme is used for the 100-km span of Fig. 2.9. The solid line shows the normal case, and the dashed line shows the increased energy required for dispersion-managed solitons when a 41-GHz FWHM Gaussian filter is inserted at the output end of the map.

several decibels greater than that in the right half. The resultant curves of pulse width versus distance are shown in Fig. 2.19. Note that without the filter, pulse breathing is also no longer symmetric, with the position of the minimum width moved toward the higher intensity region of the map, while the presence of the filter tends to restore the symmetry of the breathing [45].

The pulse width behavior of Fig. 2.19 can be understood by reference to Fig. 2.20, which shows the corresponding curves of pulse bandwidth versus distance. The explanation lies in the fact that the rate of change of bandwidth with distance increases monotonically with increasing intensity, so the rate of increase

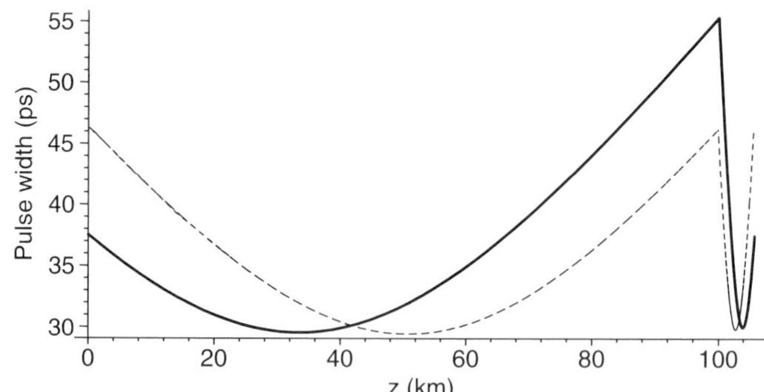

FIGURE 2.19 Pulse width vs. distance for dispersion-managed solitons in the map of Fig. 2.9 with the asymmetric intensity profile of Fig. 2.18. (The solid and dashed curves refer to the cases without and with filter, respectively, as in Fig. 2.18.) Note the resultant asymmetry in the breathing for the normal case, and the restoration of symmetry by the insertion of the filter.

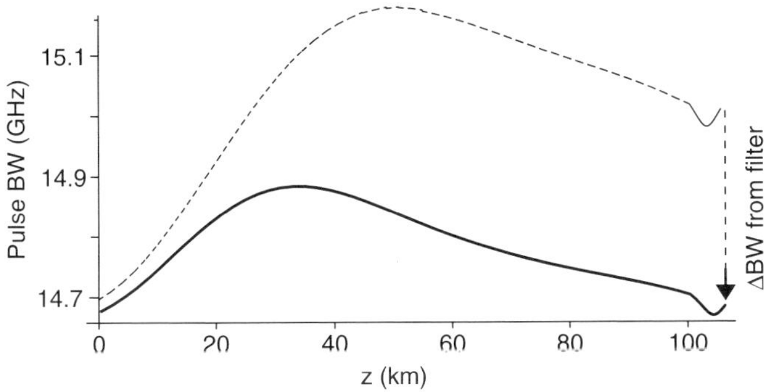

FIGURE 2.20 Pulse bandwidth vs. distance for dispersion-managed solitons in the map of Fig. 2.9 with the asymmetric intensity profile of Fig. 2.18, where, once again, the solid and dashed curves refer to the cases without and with the filter, respectively.

in the left half of the map (with greater intensity) is greater than the rate of decrease in the right half of the map (with lower intensity). Thus, in the normal case (without the frequency filter), since the bandwidth must return at the end of the map to its original value, the maximum bandwidth (and hence the minimum pulse width) must occur before the middle of the map. (Remember that the two extremes must occur in exactly the same place.) On the other hand, the step reduction in bandwidth provided by the filter forces the curve of bandwidth to tilt upward such that the position of maximum bandwidth (and correspondingly, the position of minimum pulse width) is restored to the center of the map.

Frequency filters like the one we have used here are sometimes employed to reduce noise-induced frequency, timing, and amplitude jitter in soliton transmission systems; as such they are known as "frequency-guiding filters" or simply "guiding filters." We shall discuss guiding filters at greater length in Chapter 3, Section 3.5. In the meantime, the asymmetric pulse breathing obtained in the normal case, while not ideal, usually does not have very serious consequences. This is true in part because of the fact that the maximum pulse width always corresponds to the region of lowest intensity, so that the adjacent-pulse interaction is correspondingly reduced.

2.3. Map Scaling to Higher Bit Rates

All of the real-world examples of pulse behavior we have examined in Sections 2.2.2–2.2.4, with a minimum pulse width τ_0 of $\sim$30 ps and $\tau_{max} \sim 50$ ps, refer to a bit rate of $\sim$10 Gbit/s. Since τ_0 scales inversely with the bit rate, and

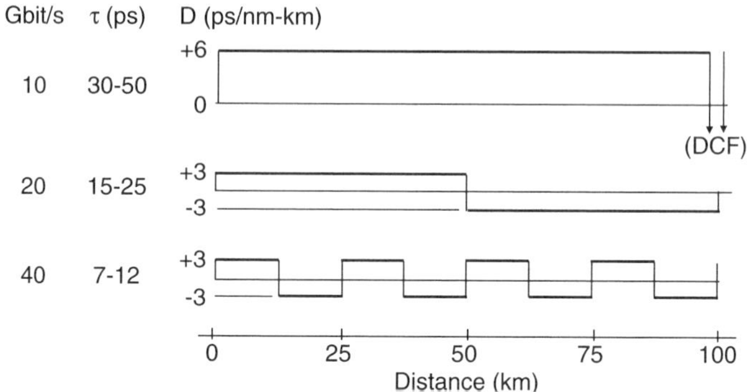

FIGURE 2.21 Illustration of map scaling to higher bit rates. The two pulse widths shown for each bit rate are τ_0 and τ_{max}, respectively (for other details, see text).

since z_c scales as $\tau_0{}^2/D_{loc}$, to maintain the same pattern of pulse breathing as the bit rate is increased, the combination of segment lengths L_{seg} and corresponding D_{loc} values must be scaled such that L_{seg}/z_c remains constant. (Note that this is the same as saying that the map strength S_{map} must be held constant.) A few examples of how such scaling might be carried out are shown in Fig. 2.21.

The map shown in Fig. 2.21 for 10 Gbit/s is just the prototypical map of Fig. 2.9. For 20 Gbit/s, the scheme shown is to use a negative D transmission fiber, with $|D^-| \cong D^+$, for compensation, rather than DCF, and to make $|D_{loc}|$ just half of the former value. Thus, z_c and L_{seg} are each reduced by the same factor (2) in achievement of the desired scaling. For a number of practical reasons, not the least of which is the fact that it is difficult to manufacture dispersion-shifted fibers without a certain degree of wander with distance in D itself, it is probably not a good idea to reduce $|D_{loc}|$ to much less than about 3 ps/nm-km, in order to keep the fractional changes in D_{loc} within reasonable bounds. Thus, for the transition from 20 to 40 Gbit/s, the necessary scaling is best achieved simply by a further reduction in L_{seg} by a factor of 4. When the dispersion parameter changes sign as often as in our example for 40 Gbit/s, the result is often referred to as "dispersion-managed cable." Since fiber cables tend to be delivered and installed in segments no longer than about 6 km each, which then have to be spliced together in the field, the installation of such dispersion-managed cables should not be at all impractical.

2.4. Dispersion-managed Solitons: Summary

By now the reader should have a rather good understanding of dispersion management and dispersion-managed solitons. Although dispersion-managed solitons cannot in general be treated analytically, as we have seen here, there are three basic principles that govern their behavior: First, the dispersive term is locally dominant, so that the pulse shape (Gaussian) and pulse breathing are accordingly very close to that expected from linear dispersion alone. Second, the pulse behavior becomes periodic with the period of the dispersion map (and hence solitons exist) when there is just enough nonlinear phase shift across the pulse to cancel the net phase shift from the dispersive term, as averaged over the entire map period. Third, periodic variation in the pulse bandwidth, though small relative to the breathing in pulse width, nevertheless tends to greatly magnify the net phase shift from the dispersive term; this in turn requires the canceling nonlinear phase shift, and hence the soliton pulse energy, to be similarly magnified for the well-known energy enhancement effect.

The detailed consequences of these principles for any particular map can be obtained quickly and easily through application of the ODE method outlined in this chapter. In particular, the ODE method yields a complete and dependable picture of the pulse behavior and an accurate value for the soliton energy.

It is perhaps useful at this point to make comparison among the three types of solitons we have studied in Chapters 1 and 2, viz., classical, path-average, and dispersion-managed solitons. The comparison is made in Table 2.2. Note that descent through the rows of Table 2.2 shows a rapid growth in pulse variation and a corresponding increase in degrees of freedom for system design. Thus, for example, while classical solitons rigidly require a fixed pulse energy and a fixed fiber dispersion parameter, both path-average and dispersion-managed solitons can tolerate considerable periodic variation in both of those parameters. The maps for dispersion-managed solitons can be based on fiber combinations that uniquely supply the nearly constant values of $\bar{D}$ required for dense WDM and the large values of D_{loc} required for thorough suppression of interchannel four-wave mixing. In comparison with classical or path-average solitons, dispersion-managed solitons

TABLE 2.2 The Three Types of Solitons Compared

Soliton type	Phase shifts from dispersion and nonlinear terms	Pulse behavior		Map strength
		Width and BW	Chirp	
Classical	True differentials	Constant	0	n.a.[a]
Path-average	Small but finite; cancel with L_{amp}	~Constant	Periodic	$S_{map} \ll 1$
Dispersion-managed	Large periodic variation; cancel with L_{map}	Large but stable periodic variation		$S_{map} > 1$ (typically ~2–4)

[a]n.a., Not applicable.

allow a given pulse energy to be obtained with much lower values of $\bar{D}$, thus affording a great reduction in timing jitter induced by frequency fluctuations. In Chapter 6 we shall see how dispersion-managed solitons allow for a much higher degree of control over soliton–soliton collisions, with a consequent suppression of unwanted effects.

Chapter 3

Spontaneous Emission and Its Effects

3.1. Some Basic Concepts

3.1.1. *Fundamental Modes of the Radiation Field*

The field in an optical fiber transmission line that arrives at some location z can usefully be resolved into Fourier components. These may be called temporal modes of the field. The word *mode* is used in a number of ways. In a fiber with a large core, there may be several transverse modes, that is, different transverse field patterns that the fiber can support. We are concerned here only with fibers that support only one transverse mode. Such fibers are called single (transverse)-mode fibers, with the word *transverse* usually omitted. In addition, in the very nearly cylindrical fibers that are primarily in use, there are two independent polarization modes, which give rise to a defect called *polarization-mode dispersion*, as discussed in Chapter 7.

Consider the temporal modes of a single polarization mode. In some appropriately long time T (see below), the field $u(t, z)$ that arrives at some location z can be resolved into a sum of orthogonal Fourier components

$$u(z,t) = \sum_n a_n u_n(t) \qquad \text{where} \qquad u_n(t) = \exp(-i2\pi \nu_n t)/\sqrt{T}, \tag{3.1}$$

where a_n is a complex coefficient and ν_n is the frequency of the nth mode. The different modes are made orthogonal by imposing periodic boundary conditions, so that $u_n(t+T) = u_n(t)$. This requirement is satisfied by $\nu_n = n/T$, so that we have

$$\int_{t_0}^{T+t_0} u_n^*(t)u_m(t)\,dt = \frac{1}{T}\int_{t_0}^{t_0+T} \exp[i2\pi(\nu_n - \nu_m)t]dt = \delta(m,n), \tag{3.2}$$

where $\delta(m,n)$ is the Kroneker delta function (zero when $n \neq m$ and one when $n = m$). The frequency separation of the modes is $1/T$, so the number of modes N in a bandwidth B is just $N = BT$. In a communications system, an important parameter is the number of modes in the effective optical bandwidth of the receiver and in the bit period, as this determines how much noise the receiver sees. This is just $B_{rec}T_{bit}$. One can think of the appropriately long time T (see above) simply as T_{bit}, or as some much longer time, in which case the number of modes per bit period is just the total number of modes in T divided by the number of bit periods in T. The result is the same.

With our usual normalization making $|u|^2$ the optical power in the fiber, the total energy W passing position z in time T is

$$W = \int_{t_0}^{t_0+T} |u(z,t)|^2 dt = \sum_n |a_n|^2. \tag{3.3}$$

Thus $|a_n|^2$ is the energy in the nth mode. The average power passing position z during the interval T is W/T, and so the average power per unit bandwidth is W/BT, which is equal to W/N. Thus the average power per unit bandwidth is equal to the average energy per mode. We note that each mode has a unique frequency. Its equation of motion, $du_n/dt = -i2\pi\nu_n u_n$, is that of a simple harmonic oscillator (SHO). Thus, the field that arrives at z can be likened to a collection of SHOs.

Another useful descriptor of the field is a degree of freedom (DOF). Each field mode is considered to comprise two independent DOFs corresponding to its two quadrature components, represented by the real and imaginary parts of a_n. In other words, $u_n(t)$ and $iu_n(t)$ are independent DOFs. The number of independent DOFs in the field is therefore twice the number of modes. More generally, we can think of any bounded function of time that is contained in the interval T as occupying one of some complete set of DOFs of the field. For example, the soliton function $u_s = \text{sech}(t)\exp(-i\Omega t)$ occupies one DOF of the field if the interval T is long enough to contain it. Mathematically this poses a problem, since u_s extends over all time, but for all practical purposes it does not. Independent DOFs u_j and u_k within the time interval T satisfy the relation $\Re\int_{t_0}^{t_0+T} u_j^* u_k\,dt = 0$. Thus iu_s occupies another of the same complete set of independent DOFs. This concept of DOFs is useful in considering the perturbations of solitons caused by the noise injected into a communications line by amplifiers. For example, suppose ϵ is a small positive real number ($\epsilon \ll 1$). If one adds ϵu_s to u_s, the energy of the soliton is increased. If one adds $i\epsilon u_s$ to u_s, the phase of the soliton is changed.

When the two polarization modes of the field are considered, the total number of modes and DOFs is doubled.

3.1.2. *Thermal Noise*

At low frequencies, where $h\nu/kT$ is small, the average noise power per unit bandwidth in a transmission line in thermal equilibrium with its surroundings is well known to be kT, where k is the Boltzmann constant and T is the absolute temperature. Where noise is concerned, the average energy per mode, which is identical with the average power per unit bandwidth, is called the equipartition energy. At room temperature, 20°C or 293°K, the ratio $h\nu/kT$ is unity at an infrared wavelength of 49 μm. At the wavelengths used in photonic communications, e.g., 1.55 μm, where $h\nu$ is considerably larger than kT, there are very few thermal photons to worry about.

According to Planck's law, the thermal equipartition energy is $h\nu/(\exp(h\nu/kT) - 1)$, which can be seen to approach kT when $h\nu/kT \ll 1$. If one carries the exponential to second order in $h\nu/kT$, the high-temperature asymptote becomes $kT - h\nu/2$. This equipartition energy may be thought of as the photon energy $h\nu$ times the average number $\bar{n}$ of photons per field mode. Thus $\bar{n} = (\exp(h\nu/kT) - 1)^{-1}$. As $h\nu/kT$ becomes greater than unity, $\bar{n}$ goes exponentially to zero. This prevents the so-called ultraviolet catastrophe, which the earlier classical theory wrongly predicted.

Note the emphasis on energy. Quantum mechanics, which is necessary to understand Planck's law, among many other things, deals relatively easily with energy levels and transitions between them, absorbing or emitting photons. It has more trouble dealing with coherence. For example, when the maser was being born in the 1950s, many atomic physicists of renown had trouble accepting the idea that the output of a maser oscillator could be a coherent field, like the output of electronic oscillators, such as klystrons and magnetrons.

The correspondence principle dictates that quantum mechanics must approach classical mechanics when the number of quanta involved gets large. In a typical communication system such as we are concerned with here, the equipartition energy of the noise corresponds to many photons, and each optical pulse has an energy corresponding to many more photons. So we would expect that a classical theory would suffice, and indeed all of the analysis in this book is based on classical theory, i.e., Maxwell's equations, and the experimental results agree well with the theory.

The appropriate classical field picture is, in essence, one that ignores the discreteness of the energy, and so mostly forgets about the photon, except as a unit of energy. It is often called the Wigner picture. It has sound foundations, which are beyond the scope of this book, as an approximation to the exact quantum field theory. It is not exact in cases where just a few photons are involved (no classical theory can be), but in the context of this book, it gives all of the correct answers.

We have noted that the field that arrives at z in time T is akin to a collection of harmonic oscillators whose number equals the number of modes of the field. The ground state of a harmonic oscillator in quantum mechanics has an energy one-half quantum above the zero of the potential energy, and its coordinate and momentum have Gaussian distributions about zero, exactly like random Gaussian noise with half a quantum of energy. To accommodate these "zero-point" fluctuations of the field, which are quite real, we must add an energy equal to the energy of half a photon to the equipartition noise energy. While this notion, if extended to ever higher frequencies, would bring back the ultraviolet catastrophe, it does no such harm in narrow frequency ranges. In every case pertinent to this book, it gives excellent approximations to whatever exact answers have been derived from quantum field theory. To be more precise, it gives results correct to first order in the quantum energy $h\nu$. For example, the first-order term $-h\nu/2$ disappears from the high-temperature limit of the thermal equipartition energy. In this regard, it is important to note that the field propagation equation, the NLS equation, is nonlinear. This makes problems for the quantum field theorists, but because the classical picture is correct to first order in the quantum energy, it still gives excellent results.

It is one of the properties of quantum mechanics that the zero-point noise fluctuations of the field cannot be seen directly by an energy detector, but they are seen by a coherent receiver such as a heterodyne detector or a laser amplifier. Furthermore, an energy detector sees the energy fluctuations caused by the interference of the zero-point noise with any other fields that may be present, such as signal fields.

3.1.3. Spontaneous Emission Noise

The most common view of spontaneous emission originated with Einstein, who introduced the concept of induced and spontaneous emission. In this (energy) view, particles in the lower states of transitions that are resonant with a field absorb photons, while particles in the corresponding upper states emit photons in two ways, by induced emission, which is the inverse of absorption, and by incoherent spontaneous emission. The ratio of spontaneous emission to induced emission and absorption is required to account for thermal noise. If we go by this picture in a fiber with absorbing and emitting particles, and consider noise, we get the relation

$$\frac{dW_{eq}}{dz} = C[(N_2 - N_1)W_{eq} + h\nu N_2], \tag{3.4}$$

where W_{eq} is the equipartition energy (noise power per unit bandwidth), C is a coupling coefficient, and N_1 and N_2 are, respectively, the number of particles in lower and upper states. The equilibrium situation $dW_{eq}/dz = 0$ requires that $W_{eq} = h\nu N_2/(N_1 - N_2)$. Boltzmann's law says that in thermal equilibrium, $N_1/N_2 = \exp(h\nu/kT)$, so we get $W_{eq} = h\nu/(\exp(h\nu/kT)-1)$, which agrees with Planck's law.

If the parameters in Eq. (3.4) are constant over some distance, say from 0 to z, its solution is

$$W_{eq}(z) = GW_{eq}(0) + \frac{N_2}{N_2 - N_1}(G-1)h\nu; \qquad G = \exp(C(N_2 - N_1)z), \tag{3.5}$$

where $W_{eq}(0)$ is the noise input at $z = 0$.

If $N_2 > N_1$ this equation describes the growth of noise in an amplifier. In the jargon of the optical amplifier field, the quantity $N_2/(N_2 - N_1)$ is called the "excess spontaneous emission factor," or n_{sp}. In more complicated circumstances, say if the parameters vary over the length of the amplifier, this result can still be described by the generic equation

$$W_{eq}(z) = GW_{eq}(0) + n_{sp}(G-1)h\nu, \tag{3.6}$$

where G is the power gain of the amplifier and n_{sp} is a numerical parameter whose minimum value is one (when $N_1 = 0$).

If $N_1 > N_2$ in Eq. (3.5), G is less than one, so it describes a loss mechanism. Many lossy parts of a photonic transmission system are passive, in which case N_2 is effectively zero. In such a case, as one can see from Eq. (3.4), the equipartition noise energy W_{eq} simply decays toward zero. If we relabel G as L, where $L < 1$ (representing loss), Eq. (3.5) becomes

$$W_{eq}(z) = LW_{eq}(0); \qquad L = \exp(-C(N_1 - N_2)z). \tag{3.7}$$

This quantum view of the spontaneous and stimulated emission of noise is quite correct. It shows precisely what a detector of energy would record. But not all detectors are energy detectors. Furthermore, if a signal field is present, things get more complicated. Following the energy in a communications system leads to a raft of unnecessary complications.

In the classical field picture, which ignores the discreteness of the energy, the spontaneous emission noise and zero-point noise are additive Gaussian fields. The zero-point noise is accounted for by adding $h\nu/2$ to the equipartition noise energy, and this zero-point noise is treated as no different from the rest of the noise. The growth of the noise energy is then described by the equation

$$\frac{dW_{eq}}{dz} = C[(N_2 - N_1)W_{eq} + \frac{1}{2}h\nu(N_1 + N_2)], \tag{3.8}$$

whose solution, on the same basis as Eq. (3.5), is

$$W_{eq}(z) = GW_{eq}(0) + \frac{1}{2}\frac{N_2 + N_1}{N_2 - N_1}(G - 1)h\nu = GW_{eq}(0) + \left(n_{sp} - \frac{1}{2}\right)(G - 1)h\nu. \tag{3.9}$$

Similarly, the equation corresponding to Eq. (3.7) is

$$W_{eq}(z) = LW_{eq}(0) + \frac{1}{2}(1 - L)h\nu. \tag{3.10}$$

Figure 3.1 attempts to summarize all we have said so far about the energy and field pictures of noise. Note that the equipartition energy W_{eq} in the field picture is always exactly $h\nu/2$ greater than that in the energy picture. What has changed is our view of the process. In the field view, random Gaussian noise is emitted into the field both by excited state particles and by ground state particles. According to Eq. (3.8), if $N_2 + N_1$ is a constant, the rate of spontaneous noise emission is independent of the ratio N_2/N_1. A passive loss mechanism (with $N_2 = 0$) now leads via Eq. (3.8) to an equilibrium vacuum state where $W_{eq} = h\nu/2$, and these zero-point noise fluctuations are continually absorbed and re-emitted. Furthermore, our view of spontaneous emission has changed. If we have an amplifier looking at a vacuum (dark) field, in the energy view the noise output comes exclusively from amplified spontaneous emission in the amplifier, whereas in the field view the noise output comes partly from the amplified zero-point input, and the rest comes spontaneously from the amplifier. For example, for an ideal amplifier ($N_1 = 0$), Eq. (3.9) gives $GW_{eq}(z) = GW_{eq}(0) + (G - 1)h\nu/2$. Here the spontaneous emission noise from

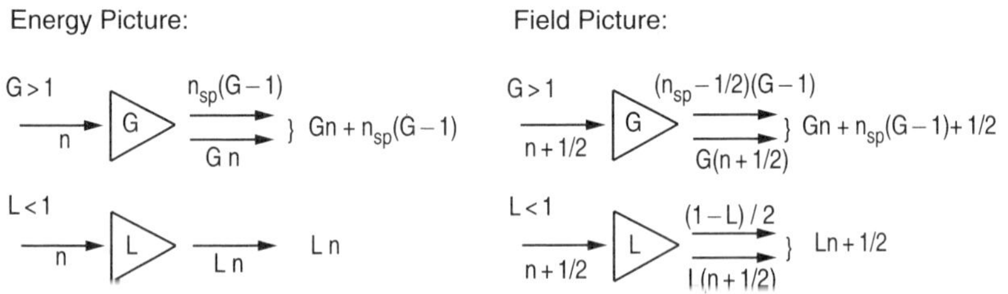

FIGURE 3.1 Illustration of the energy and field pictures of noise. On the left of each element are the noise inputs; on the right, the lower lines are amplified input noise and the upper lines are spontaneous emission noise; total noise is on the far right. (The energies here are all in units of the photon energy $h\nu$.)

the amplifier has been reduced by a factor of two from that in the energy picture. As it turns out, in computations of bit error rates and such, the optical noise at the receiver is much greater then $h\nu/2$, so the effect of adding the zero-point noise is usually negligible.

So what is the fuss all about? Do lower state particles actually emit noise into the system? The answer is yes, in a very real sense. First, consider the energy picture. If one could put exactly N photons into an attenuator ($N_2 = 0$), the output would be a binomial distribution of photons, so there would be fluctuations of the output energy. If N is large, and there is much attenuation, the output tends to a Poisson distribution. Hence the output is noisy. This kind of noise in the energy picture is called partition noise. Now, consider the field picture. The input of exactly N photons is a nonclassical situation, but one can approach it by assuming an input field with random phase and a small energy distribution, in keeping with the uncertainty product $\Delta N \, \Delta\phi \geq 1$. As the input field is attenuated, the zero-point noise field grows, resulting in almost exactly the same energy distributions, although now the distributions are continuous rather than discrete. To illustrate this point, Fig. 3.2 compares a Poisson distribution having a mean of six photons with a classical energy distribution corresponding to a signal field having the energy of six photons plus additive Gaussian noise having the mean energy of half a photon. For the Poisson distribution, the probability for n photons is placed at an energy of $n + 1/2$ to facilitate the comparison. One can see that even for a noise energy corresponding to a very few photons, the classical field picture reproduces the quantum energy picture quite well.

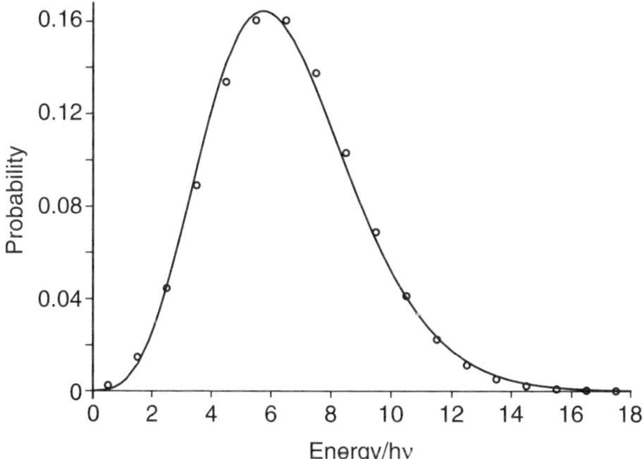

FIGURE 3.2 Probability distributions derived from energy and classical field pictures compared. Points: Poisson distribution with a mean of six photons in the energy picture. Solid line: Classical energy distribution corresponding to a signal field with the energy of six photons plus additive Gaussian noise with mean energy of half a photon.

The advantage of the field picture is that we can confidently use the wealth of science and engineering built up for communications systems at lower frequencies. The electromagnetic fields in communications are broadly broken up into signal fields, those we want, and noise fields, those we do not want. The field picture is now conventionally used in dealing with optical signals, as we do in this book. It is in dealing with noise that some of the unnecessary complications arise. If one has the case of signal plus additive Gaussian noise, the field picture simply takes $u = u_S + u_N$, where u_S is the signal and u_N is the noise. Since $|u|^2$ is the power, and

$$|u|^2 = |u_S|^2 + u_N^* u_S + u_S^* u_N + |u_N|^2, \tag{3.11}$$

the energy picture finds two separate kinds of noise. The sum of the second and third terms of this expression, which in the field picture is simply the interference of signal and noise, is known in the literature as "signal–spontaneous beat" noise, while the last term is known similarly as "spontaneous–spontaneous beat" noise. If u_N is just the zero-point field, an energy detector cannot see the $|u_N|^2$ term, but it does see the signal–spontaneous beat noise, which in the energy picture is known as *photon shot noise*.

As an example of the utility of the field picture, consider the noise figure of a linear amplifier. Conventionally, the noise figure NF of an amplifier is the ratio of signal-to-noise at the output to signal-to-noise at the input, where signal and noise are measured in power, and the noise figure is usually stated in decibels (dB). For linear amplifiers, this simply reduces to the output noise power divided by the product of the input noise power and the amplifier gain, or $NF = W_{eq}(\text{out})/(GW_{eq}(\text{in}))$. At low frequencies, the input noise is assumed to be the unavoidable minimum of thermal noise. In the field picture of photonic systems, the minimum input noise to an amplifier is the zero-point noise $W_{eq} = (1/2)h\nu$, and the corresponding output noise from the amplifier is [see Eq. (3.9)] $W_{eq}(z) = (1 + 2n_{sp}(G - 1))h\nu/2$. Thus the noise figure is $1/G + 2n_{sp}(1 - 1/G)$. If there is no amplifier gain ($G = 1$), the noise figure is unity, which translates to 0 dB. For high gain ($G \gg 1$), the noise figure approaches $2n_{sp}$. The minimum possible value of the high gain noise figure is 2 (3 dB) when $n_{sp} = 1$. The mavens of the energy picture have reached this same conclusion, albeit by a more difficult route.

As we go forward in this book, the reader will notice that the zero-point noise is mostly ignored. The energy picture is used to discuss noise generation [see Eq. (3.6)]. This makes contact with almost all of the literature. In fact, the zero-point noise is mostly ignorable—for example, where we discuss error rates—because the noise at the detector is so much greater. Also, in a typical section of a photonic transmission line consisting of a lossy section of fiber compensated by some gain mechanism, it makes little difference where the noise is generated—that is, whether the noise is generated solely by the gain (the energy picture) or by both gain and loss (the field picture). Here, because of the nonlinearity of the propagation equation, it does make some small difference, and it would be very interesting to find some situation where that difference might be measurable. It is important to realize, however, that the most accurate classical picture includes the zero-point noise, as we saw in the discussion of noise figure. Certain other situations, such as discussions of "squeezing the vacuum" using solitons, are easily explained using the classical picture, including the zero-point noise field.

3.2. Optical Amplifiers

3.2.1. *The Raman Effect and Raman Amplification*

The Raman effect in silica-glass fibers begins with a pump-induced transition to a virtual state, followed by emission from it, where the emission terminates on an excited state of the lattice; emission of an optical phonon (which typically takes place within a few femtoseconds) then completes return to the ground state (see Fig. 3.3). Because of the extremely fast relaxation, the population of the terminal state of the optical emission tends to be determined by equilibrium with the surrounding phonon bath, and hence is almost independent of the rates of optical pumping and emission. Thus, in contrast to erbium amplifiers, both the shape of the Raman gain band (see Fig. 3.4) and the excess spontaneous emission factor are essentially independent of pump and signal levels.

For gain in the neighborhood of the very broad peak of the Raman gain band, and when the fiber is at or near room temperature, the excess spontaneous emission factor $n_{sp} \cong 1.1$ (noise figure $\cong$ 3.5 dB). Nevertheless, Raman gain is prized,

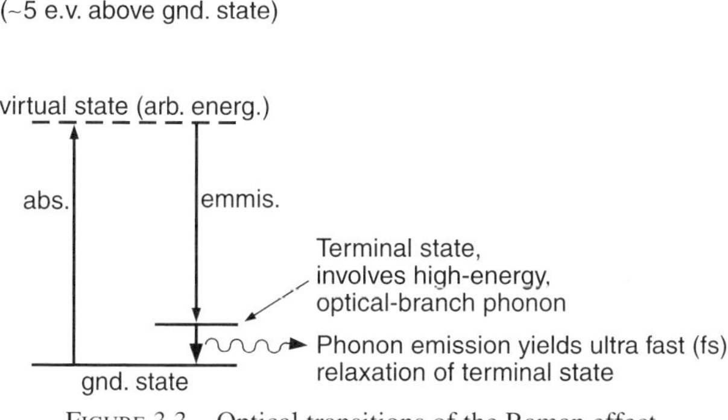

FIGURE 3.3 Optical transitions of the Raman effect.

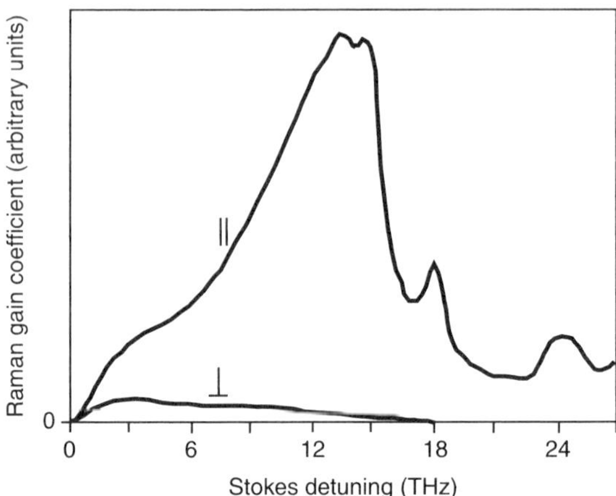

FIGURE 3.4 Relative strength of Raman gain versus the frequency difference between pump and signal photons. The parallel and perpendicular symbols refer to co-polarized and orthogonally polarized pump and signal, respectively. Reproduced with permission of Dougherty *et al.* [46].

more than anything else, for the great noise reduction afforded (as will be detailed shortly) by its ability to distribute gain over long fiber spans.

The position of the Raman gain peak depends only on the pump wavelength. Thus, it is possible to create a broad (up to ~8 THz wide), flat gain band, exactly where it is needed, just by combining the right powers at several different, carefully chosen pump wavelengths. This ease with which flat gain bands can be attained and controlled for dense WDM constitutes the second greatest advantage of Raman gain. It is so important that we shall treat it in detail later in Chapter 8.

Significant Raman gain is obtained only for that component of the signal that is polarized in the same way as the pump. Although at first this would seem to be crippling, it turns out not to be a serious problem. The modest birefringence of even the very best fibers is sufficient to cause the polarization states of the signal and pump to evolve on a scale of just a few meters. Thus, the counter-propagating signal and pump tend to undergo a thorough averaging of relative polarization states, and usually dispersion will cause a smaller, but still effective, polarization averaging when the pump and signal co-propagate. Furthermore, the pump lasers themselves can be effectively depolarized with a few simple techniques that will be detailed later.

3.2.2. *Erbium Fiber Amplifiers*

Discovered in 1987 [47, 48], erbium fiber amplifiers have been described in great detail elsewhere [49–51]. Therefore, this section sketches only the most fundamental points—essentially just that needed for the following discussion of the growth of noise in long chains of amplifiers.

A typical erbium amplifier consists of a few to some tens of meters of fiber whose core is doped with ~0.1% by weight of Er^{3+} ions and optically pumped at either 1480 nm or 980 nm (or both). Since the lengths involved are very short on the scale of typical transmission distances, they are rightfully considered as "lumped" amplifiers. The SiO_2 host fibers (typically co-doped with Al, Ge, or P) are largely compatible with ordinary transmission fibers and thus can usually be fusion spliced directly to them.

The three lowest energy levels of the Er^{3+} ion (the only ones of major significance for amplification) are diagrammed in Fig. 3.5. As shown there, the electrostatic field of the surrounding ions of the glass splits these highly degenerate levels into manifolds of 5, 4, and 3 sublevels, for the $^4I_{15/2}$, $^4I_{13/2}$, and $^4I_{11/2}$ levels, respectively. Since each of the sublevels is strongly broadened through interaction with phonons, the sublevels of each manifold tend to form an unresolved continuum. It should be further noted that since the host is a glass, the perturbing field, and hence the degree of splitting, tends to vary from site to site, thus adding a degree of inhomogeneous broadening to the mechanisms already cited. The emission and absorption cross-sections of the gain transition (that between the $^4I_{15/2}$ and $^4I_{13/2}$ manifolds) are plotted as functions of wavelength in Fig. 3.6. It should be noted that since the gain transition is first-order forbidden, the radiative time is long (~10 ms). Since, even at room temperature, kT is smaller than the total energy spread of either manifold (and since, further, phonon-induced relaxation rates among the levels of a manifold are on a much shorter time scale than 10 ms), absorption tends to take place mainly from the lower levels of the $^4I_{15/2}$ manifold, and emission from the lower levels of the $^4I_{13/2}$ manifold. Thus, the emission and

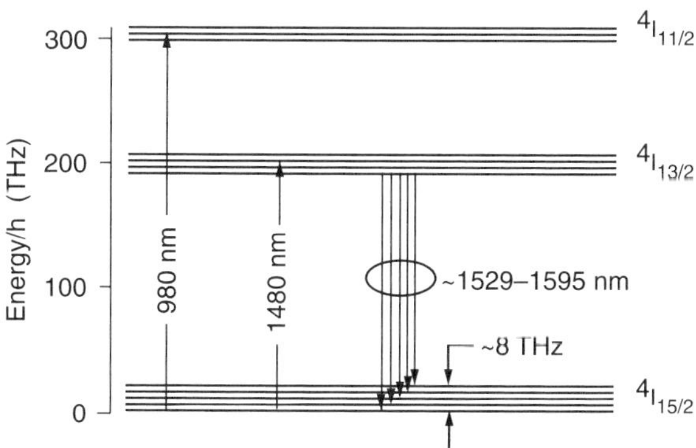

FIGURE 3.5 Simplified energy level diagram of Er.

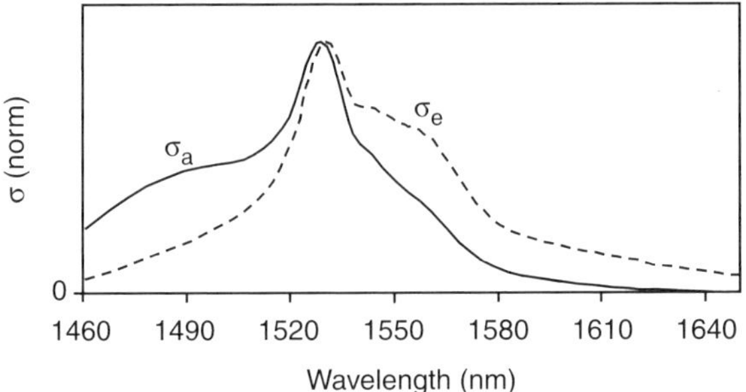

FIGURE 3.6 Emission and absorption cross-sections plotted as functions of wavelength for the gain transition of Er^{3+} in glass at room temperature. Reproduced with permission from Nufern Corporation, E. Granby, Connecticut.

absorption cross-sections are highly skewed to the high and low energy sides of the central wavelength ($\approx$1530 nm), respectively. At any wavelength, the net gain coefficient α_g (<0 for absorption) is given by

$$\alpha_g(\lambda) = \sigma_e(\lambda)n_2 - \sigma_a(\lambda)n_1, \tag{3.12}$$

where σ_e and σ_a are the emission and absorption cross-sections, and n_2 and n_1 are the population densities of the excited and ground states, respectively.

The region of high gain cross-section, extending from the central wavelength to about 1560 nm, is known as the "C" band. Note that because of the different shapes of $\sigma_e(\lambda)$ and $\sigma_a(\lambda)$, realization of a given fraction of the gain potential requires a higher degree of population inversion at the short wavelength end than at the long wavelength end of the C band. Thus, the shape of the $\alpha_g(\lambda)$ curve is highly dependent on the degree of population inversion (see Fig. 3.7). While this does not pose much of a problem when only a narrow band of signal wavelengths is involved, it becomes a very serious one for the creation of gain-flattened amplifier bands for dense WDM.

The ratio σ_a/σ_e of about 3/1 at 1480 nm (see Fig. 3.6) means that at most $\approx$75% population inversion can be produced by pumping at that wavelength, for a resultant minimum n_{sp} of 1.5, or equivalently, a best noise figure of $\sim$5 dB. Pumping at 980 nm to the $^4I_{11/2}$ manifold (always accompanied by very fast nonradiative relaxation to the $^4I_{13/2}$ manifold) avoids this problem, so that with intense pumping, one can have $n_{sp} \cong 1.0$ (noise figure $\cong$ 3.0 dB). Nevertheless, practical considerations, such as the possible need to avoid creating too great a variation in gain across the C band, often increase these numbers significantly.

Amplifiers can also be made for the so-called L band, i.e., the long-wavelength region of Fig. 3.6 extending from $\sim$1560 to $\sim$1600 nm. This is done by pumping a long, heavily doped fiber such that the amplified spontaneous emission (ASE)

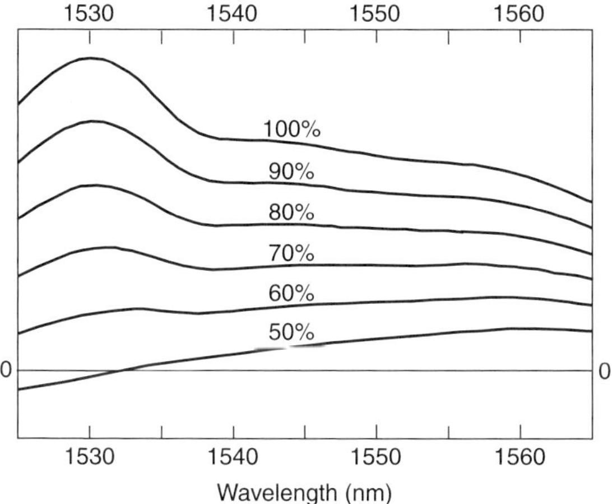

FIGURE 3.7 Gain of Er C band amplifier as a function of wavelength, with degree of inversion as a parameter.

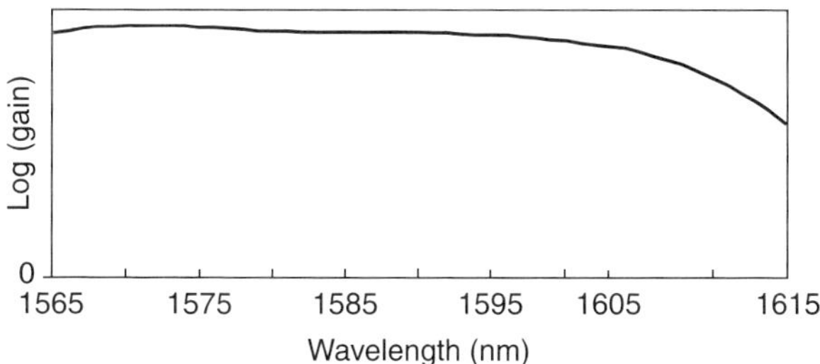

FIGURE 3.8 Gain of Er L band amplifier as a function of wavelength for pumping conditions yielding exceptionally flat gain. Reproduced with permission from Ylä-Jarkko *et al.* [52].

from the C band nearly equalizes the excited and ground state populations (thus nearly killing the C band gain). The much greater ratio of σ_e/σ_a in the L band, however, allows for the existence of significant gain there with reasonable noise figure. The resultant net gain vs. wavelength for such an L band amplifier is shown in Fig. 3.8. Since transitions in the L band terminate on elevated states that are only partially populated, the shape of the gain curve is much less dependent on the pumping rate than it is in the C band. By the same token, however, the L band shape depends rather strongly on the temperature, such that in critical applications, the erbium fiber's temperature must be regulated.

When erbium fiber amplifiers were first introduced in the late 1980s, they were much appreciated for their ability to provide up to several tens of decibels of polarization-independent power gain, with low noise figure, and with pump powers of just a few tens of milliwatts (just about the maximum available from commercial semiconductor lasers of the time). Thus, they can be relatively inexpensive and handy amplifiers for many small-signal, modest-bandwidth applications. Also, the long radiative lifetime of the gain transition enables erbium amplifiers to store considerable energy from a pump of modest power and then to release it all in a short burst of much higher peak power, as in certain pulsed laser applications. But it must also be noted that often, for application to dense WDM, separate C and L band amplifiers (usually each a complex, high-output power, multistage device) must be multiplexed together. The net result is a very complex device indeed, with many high-powered pump lasers, isolators, gain-leveling filters, WDM multiplexers, etc. The complexity rapidly becomes greater than, and the cost comparable to, the relatively simple set of multiplexed lasers used for Raman pumping.

3.3. ASE Growth in a Chain of Amplifiers and Fiber Spans

3.3.1. *Theoretical Behavior of the Model System*

We begin with a general model of a long-haul system that includes the possibility of lumped amplification. Figure 3.9 shows the prototypical system. In the system shown, each amplifier contributes a noise power per unit bandwidth, or equipartition energy, of $W_{eq} = (G-1)n_{sp}h\nu$ [Eq. (3.6)]. Since the gain from the output of each amplifier to the system output ($z = Z$) is unity, the noise at the detector is just

$$w_{eq} = N(G-1)n_{sp} = \frac{\alpha Z}{\ln G}(G-1)n_{sp}, \tag{3.13}$$

where α is the natural logarithm of the fiber loss rate, and where, to simplify appearances, we have substituted $w_{eq} = W_{eq}/h\nu$.

For perfectly uniform Raman gain, we can let N become very large, $G \to 1+\epsilon$, where $\epsilon \ll 1$, and set $n_{sp} = 1.1$, so Eq. (3.13) becomes

$$w_{eq} = \alpha Z \times 1.1. \tag{3.14}$$

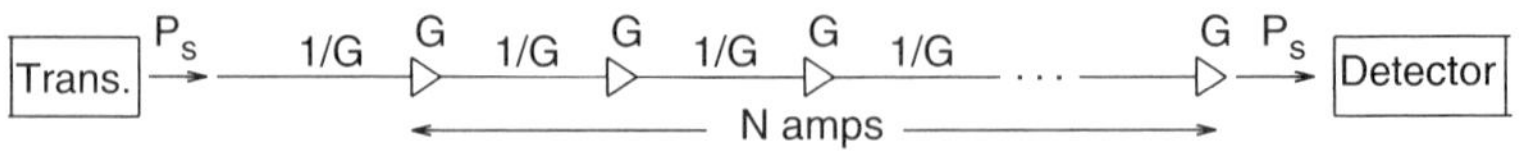

FIGURE 3.9 Prototypical ultra-long-haul system, with N amplifiers of power gain G preceded by N fiber spans of loss factor $1/G$.

From Eq. (3.14), we can immediately see that uniform Raman gain provides the lowest possible noise at the system output. Furthermore, from a comparison with Eq. (3.13), it is clear that the noise with high-gain lumped amplifiers is much higher. Consider, for example, amplifiers of 20-dB gain (as would be required for spans of length approaching 100 km); the noise at system output is then nearly 22 times, or 13.4 dB greater than with uniform Raman gain!

For injection of Raman pump power every distance L along the path (non-uniform Raman gain), each dz of path contributes $G - 1 = \alpha_g(z)\,dz$, so one has

$$w_{eq} = \alpha Z \times 1.1 \times \frac{1}{\alpha L}\int_0^L \alpha_g(z)\,g(z)\,dz, \tag{3.15}$$

where $g(z)$ is the net gain from z to L. Although the noise here is intermediate between that of Eqs. (3.13) and (3.14), as we shall soon see, for fiber spans of 100 km or less, the result is much closer to that of Eq. (3.14).

The range of acceptable signal levels is bounded on the low side by the onset of significant errors from inadequate signal-to-noise (S/N) ratio, and on the high side by the onset of significant errors from nonlinear effects. Since the most important nonlinear effects (primarily SPM and XPM) tend to scale with the path-average signal power, to facilitate comparison, we should calculate the corresponding path-average value of the noise. Thus, we must multiply the span input noise powers just calculated by the appropriate ratios of path-average to span input power. For the case of lumped amplifiers, that path-average factor is

$$\frac{\bar{P}_{sig}}{P_{sig}(0)} = \frac{1}{L}\int_0^L \exp(-\alpha z)\,dz = \frac{G-1}{G\ln G}.$$

Multiplying Eq. (3.13) by this factor, we get the path-average noise for the case of lumped amplifiers:[1]

$$\bar{w}_{eq} = \alpha Z \times \left\{ n_{sp}\frac{1}{G}\left[\frac{G-1}{\ln G}\right]^2 \right\} = \alpha Z \times \left\{ n_{sp}\,F(G) \right\}. \tag{3.16}$$

Note that the path-average noise for lumped amplifiers, while still considerable for high gains, is nevertheless substantially smaller than the noise at the detector [Eq. (3.13)] for the same gain.

For the case of Raman gain, where the pump power is injected every distance L, a similar calculation yields:

$$\bar{w}_{eq} = \alpha Z \times \left\{ 1.1 \times \frac{1}{\alpha L}\int_0^L \alpha_g(z)\,g(z)\,dz \times \frac{1}{L}\int_0^L P_{sig}(z)/P_{sig}(0)\,dz \right\}. \tag{3.17}$$

[1] The quantity $F(G)$, implicitly defined by Eq. (3.16), will be needed later in this chapter. $F(G)$ was originally defined in Reference [54].

In both Eqs. (3.16) and (3.17), the quantities inside large {} are the penalty factors, which can be interpreted equally well as follows:

1. The factor by which the path-average noise increases (over αZ) for constant path-average signal power, or
2. The factor of increase in path-average signal power (hence, increase in nonlinear penalties) required to maintain a given S/N ratio.

Although the preceding expressions for these penalty factors may not be immediately transparent, they have been evaluated numerically and are plotted in Fig. 3.10 for lumped amplifiers and for various situations of Raman pumping. Note that although the difference in penalty between the lumped and Raman amplifier curves begins with the modest difference in their n_{sp} values, and does not change much for span lengths of just a first few tens of kilometers, eventually, it becomes substantial, ≈6–8 dB in the neighborhood (≈100 km) of typical terrestrial amplifier hut spacings. To the extent that the limits of error-free transmission are governed by the growth of spontaneous emission noise, this 6- to 8-dB difference represents the factor (4–6×) by which the maximum transmission distance is increased when all-Raman amplification is substituted for lumped erbium amplifiers. It was this great increase in reach that first attracted systems developers to the all-Raman approach. Note also that although the penalty with purely backward Raman pumping is nearly 3 dB at 100 km, the use of combined forward–backward pumping

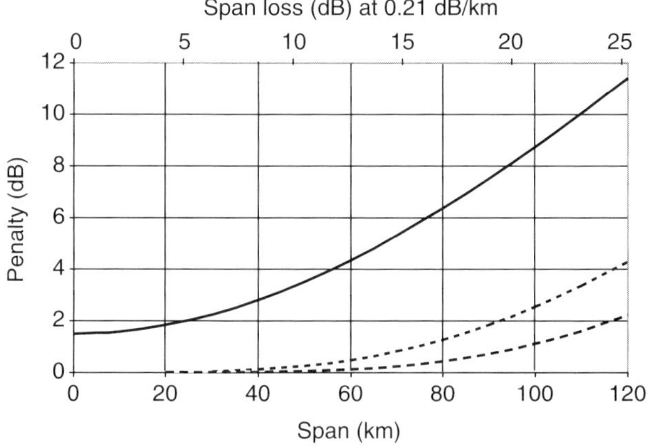

FIGURE 3.10 Noise penalty factors from Eqs. (3.16) and (3.17), plotted as a function of span loss. Solid line: Lumped (erbium) amplifiers, with noise figure 5 dB. Dotted line: Raman gain from 100% backward pumping. Dashed line: Raman gain from 50% forward/50% backward pumping. All penalty factors have been normalized to the $n_{sp} = 1.1$ of Raman gain, so that the Raman penalty curves will begin at 0 dB.

cuts that penalty to about a half. Finally, note that if the ≈100-km spans could be backward pumped at mid-span as well as at their far ends, the noise penalty is reduced to <1 dB, making it almost negligible. Such mid-span pumping also has the practical advantage of reducing the powers required of the individual pump lasers by a factor of two [53].

3.3.2. Experimental Test of ASE Growth

An accurate experimental test of the predictions of Eq. (3.17) has been made by using the recirculating loop shown schematically in Fig. 3.11. As shown there, the loop consists of six 100-km-long spans of non-zero dispersion-shifted fiber, each span properly compensated by a coil of DCF, with the DCF backward pumped, and provision made for forward as well as backward pumping of the 100-km spans. (For the experimental result reported here, the forward-pumping WDM couplers were moved to allow the normally forward pumps to be used as mid-span backward pumps.) When the loss of the DCF coils, the WDM couplers, and all the hardware (acousto-optic modulator, gain equalizer, etc.) used to close the loop on

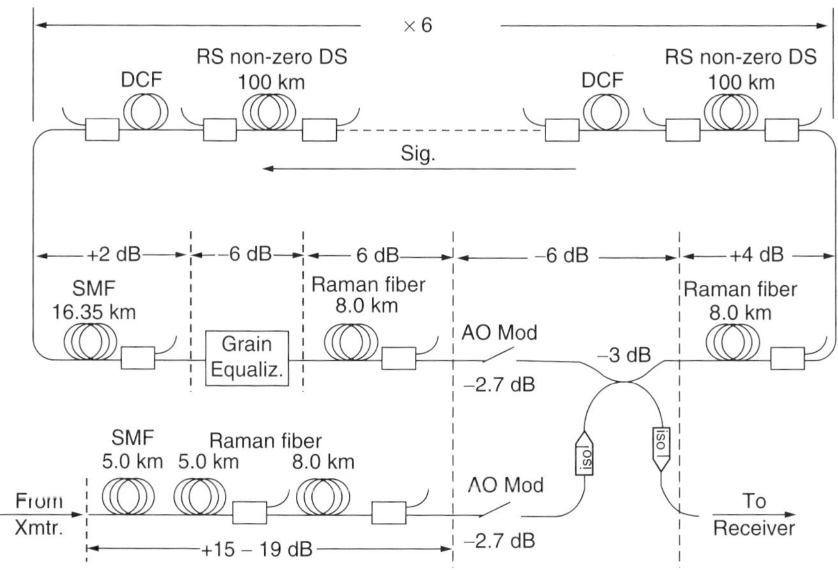

FIGURE 3.11 All-Raman amplified recirculating loop used in test of spontaneous emission noise growth. The net dispersion of the loop-closing amplifiers is essentially zero. The small rectangular boxes are WDM couplers for the introduction of Raman pump light (Xmtr, transmitter).

itself is factored in, the effective loss per 100 km of transmission fiber is about 30 dB. (The length of fiber in the DCF and loop-closing amplifiers is not included in the reported transmission distance.)

The noise measurements begin with determination, on a polarization-insensitive optical spectrum analyzer (OSA), of the ratio of the spectral intensity of a 10-Gbit/s data stream with the usual half-occupancy of bit periods, to the spectral intensity of the noise in an adjacent, empty channel, and with the OSA's spectral resolution wide enough to completely take in the entire spectrum of the pulse stream. That raw S/N ratio is then corrected by adding 6 dB to correct to just one polarization mode and for the unoccupied bit periods, and is further corrected to reflect the noise in the bandwidth (10 GHz in this case) corresponding to the bit period. (It should be noted that others often report the raw S/N ratio, but without the listed corrections and as measured with somewhat arbitrarily chosen spectral resolution.) Our corrected measurement then yields the fundamental quantity $S/N = W_{sol}/W_{eq}$. An independent determination of the signal pulse energy then lets us compute the noise itself; the result (reported as equipartition energy) is plotted in Fig. 3.12 as a function of the total transmission distance. The slope of the best-fit straight line makes almost perfect fit to the prediction of Eq. (3.17), based on the effective loss per 100 km just cited, and upon the $\approx$1-dB penalty factor from the mid-span pumping. The slight offset at the origin represents the noise contributed by the transmitter, and from the pre-amplifier shown in Fig. 3.11. In the absence of nonlinear penalties, the minimum S/N ratio required for a bit error rate (BER) $< 1 \times 10^{-9}$ is $\approx$100. Thus, the S/N ratio shown in Fig. 3.12 for 8000 km, even if degraded a decibel or so for Rayleigh double backscattering of the signal (see next section) or modest nonlinear penalties, is more than adequate for "error-free" transmission.

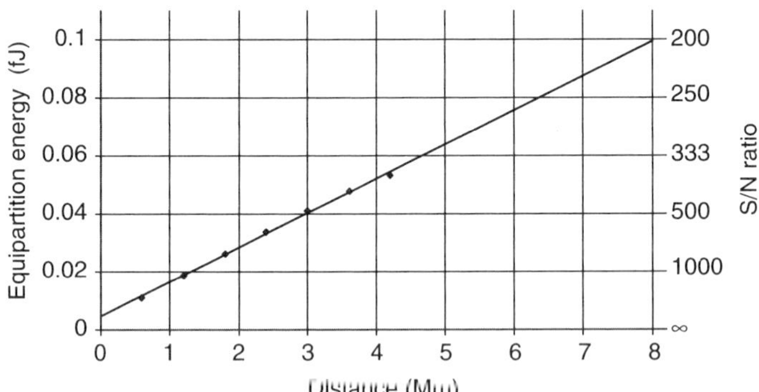

FIGURE 3.12 Experimentally measured noise as a function of distance for the loop of Fig. 3.11. The S/N ratio indicated by the vertical scale on the right is based on an assumed signal pulse energy of 20 fJ.

3.3.3. *Rayleigh Double Backscattering*

One disadvantage with Raman pumping of long spans is that Rayleigh double backscattering of the signal can add significantly to the spontaneous emission noise. The problem arises because of the fact that large Raman gain near the end of a backward-pumped span amplifies the backscattered signal just as much as it amplifies the signal itself; thus, the double backscattered signal experiences the Raman gain twice. Since the backscattered signal appears only in active channels, it is not included in the measured noise reported in Fig. 3.12 and so must be separately accounted for.

Rayleigh scattering is from index and density fluctuations that are small on the scale of the light wavelength, and thus it is isotropic. As already pointed out in Chapter 1, Rayleigh scattering tends to account for all or nearly all of the energy loss coefficient α of high-quality transmission fibers in the 1550-nm region. The Rayleigh backscattering coefficient is

$$\beta = \alpha_R \frac{\delta\Omega}{4\pi} = \frac{\alpha_R}{4}\left[\frac{N.A.}{n}\right]^2, \tag{3.18}$$

where α_R is that part of the loss coefficient attributable to Rayleigh scattering, $\delta\Omega$ is the solid angle of the scattered light captured by the fiber's core, n is the core's index of refraction, and $N.A.$, the numerical aperture, is just the sin of the half angle of the cone of light emerging from a cleaved fiber end. Table 3.1 lists β and the parameters needed for its calculation for several common fiber types.

Since isolators are always used between spans, the double backscattered noise can be computed on a per-span basis, just as with ASE noise. (The noise of either kind at the receiving end of the system is then just the per-span noise times the number of spans.) Figure 3.13 shows a representative normalized signal energy profile $H(z) = P_{sig}(z)/P_{sig}(0)$ created by backward Raman pumping of a fiber span and the setup for calculating the Rayleigh double backscattered fraction of the signal accumulated over its entire length L. As the signal travels from z to z', its energy clearly gains by the factor $H(z')/H(z)$. Note, however, that the light

TABLE 3.1 Rayleigh Scattering Loss Coefficient, Numerical Aperture, Index of Refraction n, and Backscattering Coefficient β at 1550 nm for Three Representative Fiber Types

Fiber type	$\alpha_R/0.23$ (dB/km)	$N.A.$	n	β (dB/km)	β^2 (dB/km^2)
Standard SMF	0.20	0.13	1.456	−41	−82
Dispersion shifted	0.20	0.17	1.456	−38.6	−77.2
DCF	0.35	0.2	1.456	−35	−70

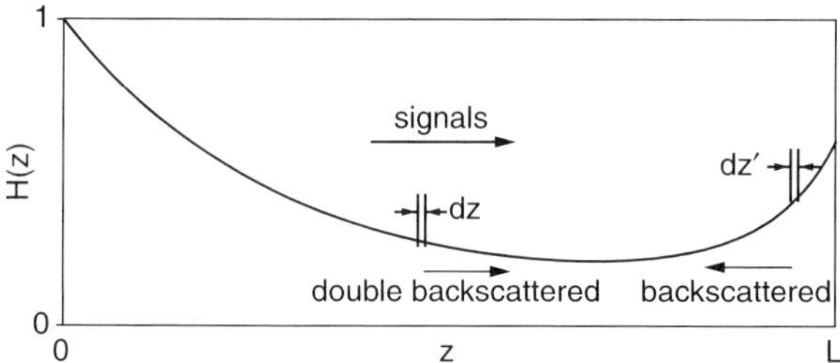

FIGURE 3.13 Normalized signal power $H(z)$ in a backward Raman pumped span of length L.

backscattered at z' experiences exactly the same gain as it travels back to point z. Thus, the net backscattered light at point z as accumulated by backscattering from points in the interval $L \longrightarrow z$ is

$$P_{bs}(z) = \int_z^L P(0)\,\beta\,H(z')\frac{H(z')}{H(z)}\,dz' = P(0)\,\beta^2\,H(L)\int_0^L H^2(z')\,dz'. \quad (3.19)$$

The double backscattered light from the entire span is then

$$P_{dbs}(L) = \beta\int_0^L P_{bs}(z)\,\frac{H(L)}{H(z)}\,dz = \frac{P(0)\,\beta^2}{H(z)}\int_z^L\left[\frac{1}{H^2(z)}\int_z^L H^2(z')\,dz'\right]dz. \quad (3.20)$$

Dividing Eq. 3.20 by the signal output power $P(0)H(L)$, and multiplying by 1/2 to account for the fact that the Rayleigh double backscattering represents a time average of the signals from many bit periods, statistically only 1/2 occupied in the original signal, we finally get the desired double backscattered noise-to-signal (N/S) pulse energy ratio:

$$(N/S)_{dbs} = \frac{1}{2}\beta^2\int_z^L\left[\frac{1}{H^2(z)}\int_z^L H^2(z')\,dz'\right]dz. \quad (3.21)$$

Figure 3.14 shows the per-span N/S ratio as calculated from Eq. 3.21, plotted as a function of span length, for various Raman gain configurations (including lossless fiber and zero Raman gain). For comparison, also shown there is the (per-span) ASE noise-to-signal ratio based on an assumed signal pulse energy of 20 fJ and the ASE noise energies reported in Fig. 3.14. The relative importance of the Rayleigh double backscattered signal noise is, of course, in direct proportion to the absolute signal level itself. The uppermost (solid) line in Fig. 3.14 plots the ASE noise-to-signal ratio for 20-fJ pulses and the W_{eq} of Fig. 3.12. Comparing that with the double backscattered noise curves of Fig. 3.14, we can calculate that for the case of 25% forward/75% backward pumping of 100-km spans, the Rayleigh

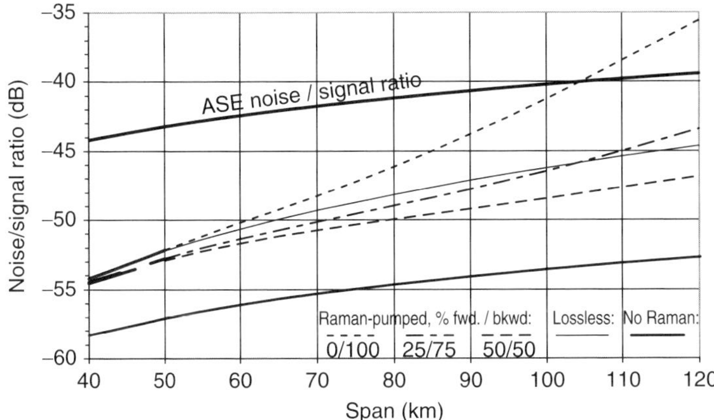

FIGURE 3.14 Lower curves: Per-span Rayleigh double backscattered noise-to-signal pulse energy ratios, as computed from Eq. (3.21), plotted as a function of span length. The various gain profile configurations are indicated. The assumed fiber core cross-sectional area is 50 μm^2 (appropriate for dispersion-shifted fiber), and for the cases with Raman gain, the spans have been pumped to unity net gain. Uppermost line: Per-span ASE noise-to-signal pulse energy ratio ($1.5 \times W_{eq}/W_{sig}$) for an assumed pulse energy of 20 fJ and the W_{eq} shown in Fig. 3.12. The factor of 1.5 is required to put the 10-GHz bandwidth of W_{eq} on an equal footing with the ≈15-GHz bandwidth of the Rayleigh double backscattered noise, which is the same as that of the original signal.

double backscattering increases the net noise by about 23% [so that it reduces the net S/N ratio to about 1/1.23 (−0.9 dB)] of the values shown in Fig. 3.10, already a significant reduction. For the case of 100% backward pumping (again of 100-km spans), on the other hand, the decrease in net S/N ratio is by (a probably intolerable) −2.6 dB. From this example, we can see why the Raman pumping of long spans to complete transparency by back pumping alone is not generally done. Rather, one either uses a combination of forward/backward pumping, or excess gain in the following coil of DCF. Finally, note that mid-span pumping (of the same 100-km spans), and insertion of an isolator there as well, tends to make the Rayleigh double backscattered signal noise almost insignificantly small.

3.4. ASE-induced Errors

3.4.1. *Amplitude or Energy Errors*

There are two main sources of error that affect the soliton system, fluctuations of the pulse energies and of their arrival times. At each amplifier, the addition of the

ASE noise changes the energy, central frequency, mean time, and phase of the solitons in statistically random ways. The changes in mean time and phase are of little importance in the present context. The other two changes can be analyzed separately. We shall focus on the energy fluctuations in this section, and on the frequency changes and the resultant jitter in arrival times in the next.

The energy fluctuations are quite similar to those that occur in a linear system. The argument is as follows: The system is effectively linear over short distances, so there is no difference in the way the noise field is injected into the system. The only difference in the soliton system is that the energy changes of first order in the noise field (the so-called signal–spontaneous noise) are captured by the solitons, which then reshape themselves as they propagate. This reshaping is done with no significant change in energy. Thus the energy fluctuations at the receiver are like those that would occur if the system were linear and dispersion free.

To evaluate the errors incurred by the energy fluctuations, some model detector must be chosen. For simplicity we shall assume that the detector consists first of an optical filter of bandwidth B_0, followed by a photodetector, followed by an integrator, so that in effect the total energy that passes the optical filter in each time slot T is measured. The detectors actually used in most systems do not work this way, but the good ones give similar results. Let $m = KB_0T$, where $K = 1$ or 2 is the number of polarization states to which the receiver is sensitive. The sampling theorem says that the detected optical field has approximately $2m$ independent degrees of freedom, and without loss of generality the soliton may be considered to occupy just one of these. The mean ASE noise energy per DOF is one-half of the equipartition energy, or $W_{eq}/2$. Thus, if we let S be the ratio of the total energy in a bit period to the equipartition energy W_{eq}, then S becomes the sum of the squares of $2m$ independent Gaussian random field variables, each with variance equal to 1/2. All but one of these have zero mean values. The exception has a mean value that is the square root of the normalized unperturbed soliton energy, S_1. One can think of S as the square of the radius to a point in a $2m$ dimensional euclidian space whose coordinates are the real amplitudes of the normalized DOF field components.

On the basis of this picture, we get the following results. First, in terms of the unit energy W_{eq}, the mean and variance of the energy of each noise mode are both unity. Thus, the mean and variance of the distribution of S for a "zero" (soliton absent) are both equal to m. For a "one" (soliton present), the mean of the energy distribution is also obtained just by adding the expected energies of the individual components, and hence is equal to $S_1 + m$. The corresponding variance is an entirely different story, however, because of the above mentioned fact that the soliton adds coherently to the noise field in its DOF. If we designate the stochastic noise field in the ith DOF by ϵ_i, the soliton field of magnitude $\sqrt{S_1}$ adds vectorially in the complex plane to the fields ϵ_1 and its quadrature complement ϵ_2 as shown

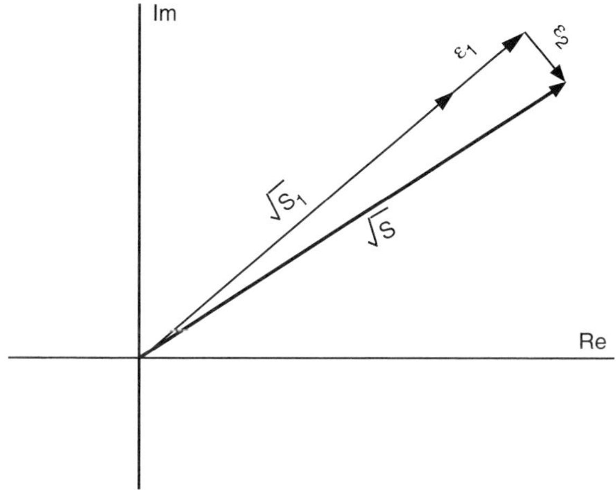

FIGURE 3.15 Illustrating vector addition of the signal ($\sqrt{S_1}$) with the noise field term ϵ_1 of its DOF and the quadrature complement ϵ_2 in the complex plane. From the geometry shown here, clearly, the field ϵ_1 has a much greater effect on the pulse energy S than does ϵ_2. Also, keep in mind that with an expected value of 0, ϵ is just as likely negative as positive, so that it can also reduce the size of the signal's field and energy. The effects of the other $2m - 2$ noise field components, not shown here, are similar to that of ϵ_2.

in Fig. 3.15. The signal energy S is the sum of the squares of the fields in the $2m$ DOFs:

$$S = \left(\sqrt{S_1} + \epsilon_1\right)^2 + \sum_{i=2}^{2m} \epsilon_i^2. \tag{3.22}$$

Expanding Eq. (3.22) and using the fact that $\langle S\rangle = S_1 + m$ (see above), we then get

$$S - \langle S\rangle = 2\sqrt{S_1}\epsilon_1 + \sum_{i=1}^{2m} \epsilon_i^2 - m. \tag{3.23}$$

To calculate the variance $\langle (S - \langle S\rangle)^2\rangle$, we must make use of the facts that the Gaussian distribution functions for the ϵ_i have second and fourth moments $\langle \epsilon_i^2\rangle = 1/2$ and $\langle \epsilon_i^4\rangle = 3/4$, respectively, and that the expected value of a product of stochastically independent terms is the product of their expected values. After a bit of manipulation, the final result is

$$\left\langle (S - \langle S\rangle)^2\right\rangle = 2S_1 + m. \tag{3.24}$$

Note that this variance is usually much bigger than the corresponding variance (m) for the zeros. Thus, spontaneous emission noise typically creates a much larger spread of energies among the ones than it does among the zeros. Incidentally, in the variances, the term m represents the "spontaneous–spontaneous beat" noise, while the term $2S_1$ represents the "signal–spontaneous beat" noise, both mentioned earlier [see Eq. (3.11) and the subsequent text].

The exact probability densities for S are given by [54]

$$P_0(S) = \frac{S^{m-1}}{(m-1)!} \exp(-S) \tag{3.25}$$

for a "zero," and by

$$P_1(S) = \left(\frac{S}{S_1}\right)^{\left(\frac{m-1}{2}\right)} \exp[-(S + S_1)] \; \mathrm{I}_{m-1}(2\sqrt{SS_1}) \tag{3.26}$$

for a "one," where I_{m-1} signifies a modified Bessel function of the first kind. It will be noted that the exponential and Bessel function terms in Eq. (3.26) tend to be very small and very large, respectively. While such numbers are not a problem for computers, the combination tends to make the expression somewhat opaque. Thus, the following approximate asymptotic form of Eq. (3.26) can give one a more immediate feel for the behavior of Eq. (3.26) in the regions where it is valid:

$$P_1(S) \approx \frac{1}{2\sqrt{\pi S}} \left(\frac{S}{S_1}\right)^{\left(\frac{m-1/2}{2}\right)} \exp\left[-(\sqrt{S} - \sqrt{S_1})^2\right] \left(1 - \frac{4(m-1)^2 - 1}{16\sqrt{SS_1}} + \cdots\right). \tag{3.26a}$$

The approximation is valid in regions where the quantity $(4(m-1)^2 - 1)/16\sqrt{SS_1} \ll 1$.

The functions $P_0(S)$ and $P_1(S)$ are often approximated by Gaussians. Gaussian distributions are determined solely by their means and variances. From the means and variances cited above, the approximate Gaussian distributions are

$$P_{0G}(S) = \frac{1}{\sqrt{2\pi\, m}} \exp\left(-(S-m)^2/(2m)\right) \tag{3.27}$$

and

$$P_{1G}(S) = \frac{1}{\sqrt{2\pi\,(2S_1+m)}} \exp\left(-(S - S_1 - m)^2/(4S_1 + 2m)\right). \tag{3.28}$$

The probability distributions P_0 and P_1 and their Gaussian approximations are shown in Fig. 3.16, for the combination $S_1 = 100$ and $m = 8$. Note that the exact probability distributions are somewhat asymmetric, with tails longer on their high-energy sides than on their low-energy sides. Note also that the P_0 and P_1 distributions are peaked at or very near $m = 8$ and $S_1 + m = 108$, respectively,

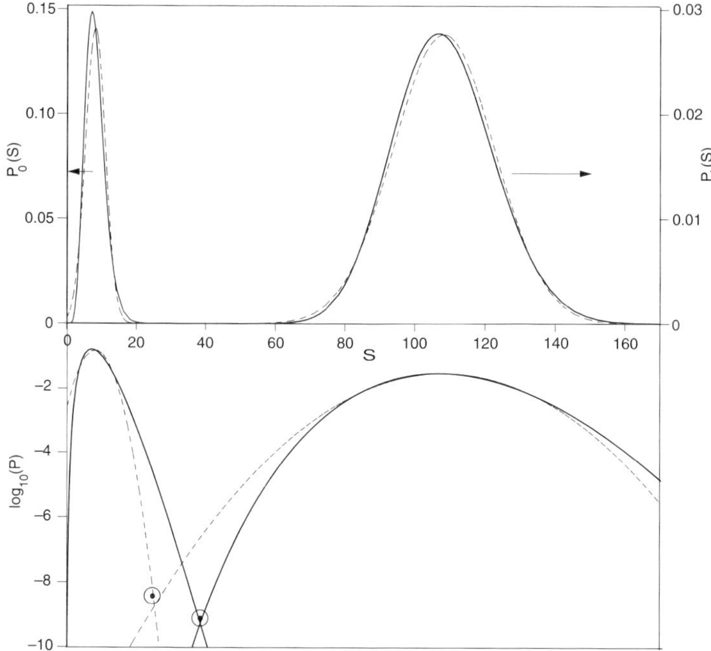

FIGURE 3.16 Solid lines: Probability densities P_0 (left) and P_1 (right) plotted as functions of S. Dashed lines: The corresponding plots of their Gaussian approximations P_{0G} and P_{1G}. Note that the plots in the upper panel are to different linear scales, in order to better display details of the P_1 and P_G curves. The far tails and their intersections are best seen in the log plots below, however. The encircled dots represent the best decision point and lowest error rate according to the exact and Gaussian distributions, respectively. [Note that the probability densities, $P(S)$ and the error rates are two different things, the latter being the integral under the tail of the first, so the vertical scale of the log plots plays two different roles here.]

reflecting the fact that the corresponding means are exactly 8 and 108, regardless of whether the distributions are exact or Gaussian approximations.

Using these probability distributions, one determines error rates by choosing a decision level S_d that equates the probability that $S > S_d$ for a "zero" with the probability that $S < S_d$ for a "one." In Fig. 3.16, these decision levels and their resulting error rates are indicated by circled dots. These points lie close to the cross-over points of their respective probability distributions on the logarithmic plot because the required integrals of the distribution tails produce multiplying factors of order one. Note that the optimum decision level is higher, but the corresponding error rate is lower, for the exact distributions (P_0 and P_1) than for their Gaussian approximations. Figure 3.17 plots the similarly determined optimum decision levels and corresponding minimum error rates as a function of the fundamental S/N ratio S_1.

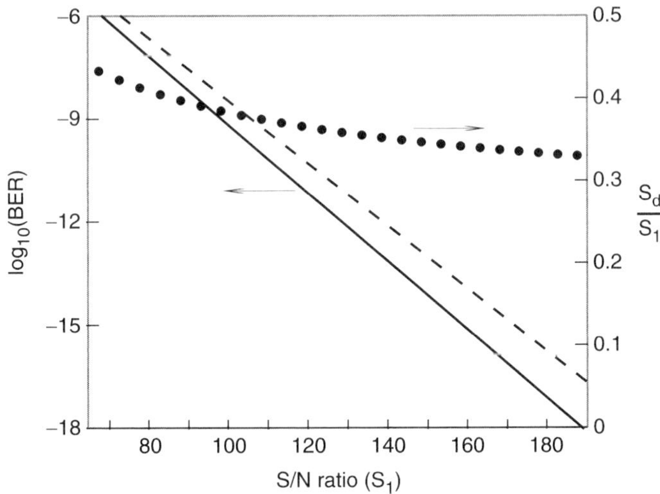

FIGURE 3.17 Bit error rate for amplitude or energy errors and the optimum decision energy level as functions of the signal-to-noise ratio for $m=8$ (S_d is the decision energy level). Also shown is the error rate estimate using the Gaussian approximation (dashed line).

Note that the more accurate computation of error rates continues to yield fewer errors than does the Gaussian approximation. For a larger value of m (larger optical bandwidth) the difference would be smaller.

3.4.2. *The Q Factor and the Gaussian Approximation*

Although the discussion thus far has centered on amplitude errors, as already noted, jitter in pulse arrival times and still other effects, such as polarization-mode dispersion (treated in Chapter 7), can also play a significant, and sometimes the dominant, role. Just as with amplitude errors, these other sources of error are also often fairly well represented by Gaussian probability distributions. The quantity Q, related to the bit error rate (BER) through the complementary error function,

$$\mathrm{BER} = \frac{1}{2}\mathrm{erfc}\left(Q/\sqrt{2}\right), \tag{3.29}$$

has become a more or less universally accepted parameter to characterize the net bit error rate in such Gaussian approximation. From the very general way that the Q factor is now used, Eq. (3.29) can be considered the *definition* of Q, i.e., $Q/\sqrt{2}$ is simply that argument of the complementary error function that yields the BER. Nevertheless, as we shall soon see, Q is very simply related to the means and variances of Gaussian distributions. Thus, it tends to be theoretically convenient and exceptionally suitable for the description of measured error rates, where it is often possible to find the means and variances of the probability distributions, but it is not easy to determine the exact form of the distribution tails. Finally, for the error rates of practical concern, the corresponding Q factors represent convenient and easily remembered numbers; for example, for a BER of 10^{-9}, the corresponding value of Q is 6. Incidentally, there is an approximate form of Eq. (3.29) that can be used when $Q \geq 3$:

$$\mathrm{BER} \approx \left(2\pi(Q^2+2)\right)^{-1/2} \exp(-Q^2/2). \tag{3.29a}$$

The error function and the complementary error function are defined as

$$\mathrm{erf}(x) = \frac{2}{\sqrt{\pi}} \int_0^x \exp\left(-t^2\right) dt \qquad \text{and} \qquad \mathrm{erfc}(x) = \frac{2}{\sqrt{\pi}} \int_x^{\infty} \exp\left(-t^2\right) dt.$$

They are defined so that $\mathrm{erf}(x) + \mathrm{erfc}(x) = 1$. Note that together they occupy only half of a complete Gaussian function. Note also that by symmetry, $(t \to -t)$, the complementary error function can also be expressed as

$$\mathrm{erfc}(x) = \frac{2}{\sqrt{\pi}} \int_{-\infty}^{-x} \exp\left(-t^2\right) dt.$$

The latter form of the erfc function applies directly to the tail below the distribution peak, while the former form applies similarly to the tail above the peak.

Let us now consider the most general case: Given two Gaussian probability distributions, $P_0(x)$ for a "zero," and $P_1(x)$ for a "one," with means respectively μ_0 and μ_1 and with variances respectively σ_0^2 and σ_1^2, we want to calculate the best decision level and the resulting error rate. Assume that $\mu_1 > \mu_0$. With a decision level μ_d somewhere between μ_0 and μ_1, the probability of making an error when reading a "zero" is $(1/2)\mathrm{erfc}((\mu_d - \mu_0)/\sqrt{2}\sigma_0)$, and the probability of making an error when reading a "one" is $(1/2)\mathrm{erfc}((\mu_1 - \mu_d)/\sqrt{2}\sigma_1)$. As already noted in the previous section, errors are minimized when the two error rates are equal, that is, when $(\mu_d - \mu_0)/\sigma_0 = (\mu_1 - \mu_d)/\sigma_1$. From this, we find that the optimum decision level, and the resulting error rate for both "ones" and "zeros", are

$$\mu_d = \frac{\mu_0\sigma_1 + \mu_1\sigma_0}{\sigma_1 + \sigma_0} \tag{3.30a}$$

and

$$\mathrm{BER} = \frac{1}{2}\mathrm{erfc}\left(\frac{1}{\sqrt{2}} \frac{\mu_1 - \mu_0}{\sigma_0 + \sigma_1}\right). \tag{3.30b}$$

Thus, from Eq. (3.30b), the value of Q needed to determine error rates from Eq. (3.29) is simply

$$Q = \frac{\mu_1 - \mu_0}{\sigma_1 + \sigma_0}. \tag{3.31}$$

Let us now return to the specific case of amplitude error. Substituting m, $S_1 + m$, $\sqrt{m}$, and $\sqrt{S_1 + m}$ for μ_0, μ_1, σ_0, and σ_1, respectively, into Eqs. (3.30a) and (3.30b), we get a decision level and a BER given by

$$S_d = \frac{(S_1 + m)\sqrt{m} + m\sqrt{2S_1 + m}}{\sqrt{m} + \sqrt{2S_1 + m}} \tag{3.32a}$$

and

$$\mathrm{BER} = \frac{1}{2}\mathrm{erfc}\left(\frac{S_1}{\sqrt{2m} + \sqrt{4S_1 + 2m}}\right). \tag{3.32b}$$

For the combination $S_1 = 100$ and $m = 8$ used in Fig. 1.16, these equations yield $S_d = 24.396$ and $\mathrm{BER} = 3.378 \times 10^{-9}$, respectively. The data for the Gaussian BER curve of Fig. 3.17 were calculated from Eq. (3.32b).

From the approximate form of Eq. (3.29), it is evident that the logarithm of the BER is almost linearly dependent on Q^2. Thus, in the engineering world, the value of Q^2 is cited more often than that of Q itself; furthermore, it is common practice to quote the quantity $10\log_{10}(Q^2)$, i.e., Q^2 is quoted in "dB." Figure 3.18 plots $\log_{10}(\mathrm{BER})$ versus Q^2 in decibels.

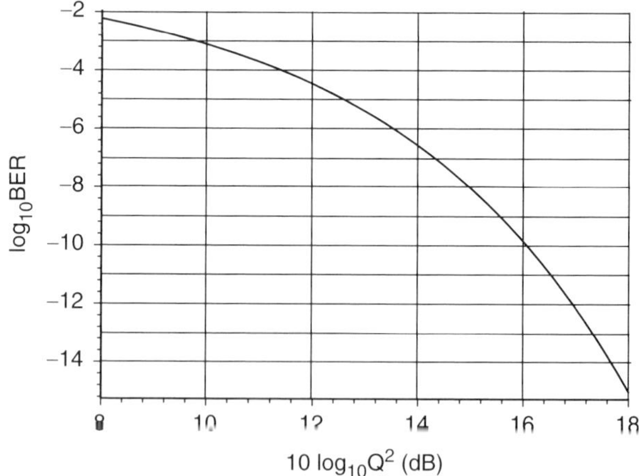

FIGURE 3.18 Logarithm of bit error rate for amplitude errors (in the Gaussian approximation) plotted as a function of Q^2 in decibels.

3.4.3. *The Gordon–Haus Effect*

The ASE noise also acts to produce random variations of the solitons' central frequencies. The fiber's chromatic dispersion then converts these variations in frequency to a jitter in pulse arrival times, known as the Gordon–Haus effect [55]. Such timing jitter can move some pulses out of their proper time slots. Thus, the Gordon–Haus effect is a fundamental and potentially serious cause of errors in soliton transmission.

The calculation of the jitter can be summarized as follows: Recall that each DOF of the noise field produced by an amplifier has a mean path-average energy of $(1/2)\bar{W}_{amp}(\nu)$. The field of one such DOF shifts the frequency of the solitons. From perturbation theory one can deduce that the effective noise field component has the form $\delta u = iau_{sol}\tanh(t)$ (see Fig. 3.19), and that it shifts the solitons' frequency by an amount $\delta\Omega = 2a/3$. Here a is a real random Gaussian variable whose variance $(3/4)\bar{W}_{amp}(\nu)$ is determined by the DOF's mean energy requirement. The solitons' random frequency shift therefore has a variance of

$$\langle \delta\Omega^2 \rangle_{amp} = \frac{1}{3}\bar{W}_{amp}(\nu). \tag{3.33}$$

Since in soliton units (see Chapter 1, Section 1.3.3) the inverse velocity shift is numerically just -1 times the frequency shift, the net time shift of a given pulse is

$$\delta t = -\sum_{amps} \delta\Omega_n z_n, \tag{3.34}$$

where z_n is the distance from the nth amplifier to the end. On the right side of Eq. (3.34) we have the sum of N independent variables, each of which has a Gaussian distribution. In such a case the sum also has a Gaussian distribution whose variance is the sum of the variances of the individual terms. Thus, the

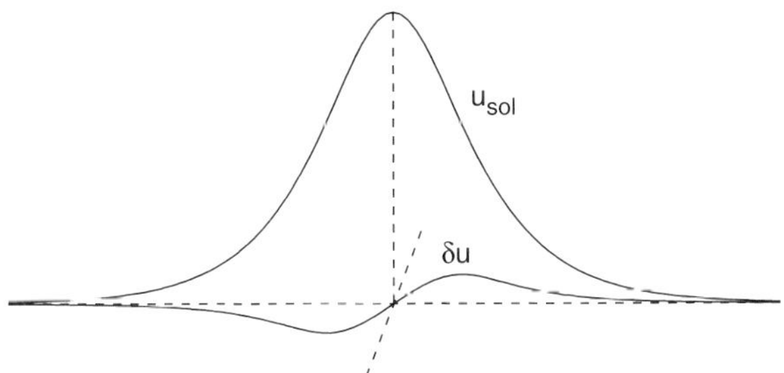

FIGURE 3.19 Noise component that modifies the frequency of a soliton in relation to the field envelope of the soliton. The two field components are in quadrature.

variance of δt is

$$\langle \delta t^2 \rangle = \langle \delta \Omega^2 \rangle_{amp} \sum_{amps} z_n^2 = \frac{\bar{W}_{amp}(\nu)}{3} \frac{Z^3}{3L_{amp}}, \tag{3.35}$$

where, for the second step in Eq. (3.35), Z represents the total system length, and we have approximated the discrete sum over the (many) amplifiers by an integral. Now, substituting $\ln G/\alpha$ for L_{amp} and using the path-average form of Eq. (3.13), we obtain

$$\langle \delta t^2 \rangle = \frac{1}{9} n_{sp} F(G) h\nu Z^3. \tag{3.36}$$

(From now on in this chapter, we shall write the variance in arrival times, $\langle \delta t^2 \rangle$, as σ^2, so σ then becomes the corresponding standard deviation.) Translated from soliton units into practical units, Eq. (3.36) becomes

$$\sigma_{gh}^2 = 3600 n_{sp} F(G) \frac{\alpha}{A_{eff}} \frac{D}{\tau} Z^3, \tag{3.37}$$

where σ_{gh} is in ps, $F(G)$ is as defined by Eq. (3.16), the fiber loss factor α is in km^{-1}, the effective fiber core area A_{eff} is in μm^2, the group delay dispersion D is in ps/nm-km, τ is the soliton full width at half maximum intensity in ps, and Z is the total system length in Mm (1 Mm = 1000 km). [The numerical constant in Eq. (3.37) is not dimensionless.] We can deduce from Eq. (3.37) that σ_{gh}^2 is proportional to the energy of the solitons, since the latter is also proportional to D/τ.

To get a feeling for the size of the effect, consider the example $Z = 9$ Mm (trans-Pacific distance), $\tau = 20$ ps, $D = 0.5$ ps/nm-km, $A_{eff} = 50\ \mu\text{m}^2$, $\alpha = 0.048\ \text{km}^{-1}$, $n_{sp} = 1.4$, and $F = 1.19$ (~30-km amplifier spacing); Eq. (3.37) then yields $\sigma_{gh} = 11$ ps.

The bit error rate from the Gordon–Haus effect is the probability that a pulse will arrive outside the acceptance window of the detection system. If the window width is $2w$ and we assume that these errors only affect the "ones," then the BER has a Q value of w/σ [see Eq. (3.31)]. For example, this implies that for an error rate no greater than 1×10^{-9}, $2w = 12\sigma_{gh}$. Now, the upper bound on $2w$ is just the bit period, although practical considerations may make the effective value of $2w$ somewhat smaller. Note, therefore, that for the example given above, where $\sigma_{gh} = 11$ ps, the quantity $12\sigma_{gh}$ corresponds to a maximum allowable bit rate of about 7.5 Gbit/s.

3.4.4. Gordon–Haus Effect for Dispersion-managed Solitons

For dispersion-managed solitons, the standard deviation, σ, of the Gordon–Haus jitter can be computed as

$$\sigma\,(\text{ps}) = 1.005\ldots(\text{ps/nm})\sqrt{\frac{W_{eq}}{W_{sol}}}\sqrt{\frac{Z}{L}}\frac{\bar{D}(\text{ps/nm-km})}{\tau\,(\text{ps})}Z\,(\text{km}), \tag{3.38}$$

where W_{eq} is the equipartition energy per span, L is the span length, and τ is the unchirped pulse width. Equation (3.38) is just the square root of Eq. (3.37), rewritten to display the quantities W_{sol} and $\bar{D}/\tau$ explicitly, as those two quantities are not rigidly coupled for dispersion-managed as they are for ordinary solitons (for ordinary solitons, $W_{sol} \propto \bar{D}/\tau$). Now, by virtue of the energy enhancement effect discussed earlier, for the same W_{sol}, the value of $\bar{D}/\tau$ is several times smaller for dispersion-managed than it is for ordinary solitons. Then σ [by virtue of Eq. (3.38)] is also smaller for dispersion-managed solitons by the same factor. Thus, while the Gordon–Haus effect tends to impose a serious penalty in ultra-long-haul transmission with ordinary solitons at 10 Gbit/s, it poses much less of a threat for the same with dispersion-managed solitons (DMSs), at least when it is the sole source of timing jitter. That is, note from the data of Fig. 3.20 that the total spread in arrival times out to the 10^{-9} probability level ($\approx 13\sigma$) is a small fraction of the bit period at 10 Gbit/s, especially when the optimal post-dispersion compensation is used. (On the other hand, note that the same spread is comparable to the bit period at 40 Gbit/s.)

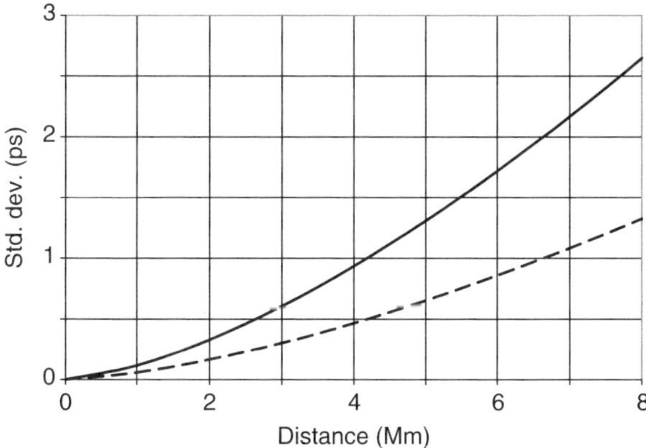

FIGURE 3.20 Standard deviation of the Gordon–Haus jitter vs. distance for a DMS system like that shown in Fig. 3.11. Solid curve: Without post-dispersion compensation [the full result from Eq. (3.38)]. Dashed curve: With optimum post-dispersion compensation [half the result of Eq. (3.38)].

3.4.5. *The Acoustic Effect*

Traditionally, the Gordon–Haus effect is considered to be the dominant source of timing jitter. There is, however, another contribution, one arising from an acoustic interaction among the pulses. Unlike the bit-rate-independent Gordon–Haus jitter, the acoustic jitter increases with bit rate, and as we shall soon see, it also increases as a higher power of the distance. Thus the acoustic jitter tends to becomes important for the combination of great distance and high bit rate. In this section we briefly review what is known about the acoustic effect.

The acoustic effect appeared in the earliest long-distance soliton transmission experiments [56] as an unpredicted "long-range" interaction: one that enabled pairs of solitons separated by at least several nanoseconds (and which were thus far beyond the reach of direct nonlinear interaction) to significantly alter each other's optical frequencies, and hence to displace each other in time. Shortly thereafter, Dianov *et al.* [57] correctly identified the source of the interaction as an acoustic wave, generated through electrostriction as the soliton propagates down the fiber (see Fig. 3.21).

Other pulses, following in the wake of the soliton, experience effects of the index change induced by the acoustic wave. In particular, they suffer a steady "acceleration," or rate of change of inverse group velocity with distance, dv_g^{-1}/dz, proportional to the local slope of the induced index change (again, see Fig. 3.21).

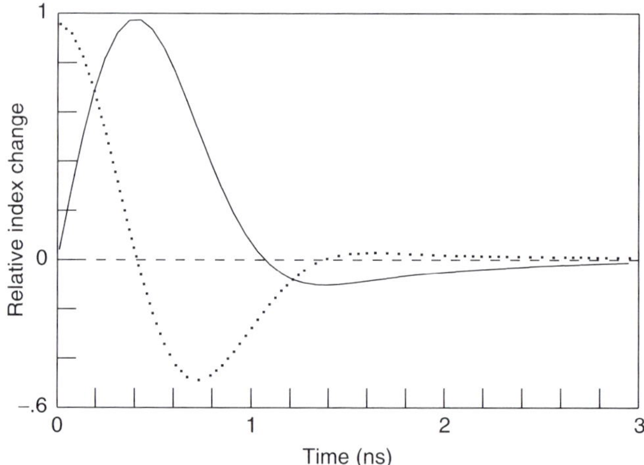

FIGURE 3.21 Solid curve: Relative index change due to the acoustic effect following passage of a soliton at $t = 0$. Dotted curve: Relative force acting on the following soliton at t; this curve is proportional to the time derivative of the relative index curve. For these curves, the interacting solitons are assumed to have a common state of polarization. The effect is only weakly dependent on their relative polarizations, however.

In a broadband transmission line, when this steady acceleration is integrated over z, it yields $\delta v_g^{-1} \propto z$, and a second such integration yields a time displacement $\delta t \propto z^2$. It can be shown that the standard deviation of the acoustic effect for a fiber with $A_{eff} = 50\,\mu\text{m}^2$ is approximately [58, 59],

$$\sigma_a \approx 8.6 \frac{D^2}{\tau} \sqrt{R - 0.99} \frac{Z^2}{2}, \tag{3.39}$$

where σ_a is in ps, D is in ps/nm-km, τ is in ps, R is in Gbit/s, and Z is in Mm. Comparing Eq. (3.39) with the square root of Eq. (3.37), note the different power dependencies on D, τ, R, and Z, most of which have already been discussed for both the Gordon–Haus and the acoustic effects. For the acoustic jitter, the scaling of σ_a as D^2/τ is easily understood, since it is clearly in direct proportion to the soliton energy, or D/τ, and the extra factor in D is required for the conversion of frequency shifts into timing shifts.

3.4.6. Phase-shift Keying and the Gordon–Mollenauer Effect

Differential Phase-shift Keying

Thus far in this book, we have discussed the representation of digital "ones" and "zeros" in an optical data stream as the simple presence or absence of a pulse, respectively, in the most commonly used encoding method known as "amplitude-shift keying" (ASK), or more recently as "on–off keying" (OOK). Recently, however, there has been much interest in a phase-shift encoding method known as "differential phase-shift keying," or DPSK, where ones and zeros are represented by a π or 0 phase-shift difference between successive pulses (see Fig. 3.22). There are two reasons for this interest. First, as with any phase- or frequency-shift keying, each bit slot contains a nominally standard pulse. Thus, absent the effects of noise, each pulse should experience exactly the same set of nonlinear interactions as every other pulse, so that effects such as timing jitter from collisions in WDM should disappear. Second, in a sufficiently linear system, DPSK provides a nearly 3-dB improvement in effective signal-to-noise ratio [60]; (see Fig. 3.23). The improvement stems from the doubled output swing between ones and zeros at the output of the balanced receiver that results from coherent addition of the pulse fields in the Mach–Zehnder (MZ) interferometer (for a more complete explanation, see Xu *et al.* [60]).

As the system becomes more nonlinear, however, the effective S/N ratio of a DPSK system is rapidly degraded [61, 62] as the nonlinear term begins to convert amplitude noise into phase noise. This phenomenon, known as the "Gordon–Mollenauer" effect [63], is the subject of the next section.

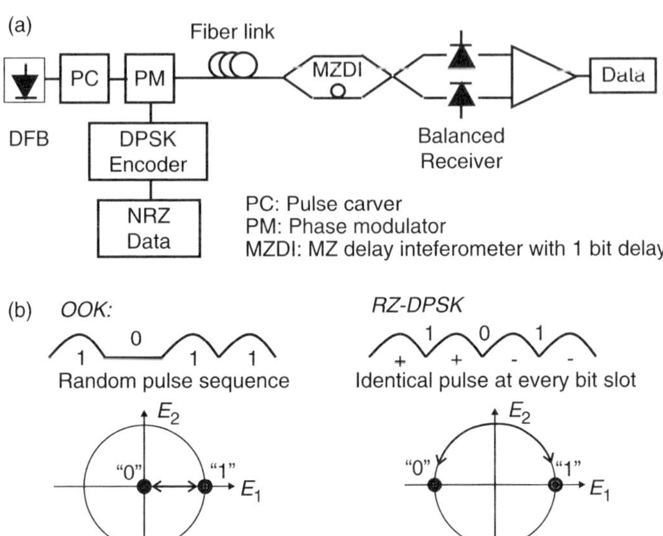

FIGURE 3.22 (a) Basic scheme for a return-to-zero (RZ), DPSK transmission system. At the transmitter, a phase modulator ("encoder") imposes relative phase shifts of 0 or π uniformly across each RZ pulse. At the receiving end, transit times through the two arms of the Mach–Zehnder (MZ) interferometer differ by exactly one bit period, such that successive incoming pulses precisely overlap in time at its output; interference then causes all of the energy of each pulse pair to appear at one or the other of its two output ports. The electrical output of the balanced receiver reflects the algebraic difference between its two inputs. (b) Phasor diagrams of the received optical E fields for OOK and RZ-DPSK, assuming the same peak pulse power for each. DFB, Distributed feedback; NRZ, non-return-to-zero. Reproduced with permission from Xu *et al.* [60].

The Gordon–Mollenauer Effect

Just as in the theory of the Gordon–Haus effect, we are concerned with a particular component of the noise field produced by each amplifier in a chain like that of Fig. 3.9. The phase of a pulse is affected by that part of the noise field that looks like the pulse itself, and both of its degrees of freedom play a role. That part in quadrature with the pulse, which we shall call E_1, is responsible for generating the linear contribution $\delta\phi_L$, and that part in phase with the pulse, E_2, generates the nonlinear part $\delta\phi_{NL}$ that is responsible for the Gordon–Mollenauer effect (see Fig. 3.24). Once again, the expected energies associated with these degrees of freedom, $\langle W_1\rangle_{amp}$ and $\langle W_2\rangle_{amp}$, are each equal to $(W_{eq})_{amp}/2$. From Fig. 3.24, and using the small angle approximation, we have $\delta\phi_L \cong E_1/E_{sig}$. Since each of the field quantities is proportional to the square root of its associated energy,

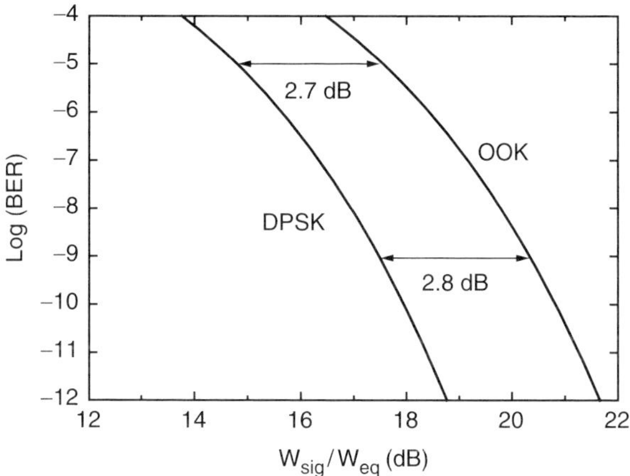

FIGURE 3.23 Log_{10}(BER) vs. fundamental S/N ratio for both RZ-DPSK and OOK; note the nearly 3-dB reduction in required S/N ratio for a given BER with DPSK. Adapted from Xu *et al.* [60].

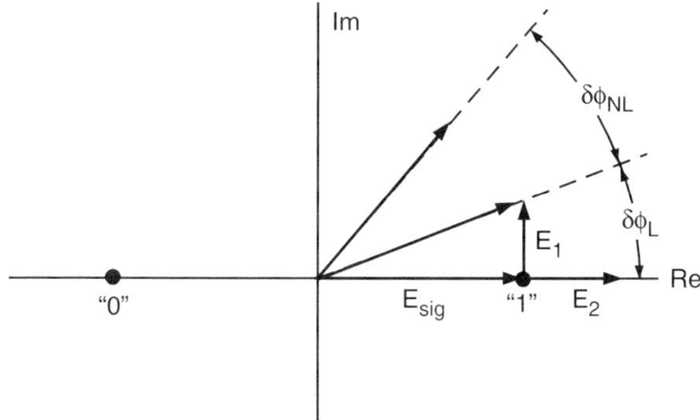

FIGURE 3.24 Phasor diagram showing addition of noise fields E_1 and E_2 at the output of an amplifier to E_{sig}, and the resultant phase shifts $\delta\phi_L$ and $\delta\phi_{NL}$ produced. The relative sizes of the noise fields to the signal field have been greatly exaggerated here for clarity.

the variance $\langle\delta\phi_L^2\rangle_{amp} = (W_{eq})_{amp}/(2W_{sig})$. Thus, after transmission over a net distance Z, we have

$$\langle\delta\phi_L^2\rangle_Z = \sum_{amps}\langle\delta\phi_L^2\rangle = \frac{W_{eq}}{2W_{sig}} = \frac{1}{2S}, \tag{3.40}$$

where S is the fundamental signal-to-noise ratio.

For the nonlinear effect, we must first calculate the noise-induced changes in pulse energy. Since the additions at each amplifier are small, the only significant change comes from addition of the E_2 noise field. Keeping only the cross term from the quantity $(E_{sig}+E_2)^2$ and expressing the result in terms of equivalent energies, we get $\delta W_{sig}=2\sqrt{W_{sig}}\sqrt{W_2}$. Squaring again and substituting the expected value $(W_{eq}/2)_{amp}$ for W_2, we get

$$\left\langle \delta W_{sig}^2 \right\rangle_{amp} = 2W_{sig}(W_{eq})_{amp}. \tag{3.41}$$

For the nonlinear phase shift, we shall need the appropriate path-averages over the following fiber spans, so Eq. (3.41) can be rewritten as

$$\left\langle \delta \bar{W}_{sig}^2 \right\rangle_{amp} = 2\bar{W}_{sig}(\bar{W}_{eq})_{amp}. \tag{3.41a}$$

The change in nonlinear phase shift produced by the nth amplifier, counting back from the receiver, is then

$$(\delta\phi_{NL})_n = k_{NL}\frac{\delta \bar{W}_{sig}}{\tau} nL, \tag{3.42}$$

where the nonlinear coefficient $k_{NL}=(2\pi n_2)/(\lambda A_{eff})$ is as defined by Eq. (1.11), and where τ is an effective pulse width, such that the fraction $\delta\bar{W}_{sig}/\tau$ yields the correct path-average power. [Thus, for dispersion-managed solitons, τ is not the unchirped (minimum) pulse width; rather, it is a value lying somewhere between the minimum and maximum widths.] Squaring Eq. (3.42) and using the energy variance from Eq. (3.41a), we obtain

$$\langle \delta\phi_{NL}^2 \rangle_n = 2\left(\frac{k_{NL}\, nL}{\tau}\right)^2 \bar{W}_{sig}\bar{W}_{eq}. \tag{3.43}$$

Since the $\delta\phi_{NL}$ contributed by each amplifier is independent of all the others, the net variance for transmission over the distance Z is

$$\left\langle \delta\phi_{NL}^2 \right\rangle = \sum_{n=1}^{(Z/L)} \left\langle \delta\phi_{NL}^2 \right\rangle_n = 2\left(\frac{k_{NL}L}{\tau}\right)^2 \bar{W}_{sig}\bar{W}_{eq}\sum_{n=1}^{(Z/L)} n^2. \tag{3.44}$$

Using the fact that the final sum in Eq. (3.44) is essentially equal to $(Z/L)^3/3$, regrouping terms and then using the substitutions $\phi_{NL}=k_{NL}(\bar{W}_{sig}/\tau)Z$ and $\bar{W}_{eq}=(Z/L)(\bar{W}_{eq})_{amp}$, we finally obtain

$$\left\langle \delta\phi_{NL}^2 \right\rangle = \frac{2}{3}\phi_{NL}^2\frac{\bar{W}_{eq}}{\bar{W}_{sig}} = \frac{2}{3}\phi_{NL}^2/S, \tag{3.44a}$$

the principal result of Gordon and Mollenauer [63]. For DPSK, which involves the difference in phase between two independent terms, each of which has a variance given by Eq. (3.44a), the effective variance is twice the result of Eq. (3.44a).

Note from Eqs. (3.40) and (3.44) that the variances of $\delta\phi_L$ and $\delta\phi_{NL}$ vary as $\bar{W}_{sig}^{-1}$ and $\bar{W}_{sig}$, respectively. Thus the total variance is minimized when the two terms are equal, and this occurs when ϕ_{NL} is approximately 1 radian (rad). Note also that the variance of $\delta\phi_{NL}$ scales as Z^3, just as in the Gordon–Haus effect, while that of $\delta\phi_L$ scales only as Z. Thus, as Z is increased in ultra-long-haul transmission, DPSK is eventually defeated by the Gordon–Mollenauer effect, while OOK continues to function. This important scaling can perhaps be best appreciated through a specific example. Figure 3.25 shows some results from Xu *et al.* [61], viz., BER measurements, carried out in the recirculating loop of Fig. 3.11, comparing transmission with DPSK vs. that for single-channel OOK. For DPSK, $\bar{W}_{sig} = 5.1$ fJ yielded the best BER performance (see Fig. 3.25). For that energy, the pulses are definitely not solitons, and hence tend to suffer considerable net dispersive broadening over 5 Mm, and $S \approx 81$ at that distance. Thus, we can only make the rough estimate $\phi_{NL} \approx 1.1$ rad. Nevertheless, it is clear that in this case, the resultant $\langle \delta\phi_{NL}^2 \rangle$ is small enough that the BER performance is dominated by the direct effects of amplitude jitter and not by the Gordon–Mollenauer effect. For the higher DPSK energy ($\bar{W}_{sig} = 8.9$ fJ), however, the situation is reversed. At 5 Mm, for that increased energy, $\phi_{NL} \approx 2.76$ rad and $S \approx 142$. Using those quantities in Eqs. (3.44) and (3.40), we get $\langle \delta\phi_{NL}^2 \rangle = 0.036$ and $\langle \delta\phi_L^2 \rangle = 0.0035$, respectively. Summing the two variances, multiplying by the factor of two required for DPSK, and taking

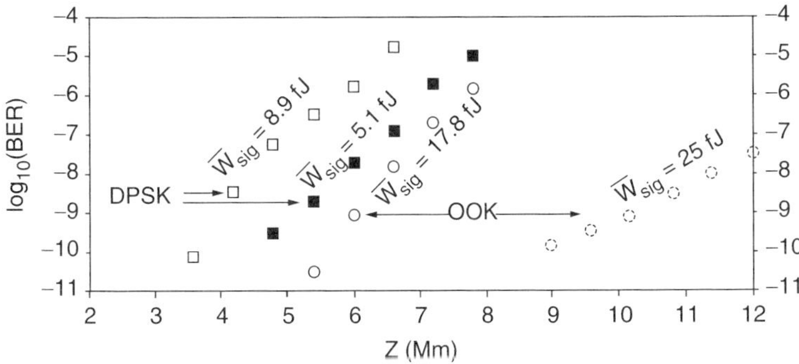

FIGURE 3.25 BER vs. distance, as measured in the recirculating loop of Fig. 3.11, for DPSK and single-channel OOK transmissions. Black squares: BER for single-channel DPSK at the optimal path-average pulse energy ($\bar{W}_{sig} = 5.1$ fJ). Open squares: WDM DPSK measurement, with $\bar{W}_{sig} = 8.9$ fJ. (With DPSK, however, the difference between WDM and single-channel performance is small; hence, the important difference here is in the pulse energies.) All of the data mentioned thus far are from Xu *et al.* [61]. Open circles: BER for OOK at a pulse energy of 17.8 fJ. Dashed circles: The same, but for 1.4× greater soliton pulse energy; this last set of data is from Chapter 6, Fig. 6.22.

the square root, we get an effective standard deviation $\sigma_{eff} = 0.28$. Corresponding to the $\approx 10^{-7}$ BER rate observed at 5 Mm, $\pm 5 \times \sigma_{eff} = \pm 1.4$ rad, which is remarkably close to the theoretical threshold for errors of $|\delta\phi| = \pi/2$; at the same time, increased S drastically reduces the BER directly from amplitude jitter. Thus, for this latter case, the Gordon–Mollenauer effect is clearly the dominant source of errors. By contrast, since the only nonlinear penalty in single-channel DMS transmission is from the Gordon–Haus effect (almost negligible for the range of parameters in Fig. 3.25), the distance for a given BER tends to scale in direct proportion to $\bar{W}_{sig}$, and considerably higher values of $\bar{W}_{sig}$ are allowed.

3.4.7. Optimization of the Pulse Energy for Best BER Performance

It should be clear from the discussion in Sections 3.2 and 3.3 that energy errors decrease, while errors from the Gordon–Haus jitter increase, with increasing soliton pulse energy W_{sol}. Thus there will be an optimum value of W_{sol} for which the combined error rates are a minimum. Since $W_{sol} \propto D/\tau$, one can hope to attain that optimum value of W_{sol} by adjusting D and τ. There are, of course, certain limitations on the practical ranges for both parameters. For example, lack of perfect uniformity of fiber preforms and other factors tend to limit the smallest values of D that can be produced reliably. To avoid significant interaction between nearest-neighbor soliton pulses, τ can be no more than ~20–25% of the bit period. Nevertheless, D/τ can usually be adjusted over a considerable range.

The optimum value of W_{sol} can be most efficiently found from a diagram [54] like that shown in Fig. 3.26, where, for a fixed value of the transmission distance, the rates for both energy and timing errors are plotted as a function of the parameter τ/D. Proceeding from the far right, where W_{sol} is smallest, note that at first, only energy errors are significant, but as W_{sol} decreases, those errors fall off exponentially. Eventually, timing errors become significant and then dominate. Also note that although the energy errors are bit-rate independent, the timing errors are not, since the allowable size of the acceptance window in time is determined by the bit period. Note that for transmission at 5 Gbit/s, the optimum value of $\tau/D \approx 70$ nm-km. If we choose $D = 0.5$ ps/nm-km, a value large enough to be reproducible, then we have $\tau = 35$ ps, a value short enough relative to the (200 ps) bit period to allow for negligible pulse interactions.

The curves of Fig. 3.26 would seem to imply a maximum allowable bit rate not much greater than 5 Gbit/s for trans-Pacific soliton transmission over a broadband transmission line (at least not for the specific choice of parameters reflected there). As is thoroughly explored in the section following this one, however, the technique of passive regeneration known as "guiding filters" has enabled that limit to be

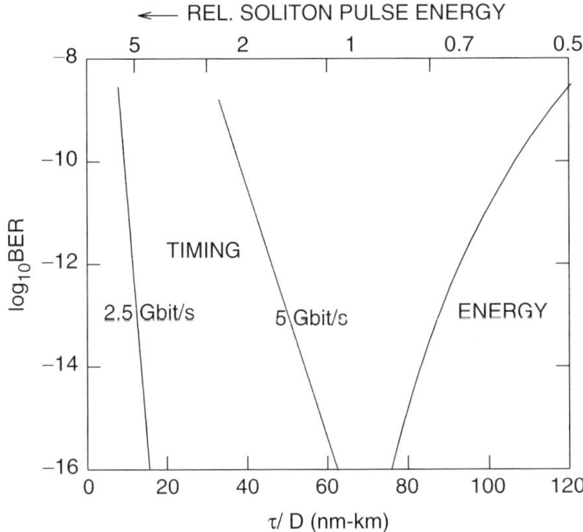

FIGURE 3.26 Bit error rates for energy and timing errors, as a function of the parameter τ/D, for a transmission distance of 9 Mm (the trans-Pacific distance). The other assumed parameters are fiber loss rate, 0.21 dB/km; $L_{amp} = 30$ km; $n_{sp} = 1.5$; $A_{eff} = 50\,\mu\text{m}^2$; $m = 8$ (see Section 3.3.1).

surpassed by a large factor. Thus, the theory in this section and its predictions are largely of interest as background for the understanding of transmission using filters. Nevertheless, for the record, we close this section by citing results of an experimental test, made a number of years ago [64], of single-channel transmission at 5 Gbit/s. Although made with guiding filters, the filters were only of the weak, fixed, tuned type, so the results that would have been obtained without filters may reasonably be projected from them (see Fig. 3.27).

3.5. Frequency-guiding Filters

3.5.1. *Introduction*

In mid-1991, two groups independently suggested the idea that the Gordon–Haus jitter and other noise effects could be significantly suppressed in soliton transmission systems simply through a narrowing of the amplifier gain-bandwidth [65,66]. In practice this means the use of narrow-band filters, typically one per amplifier. Figure 3.28 shows appropriate filter response curves in comparison with the spectrum of a 20-ps-wide soliton. The fundamental idea is that any soliton whose

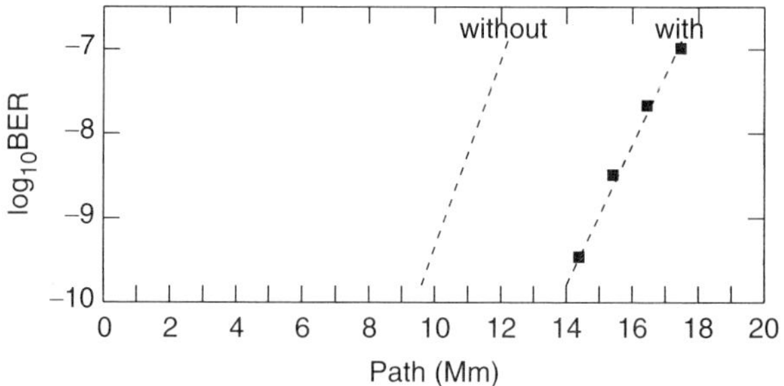

FIGURE 3.27 Experimentally measured bit error rate for a single-channel rate of 5 Gbit/s, as a function of transmission distance, and with a 2^{13}-bit-long random sequence. Curve labeled "with": Weak, fixed frequency-guiding filters used. Curve labeled "without": scaled-back projection with no filters. The other parameters are fiber loss rate, 0.21 dB/km; $L_{amp} = 28$ km; $D = 0.7$ ps/nm-km; $n_{sp} = 1.6$; $\tau = 40$ ps; $A_{eff} = 35\,\mu\text{m}^2$. At the receiver, the effective window width was about 170 ps.

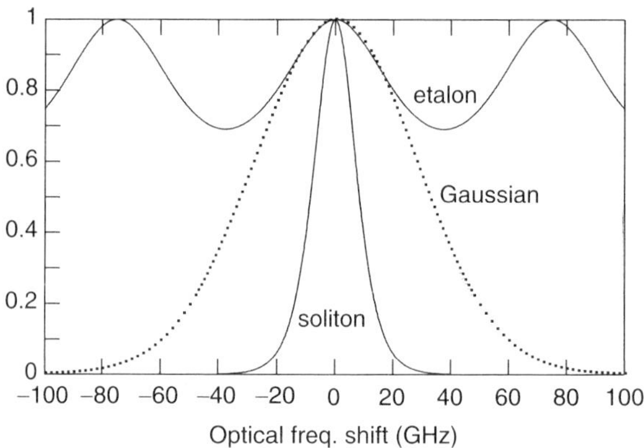

FIGURE 3.28 Intensity response curves of practical etalon guiding filter, Gaussian with same peak curvature, and spectrum of 20-ps soliton. The etalon mirrors have $R = 9\%$ and their 2.0-mm spacing creates the 75-GHz free spectral range.

central frequency has strayed from the filter peak will be returned to the peak, in a characteristic damping length Δ, by virtue of the differential loss the filters induce across its spectrum. The resultant damping of the frequency jitter leads in turn to a corresponding damping of the jitter in pulse arrival times. For example, in Eq. (3.35) for the variance of the Gordon–Haus jitter, when guiding filters are used, the quantities z_n^2 in the sum are all replaced with the common factor Δ^2.

Thus, the factor $Z^3/3$ in the final expression is replaced by the (potentially much smaller) factor $Z \times \Delta^2$.

The filters also cause a reduction in amplitude jitter. Consider, for example, a pulse with greater than normal power; that pulse will be narrower in time, have greater bandwidth, and hence experience greater loss from passage through the filter than the normal pulse, and, of course, the opposite will occur for a pulse of less than standard power. Thus, amplitude jitter also tends to be dampened out, as will be detailed later, in essentially the same characteristic length Δ as is the frequency jitter.

Since the major benefit comes from the filter response in the neighborhood of its peak, the etalon filter whose shallow response is shown in Fig. 3.28 provides almost as much benefit as a complete Gaussian filter. But the etalon, with its multiple peaks, has the great fundamental advantage that it is compatible with extensive WDM. The etalons also have the practical advantage that they are simple, cheap, and can be easily made in a rugged and highly stable form.

It should be understood that linear pulses cannot traverse a long chain of such filters: After a sufficient distance, their spectra will be greatly narrowed, and the pulses correspondingly spread out in time. The solitons survive because they can regenerate the lost frequency components, more or less continuously, from the nonlinear term of the NLS equation. On the other hand, the amplifiers must supply a certain excess gain to compensate for the net loss the solitons suffer from passage through the filters. As a result, noise components at or near the filter peak grow exponentially with distance. In order to keep the noise growth under control, the filters can be made only so strong, so the maximum possible benefit from them tends to be somewhat limited. For example, Fig. 3.29 shows the standard deviation of timing jitter, as a function of distance, for systems with the optimum strength filters (those experimentally observed [64] to produce the best BER performance), and for those with no filters. Note that at the trans-Pacific distance of 9 Mm, the filters reduce the standard deviation of the jitter by a factor somewhat less than two times.

3.5.2. Sliding-frequency Guiding Filters

There is a simple and elegant way [67] to overcome the noise growth, and hence the limited performance, of a system of fixed-frequency filters. The trick is to "slide," i.e., translate, the peak frequency of the filters with distance along the transmission line (see Fig. 3.30). As long as the sliding is gradual enough, the solitons will follow, in accord with the same "guiding" principle that dampens the jitter. On the other hand, the noise, being essentially "linear," can follow only the horizontal path in Fig. 3.30. Thus, the sliding creates a transmission

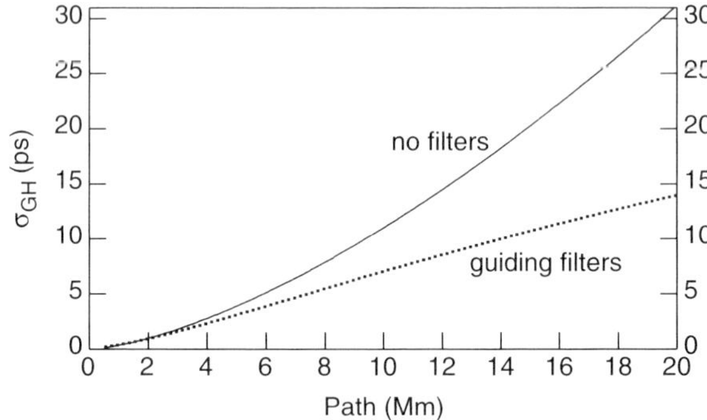

FIGURE 3.29 Computed standard deviations, σ_{GH}, of pure Gordon–Haus jitter as a function of total path, for a broadband transmission line ("no filters"), and for one having optimum strength, fixed-frequency guiding filters. The optimum filter strength, determined experimentally, corresponds to one uncoated, 1.5-mm-thick solid quartz etalon filter every 78 km. The other pertinent transmission line and soliton parameters in the strength-determining experiment were $D = 0.7$ ps/nm-km, $n_{sp} = 1.4$, and $\tau = 40$ ps. (The BER data of Fig. 3.17 were obtained in the same experiments [64].)

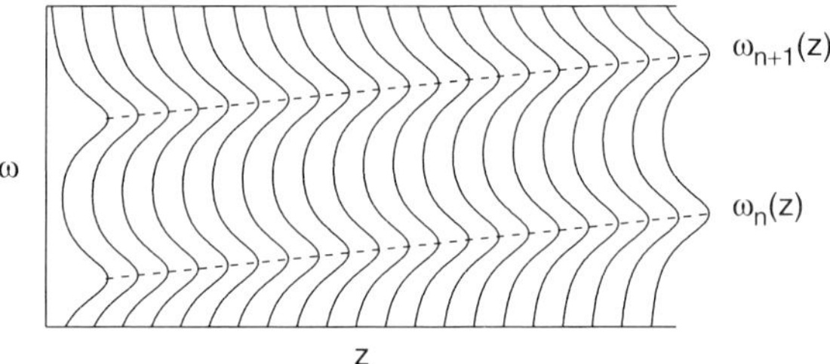

FIGURE 3.30 Transmission of sliding-frequency guiding filters versus z.

line that is opaque to noise for all but a small, final fraction of its length, yet remains transparent to solitons. In consequence, the filters can be made many times stronger, and the jitter reduced by a correspondingly large factor, with the final result that the maximum bit rate can be increased at least severalfold over that possible without sliding. The sliding-frequency filters provide many other important benefits, beyond the simple suppression of timing and amplitude jitter. Note, for example, that they suppress all noise-like fields, whatsoever the source, such as dispersive wave radiation from imperfect input pulses or other perturbations. They provide tight regulation of all of the fundamental soliton properties, such as energy, pulse width, and optical frequency. As will be detailed later, in WDM the filters suppress timing shifts and other defects from soliton–soliton collisions, and they provide a powerful regulation of the relative signal strengths among the channels in the face of wavelength-dependent amplifier gain. Thus, in short, the sliding-frequency guiding filters can be regarded as an effective form of passive, all-optical regeneration, and one that is uniquely compatible with WDM.

Finally, it should be noted that the required set of hundreds of sliding-frequency filters can be more easily and cheaply supplied than the corresponding set of fixed-frequency filters! That is, unlike the fixed-frequency mode, in which all filters must be carefully tuned to a common standard, for the sliding-frequency mode, no tuning at all is required. Rather, only a statistically uniform distribution of frequencies is needed. Provided the distribution is over at least one or more free spectral ranges, simple ordering of the filters should be able to provide any reasonable desired sliding rate.

3.5.3. *Analytic Theory of Guiding Filters*

In the mathematical representation of a transmission line with filters, in general, only numerical solutions are possible when the exact response functions of real filters are used. Nevertheless, analytic solutions are possible when the filter response is approximated by a truncated series expansion [67]. When expanded in a Taylor series, the logarithm of the filter response function F takes the general form:

$$\ln F(\omega-\omega_f)=i\zeta_1(\omega-\omega_f)-\zeta_2(\omega-\omega_f)^2-i\zeta_3(\omega-\omega_f)^3+\cdots \tag{3.45}$$

where ω_f is the filter peak frequency, and where the constants ζ are all real and positive. The first-order term can be ignored, since the linear phase shift it provides serves only to translate the pulses in time. While higher order filter terms can have important effects, the most fundamental features are revealed by analytic solutions to the simplified propagation equation employing only the second-order term,

$$\frac{\partial u}{\partial z}=i\left[\frac{1}{2}\frac{\partial^2 u}{\partial t^2}+u^*u^2\right]+\frac{1}{2}\left[\alpha-\eta\left(i\frac{\partial}{\partial t}-\omega_f\right)^2\right]u, \tag{3.46}$$

where α is the gain required to overcome the loss imposed on the solitons by the filters, and where $\eta=2\zeta_2$. (Both continuously distributed quantities α and η are easily converted into lumped, periodic equivalents.) Without filter sliding ($d\omega_f/dz\equiv\omega_f'=0$), and where, for convenience, we set $\omega_f=0$, the exact stationary solution is

$$u=\sqrt{P}\,\mathrm{sech}(t)\exp(i\phi), \tag{3.47}$$

where

$$\phi=Kz-\nu\ln\cosh(t),$$

and where the parameters $(\alpha, \eta, P, \nu, K)$ must satisfy

$$\nu = \frac{3}{2\eta}\left[\left(1+\frac{8\eta^2}{9}\right)^{1/2} - 1\right] = \frac{2}{3}\eta - \frac{4}{27}\eta^3 + \cdots, \tag{3.48a}$$

$$\alpha = (\eta/3)\left(1+\nu^2\right), \tag{3.48b}$$

$$P = \left(1+\eta^2\right)\left(1-\nu^2/2\right), \tag{3.48c}$$

$$K = (1/2)\left(1-\nu^2\right) + \left(\nu^2/3\right)\left(2-\nu^2\right). \tag{3.48d}$$

Note from Eq. (3.47) that the pulse's frequency is chirped, i.e., $\partial\phi/\partial t = -\nu\tanh(t)$. Because of this chirp, the root-mean-square (rms) bandwidth is increased by the factor $(1+\nu^2)^{1/2}$.

Numerical simulation involving real filters shows that the principal features (the chirp, extra bandwidth, and increased peak power) of the above solution are approximately preserved. The major differences lie in slight asymmetries induced in $u(t)$ and in its spectrum by the third-order filter term, and in the fact that, through a complex chain of events, those asymmetries cause the soliton mean frequency to come to rest somewhat above the filter peak.

Numerical simulation has also shown that sliding (within certain limits, given below) does not significantly alter this solution. Sliding does, however, have the potential to alter the damping of amplitude and frequency fluctuations. To get some notion of the effects of sliding on damping, we introduce the general form for the soliton $u = A\,\mathrm{sech}(At-q)\exp(-i\Omega t + i\phi)$ into Eq. (3.46). We then obtain the following pair of coupled, first-order perturbation equations:

$$\frac{1}{A}\frac{dA}{dz} = \alpha - \eta\left[(\Omega-\omega_f)^2 + \frac{1}{3}A^2\right], \tag{3.49a}$$

$$\frac{d\Omega}{dz} = -\frac{2}{3}\eta(\Omega-\omega_f)A^2. \tag{3.49b}$$

According to Eq. (3.49b), equilibrium at $A=1$ and at constant ω_f' requires that the lag $\Delta\Omega \equiv (\Omega-\omega_f)$ of the soliton mean frequency behind the filter frequencies is

$$\Delta\Omega = -\frac{3}{2\eta}\omega_f'. \tag{3.50}$$

Equation (3.50), as written, correctly predicts the difference in lag frequencies for up versus down sliding, $\Delta\Omega_u - \Delta\Omega_d$. To account for the offset in $\Delta\Omega$ produced by the third-order filter term (mentioned above for the case of no sliding), one must add a positive constant (as determined empirically from numerical simulation) to the right-hand side of Eq. (3.50). Equation (3.50) then correctly predicts $\Delta\Omega$ for all sliding rates. For etalon filters, the offset in $\Delta\Omega$ has been estimated from the third-order filter term [68]. That is, it has been shown from perturbation theory that

$$\Delta\Omega_{offset} \approx \frac{6}{5}\zeta_3, \tag{3.51}$$

where ζ_3 is computed as

$$\zeta_3 = \frac{1.762\ldots}{6}\frac{(1+R)}{(1-R)}\frac{\eta}{\tau F} \tag{3.51a}$$

where F is the free spectral range of the etalon filters. Combining Eqs. (3.50), (3.51), and (3.51a), one obtains

$$\Delta\Omega = \left[\frac{1.762\ldots}{5}\frac{(1+R)}{(1-R)}\frac{\eta}{\tau F}\right] - \frac{3}{2\eta}\omega_{f'}. \tag{3.50a}$$

Note that the two terms in Eq. (3.50a) tend to cancel for up-sliding ($\omega_f' > 0$), while they add for down-sliding. Since the damping of the filters is best for the smallest $|\Delta\Omega|$ [see Eq. (3.52)] below, up-sliding is definitely preferable to down-sliding.

Equations (3.49a) and (3.49b), when linearized in small soliton frequency and amplitude displacements δ and a, respectively, yield two eigenvalues (damping constants),

$$\gamma_1 = \frac{2}{3}\eta(1+\sqrt{6}\Delta\Omega) \text{ and } \gamma_2 = \frac{2}{3}\eta(1-\sqrt{6}\Delta\Omega), \tag{3.52}$$

with corresponding normal modes $x_1 = \delta + \sqrt{2/3}a$ and $x_2 = \delta - \sqrt{2/3}a$. This implies a monotonic decrease of damping for both frequency and amplitude fluctuations with increasing $|\Delta\Omega|$ and, through Eq. (3.50), the existence of maximum allowable sliding rates for stability. The numerical simulations performed to date with real filters are at least qualitatively consistent with these predictions.

In principle, based on the damping constants of Eq. (3.52), one can go on to write expressions for the variances in soliton energy and arrival time. It is not at all clear, however, how accurate such expressions would be in predicting the effects of strong, real filters. Nevertheless, since we are primarily interested in the behavior for $\gamma z \gg 1$, where the energy fluctuations have come to equilibrium with the noise, the energy variance can be written as

$$\frac{\langle \delta E_{sol}^2 \rangle}{E_{sol}^2} \approx \frac{N}{\gamma_E E_{sol}} = \frac{N\Delta_E}{E_{sol}}, \tag{3.53}$$

where N is the spontaneous emission noise spectral density generated per unit length of the transmission line, E_{sol} is the soliton pulse energy, and where the effective damping length, $\Delta_E \equiv 1/\gamma_E$, is expected to increase monotonically with increasing $|\Delta\Omega|$. Note that Eq. (3.53) implies that as far as the noise growth of "ones" is concerned, the system is never effectively longer than Δ_E. Since the characteristic damping lengths with sliding frequency filters are typically ~500 km or less, this means a very large reduction of amplitude jitter in transoceanic systems.

As far as the variance in timing jitter is concerned, we have already seen that, for $\gamma z \gg 1$, the factor $Z^3/3$ is replaced by $Z\Delta_t^2$. In other words, the variance in timing jitter is subject to a reduction factor

$$f(\gamma_t, z) \approx \frac{3}{(\gamma_t Z)^2} = 3\left(\frac{\Delta_t}{Z}\right)^2, \tag{3.54}$$

where Δ_t is also expected to increase monotonically with increasing $|\Delta\Omega|$. Although Δ_E and Δ_t are in general different, nevertheless, for Gaussian filters, both are expected to be approximately equal to $3/(2\eta)$ in the neighborhood of $\Delta\Omega = 0$.

For a transmission line using Fabry–Perot etalon filters with mirror spacing d and reflectivity R, the parameters η, ω_f', and α, in soliton units, are computed from the corresponding "real-world" quantities as follows:

$$\eta = \frac{8\pi R}{(1-R)^2}\left(\frac{d}{\lambda}\right)^2 \frac{1}{cDL_f}, \tag{3.55a}$$

$$\omega_f' = 4\pi^2 f' c t_c{}^3 / \left(\lambda^2 D\right), \tag{3.55b}$$

$$\alpha = \alpha_R t_c{}^2 2\pi c / \left(\lambda^2 D\right). \tag{3.55c}$$

Here f' and α_R are just $\omega_f'/2\pi$ and α, respectively, but as expressed in "real" units (such as GHz/Mm, for example), $t_c \equiv \tau/1.763\ldots$ [Eq. (1.17)], and L_f is the filter spacing.

We can now illustrate the power of sliding-frequency filters through a specific numerical example. Anticipating a bit from the next section, where we discuss the optimum choice of filter parameters, we choose the following soliton and fiber parameters: $D = 0.5$ ps/nm-km, $\tau = 16$ ps, so $z_c = 128$ km. The sliding rate will be 13 GHz/Mm (note that this means that the total sliding will be just about 1 nm in the trans-Pacific distance), so by Eq. (3.55b), $\omega_f' = 0.095$. For the filters, we choose $R = 8\%$, 2-mm air-gap etalons, with $L_f = 50$ km. By Eq. (3.55a), $\eta = 0.52$. Thus, we have $\alpha = 0.185$, and by Eq. (3.55c), $\alpha_R = 1.4$/Mm. For the damping constants of shallow etalons, however, η has a certain functional dependence on τ, and must be degraded to about $\eta_{eff} = 0.4$ for the 16-ps pulses to be used here.

The relative noise growth with sliding-frequency filters is easily simulated. Figure 3.31 shows the results of such a simulation for the conditions of our example: Note that while the sliding keeps the peak spectral density clamped to a value less

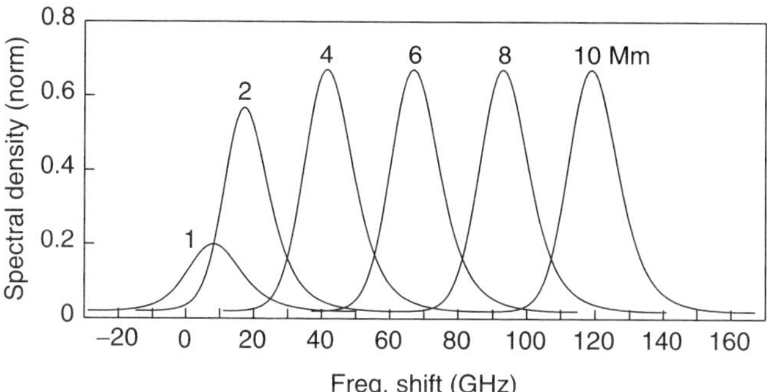

FIGURE 3.31 Noise spectral density, normalized to value at 10 Mm with no filtering, as a function of frequency and distance, for the conditions of our example (one $R = 9\%$, 75-GHz free spectral range etalon filter per 50 km, sliding rate = 13 GHz/Mm, $\alpha_R = 1.4$/Mm).

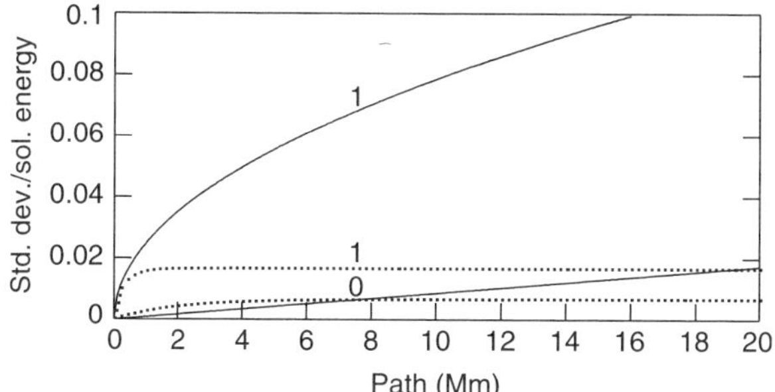

FIGURE 3.32 Standard deviations of the soliton energy ("ones") and of the noise in empty bit periods ("zeros") vs. distance, both normalized to the soliton energy, for the sliding-filter scheme of Fig. 3.30 (dotted curves), and with no filtering at all, save for a single filter at z, passing only 8 noise modes (solid curves). Assumed fiber loss rate and effective core area are 0.21 dB/km and 50 μm^2, respectively; amplifier spacing and excess spontaneous emission factor are ≈30 km and 1.4, respectively.

than would be obtained at 10 Mm without filtering, without the sliding, the noise would potentially grow by e^{14}, or about 1.2 million times, in the same distance! (Long before that could happen, however, the amplifiers would saturate.) Also note the spectral narrowness of the noise.

In Fig. 3.32 normalized standard deviations of the soliton energy ("ones") and of the noise energy in empty bit periods ("zeros") are shown as functions of distance. These curves are obtained from Eq. (3.53) and the estimate $\Delta_E \approx 600$ km [from $\eta_{eff} \approx 0.4$ and Eq. (3.52), data of Fig. 3.31, and the analysis of Section 3.3]. Note that with the filtering, both standard deviations soon become clamped to small,

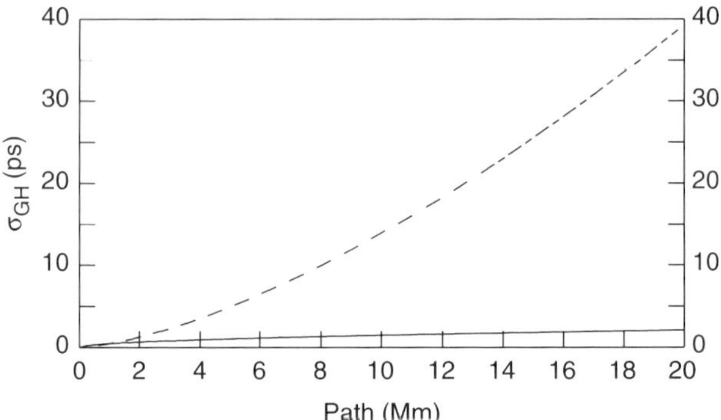

FIGURE 3.33 Standard deviation of Gordon–Haus jitter, σ_{GH}, as a function of total path. Solid curve: With strong, sliding-frequency guiding filters. Dashed curve: No filters at all. Conditions: $\tau = 16$ ps, $D = 0.5$ ps/nm-km; $n_{sp} = 1.4$; $F(G) = 1.1$; filter strength parameter $\eta = 0.5$; damping length $\Delta \approx 600$ km.

indefinitely maintained values, corresponding to immeasurably small bit error rates. Finally, in Fig. 3.33, the standard deviation of the Gordon–Haus jitter is plotted vs. distance, both for when the sliding filters are used and for when there are no filters at all. Note the nearly 10× reduction in σ_{GH} at 10 Mm, and compare with the same factor from Fig. 3.29.

3.5.4. Experimental Confirmation

Measurement of Noise and Amplitude Jitter

The following two figures refer to experimental transmission with sliding-frequency guiding filters, where the parameters are at least similar, if not identical, to those in the example just cited. Figure 3.34 shows the signal and noise levels

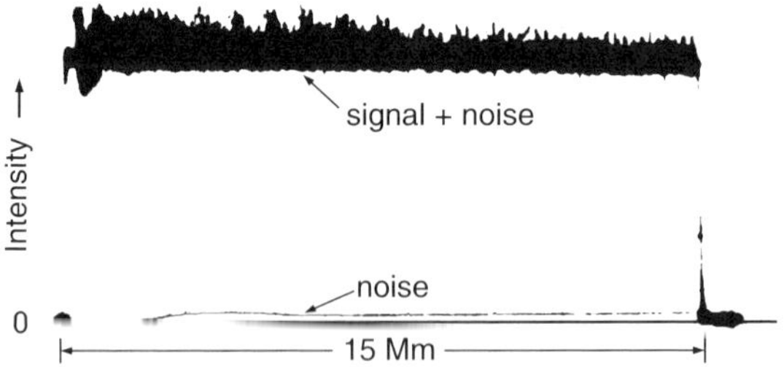

FIGURE 3.34 Noise and signal levels during a transmission using strong, sliding-frequency filters. The signal level is represented by the thick, upper line, while the noise level is represented by the fine line immediately above the zero signal level.

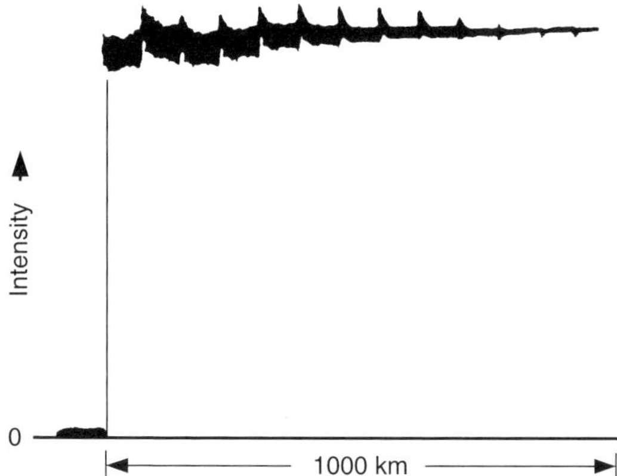

FIGURE 3.35 Observed amplitude jitter reduction with successive round trips in a transmission using sliding-frequency filters.

during a 15-Mm-long transmission, where the signal train was purposely made not quite long enough to fill the recirculating loop. Thus, once each round trip, for a period too brief (a few microseconds) for the amplifier populations to change significantly, one sees only the noise. Note that, just as in the theoretical model (Fig. 3.32), the noise grows only for a few megameters and then saturates at a steady, low value. In another early experiment with sliding-frequency filters, the pulse source was a mode-locked, erbium fiber ring laser, which had been purposely maladjusted to produce a substantial amplitude jitter at a few tens of kilohertz. Figure 3.35 shows the very rapid reduction in that amplitude jitter with successive round trips in the recirculating loop. The data shown there imply a damping length of about 400 km, which is consistent with the filter strength parameter of $\eta \approx 0.6$ and the known dispersion length $z_c = 160$ km [see Eq. (3.52)].

3.5.5. *Measurement of Timing Jitter*

The timing jitter in a transmission using sliding-frequency filters has been measured accurately by observing the dependence of the BER on the position, with respect to the expected pulse arrival times, of a nearly square acceptance window in time [69]. The scheme, which involved time division demultiplexing, is shown in Fig. 3.36. The fundamental measurement is of the time span (inferred from the precision phase shifter in Fig. 3.26), for which the BER is $\leq 10^{-10}$ for each distance. The resultant spans, or time-phase margins, are plotted in Fig. 3.37, as a function of distance, for three different cases: (1) a 2.5-Gbit/s data stream (which, since it also passes through the loop mirror, can be thought of as a 10-Gbit/s

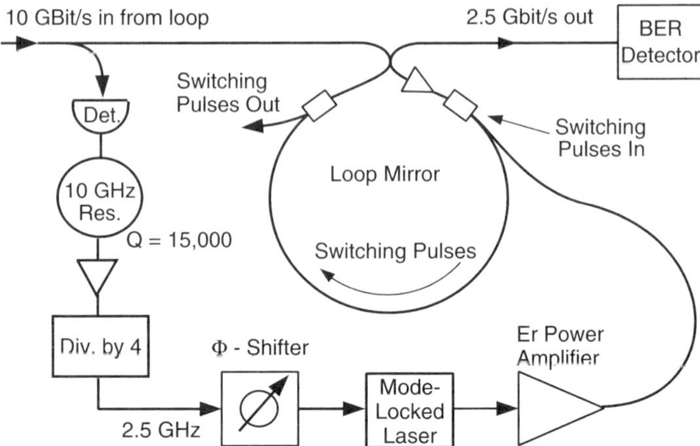

FIGURE 3.36 Scheme for the pulse timing measurements. Main elements of the clock recovery are the detector, the high-Q, 10-GHz resonator, and the divide-by-4 chip. The wavelength-dependent couplers in the loop mirror (small rectangular boxes) each contain an interference filter that transmits at the signal wavelength ($\approx$1557 nm) and reflects the $\lambda = 1534$-nm switching pulses.

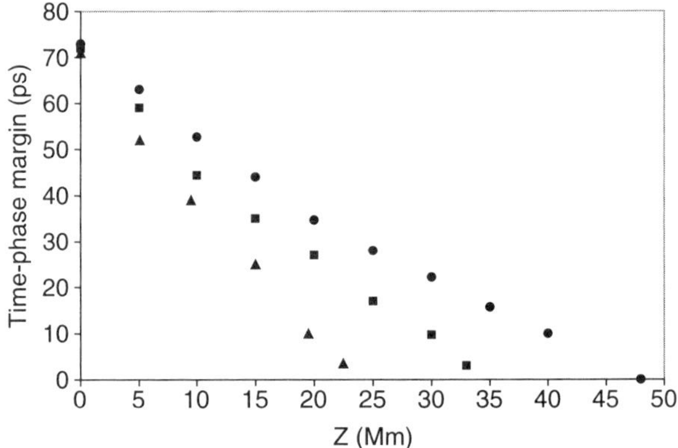

FIGURE 3.37 Time-phase margin vs. distance. Bullets: 2.5 Gbit/s. Squares: 10 Gbit/s, adjacent pulses orthogonally polarized. Triangles: 10 Gbit/s, all pulses co-polarized.

data stream for which only every fourth 2.5-Gbit/s subchannel is occupied); (2) a true 10-Gbit/s data stream with adjacent pulses orthogonally polarized; (3) a 10-Gbit/s data stream with all pulses co-polarized. Note that the error-free distances (for which the phase margin first becomes zero) are 48, 35, and 24 Mm, respectively.

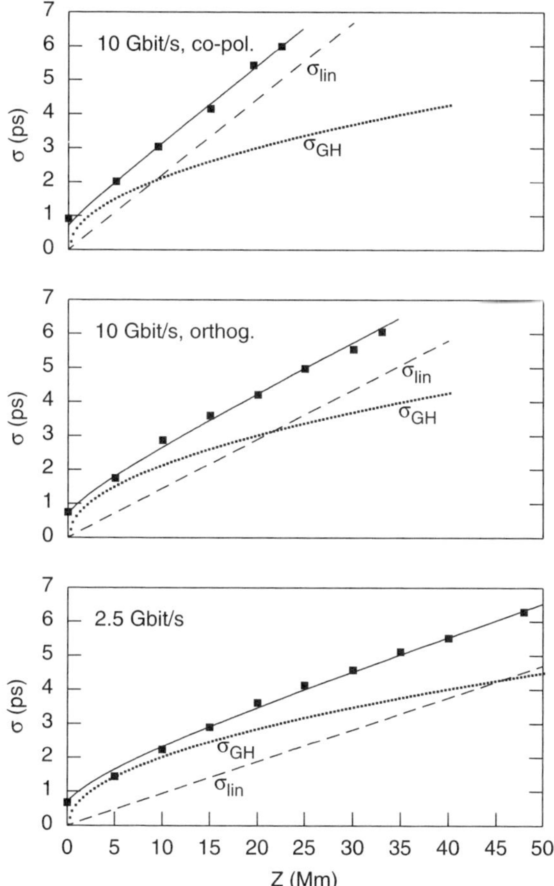

FIGURE 3.38 Standard deviation of jitter vs. distance for (bottom to top) 2.5 Gbit/s, 10 Gbit/s with adjacent pulses orthogonally polarized, and 10 Gbit/s with adjacent pulses co-polarized. Squares: Experimental points, as extrapolated from the data of Fig. 3.27. Solid curve: Best fit to theoretical curve of form $\sigma = \sqrt{\sigma_0^2 + \sigma_{GH}^2 + \sigma_{lin}^2}$. Dotted curve: σ_{GH}. Dashed line: σ_{lin}.

From the known properties of the error function, the difference between the effective width (here 82 ps) of the acceptance window and the measured time-phase margin at 10^{-10} BER should be $\approx 13\sigma$ of the Gaussian distribution in pulse arrival times. From this fact, one can then obtain the plots of σ shown in Fig. 3.38. Note that in all three cases shown there, the data make good fit to a curve of the form

$$\sigma = \sqrt{\sigma_0^2 + \sigma_{GH}^2 + \sigma_{lin}^2}, \tag{3.56}$$

where σ_0, σ_{GH}, and σ_{lin} represent the standard deviations of the source jitter (a constant), filter-damped Gordon–Haus jitter (varies as $z^{1/2}$), and jitter whose σ varies linearly with z, respectively. The best-fit Gordon–Haus term (σ_{GH}) is always about two times greater than expected for the known parameters of the experiment, and for the damping length of ≈400 km as calculated, and as confirmed by the independent measurement of the damping of amplitude jitter (Fig. 3.35). The frequency offset of the pulse spectra from the filter peaks is very small and can thus at best account for only a small part of the discrepancy, despite speculation to the contrary [70]. Thus, the only likely explanation here is in terms of the noise-like fields of dispersive wave radiation. Among the perturbations that may be responsible for significant amounts of such noise are the fiber's birefringence (see Chapter 7) and the periodic intensity associated with the use of lumped amplifiers (Section 1.4.2).

There are two contributions to the linear term, polarization jitter and the acoustic effect (Section 3.4.5). (The polarization jitter arises from a noise-induced spread in the polarization states of the solitons, and the conversion of that spread by the fiber's birefringence into a timing jitter. This effect will be discussed further in Chapter 7.) Since both contributions have essentially Gaussian distributions, the effects add as $\sigma_{lin} = \sqrt{\sigma_{pol}^2 + \sigma_a^2}$. For a filtered transmission line, the factor $Z^2/2$ in Eq. (3.39) is replaced with $Z \times \Delta$. Thus modified, Eq. (3.39) becomes

$$\sigma_a = 8.6 \frac{D^2}{\tau} Z\Delta\sqrt{R - 0.99}. \tag{3.39a}$$

Using the bit-rate dependence of σ_a and the slopes of σ_{lin} from the two lower plots of Fig. 3.37, one can easily extract values for σ_{pol} and σ_a. At $Z = 10$ Mm, those values are $\sigma_{pol} = 0.80$ ps, $\sigma_{a,2.5} = 0.50$ ps, and $\sigma_{a,10} = 1.21$ ps. The experimental values for σ_a are just 7% less than predicted by Eq. (3.39a), a remarkable degree of agreement.

Bit error rate measurements have been made at 12.5 and 15 Gbit/s, as well as at the 10 Gbit/s already cited. Figure 3.39 summarizes those results. Finally, it should be noted that by using sliding-frequency guiding filters, LeGuen *et al.* [71] achieved error-free transmission at 20 Gbit/s over more than 14 Mm.

3.5.6. *Stability Range*

The range of soliton pulse energies for which the transmission with sliding-frequency filters is stable and error-free will henceforth be simply referred to as the "stability range." It is important for the stability range to be large enough (at least several decibels) to allow for the aging of amplifier pump lasers, and other factors that may tend to degrade the signal strength with time, in real systems.

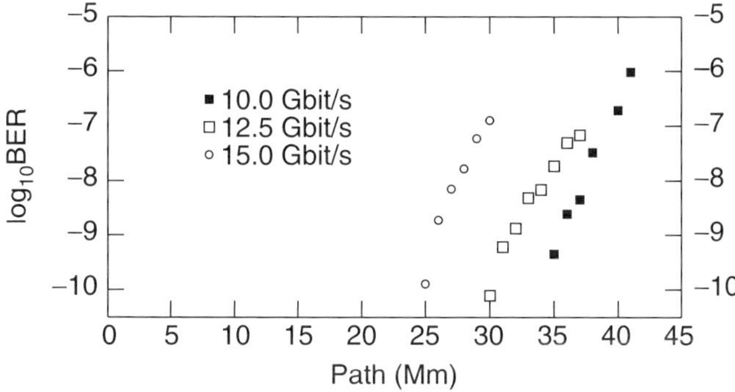

FIGURE 3.39 Measured bit error rate as a function of distance, at 10, 12.5, and 15 Gbit/s. In all cases, adjacent pulses were orthogonally polarized, and the data stream was a repeated, 2^{14}-bit random word.

Perhaps not surprisingly, the stability range is a function of both the filter strength parameter (η) and the sliding rate (ω_f'). The following brief summary is of an experimental determination of that dependence, and of the optimum values for those parameters [72].

The experiment was carried out in a small recirculating loop with piezo-driven etalon sliding-frequency filters having fixed reflectivity ($R = 9\%$) and fixed mirror spacing ($d = 1.5$ mm), and where $L_f = 39$ km. Because of the fact that η is inversely proportional to D [see Eq. (3.55a)], the fiber's third-order dispersion ($\partial D/\partial\lambda = 0.7$ ps/km-nm^2) enabled η to be varied simply through change of the signal wavelength itself. At the same time, the signal power at equilibrium, hence the soliton pulse energies, could be controlled by way of the pump power supplied to the loop amplifiers. Thus, the experiment consisted simply of measuring, for each signal wavelength, and for a fixed sliding rate, the maximum and minimum signal power levels for which a transmission over 10 Mm was stable and error-free. The results are shown in Fig. 3.40. Note that while error-free propagation ceases for $\eta \geq 0.8$, the stability range reaches a maximum, of nearly two to one, for $\eta \approx 0.4$. Figure 3.41 shows the complementary data, i.e., the measured stability range as a function of sliding rate, for fixed $\eta = 0.4$. Note that here, too, there is an optimum rate, of about 13 GHz/Mm. Essentially the same results as in Figs. 3.40 and 3.41 were obtained for two other values of L_f (26 and 50 km, respectively) and for etalons having a free spectral range (FSR) of 75 GHz (as opposed to 100 GHz).

The existence of E_{min} is easily predicted from the analysis of Section 3.5.3. That is, for stability, neither of the damping constants can be negative, so from Eq. (3.52), one has $|\Delta\Omega| \leq 1/\sqrt{6}$. From Eq. (3.50) or (3.50a), one then gets a

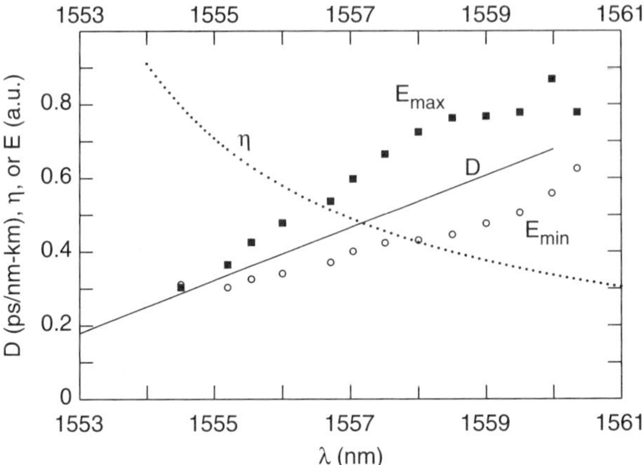

FIGURE 3.40 Fiber dispersion D (solid line), filter strength parameter η (dotted line), and experimentally determined allowable soliton pulse energy limits E_{max} and E_{min} (filled squares and open circles, respectively), as functions of the signal wavelength λ, for a fixed sliding rate of 13 GHz/Mm.

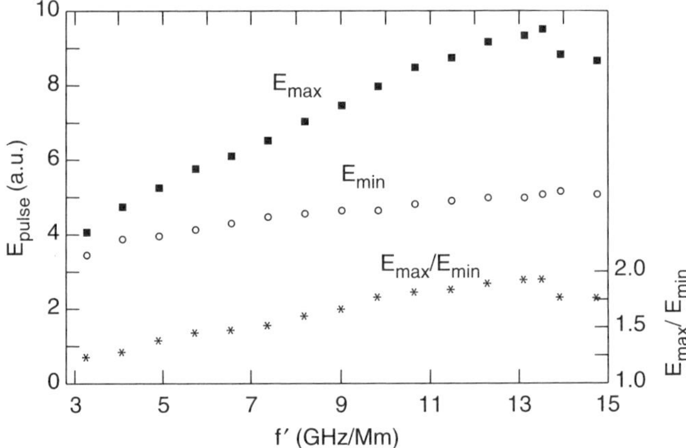

FIGURE 3.41 Experimentally determined allowable soliton pulse energy limits E_{max} and E_{min} (filled squares and open circles, respectively), and their ratio (stars), as functions of the frequency sliding rate f', for fixed, optimum filter strength parameter, $\eta = 0.4$.

maximum allowed sliding rate, ω_f', in soliton units. Finally, from Eq. (3.55b), one sees that, for fixed real sliding rate f' and for fixed D, ω_f' increases as the third power of the pulse width, and hence inversely as the third power of the pulse energy. Perhaps less abstractly, one can easily see that as the pulse energy is lowered, the rate at which the nonlinear term can alter the soliton's frequency will eventually become so low that it can no longer keep up with the filter sliding.

The existence of the upper limit, E_{max}, may be somewhat less obvious, but it has to do with the fact that eventually, as its energy is raised, the soliton's bandwidth, hence its loss from the filters, becomes too great. Here numerical simulation was helpful in elucidating the precise failure mechanism. Figure 3.42 shows the simulated pulse intensity evolution at $\eta = 0.4$ for different values of α_R. Note that for α_R below some critical value (here $\approx$1.45/Mm), there is no stable solution, and the pulse disappears after some distance of propagation; this corresponds to the lower energy limit already discussed. Above this lower limit, there is a range of allowable values of α_R (between 1.5/Mm and 3.5/Mm in Fig. 3.42). Nevertheless, one can see nondecaying oscillations in the pulse intensity evolution for the higher values of α_R. These oscillations are due to a nonsoliton component, not completely removed by the sliding, and generated by the perturbing effects of the filtering and sliding. If the excess gain is further increased, this nonsoliton component evolves into a second soliton. Clearly, this process determines the upper limit of the excess gain and of the soliton energy.

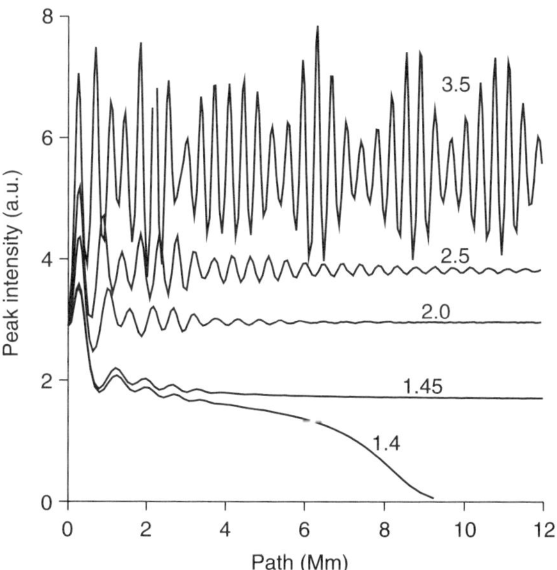

FIGURE 3.42 Soliton peak intensities as a function of distance, as determined by numerical simulation, for filter strength $\eta = 0.4$ and for various values of excess gain. The number next to each curve represents the excess gain parameter, α_R, in units of Mm^{-1}.

3.5.7. Filtering in Time

Another form of optical regeneration for solitons involves the use of intensity modulators periodically placed along the transmission line, and timed to open only during the middle of each bit period. The mean position in time of a pulse that is either early or late is thus guided back to the center of its bit period in a manner that is analogous to the guiding in the frequency domain provided by the etalon filters. Note, for example, that like frequency filtering, this "filtering in time" requires excess gain to overcome the loss imposed by the modulators, even to those arriving exactly on time. Unlike frequency filtering, however, filtering in time is not stable by itself; rather, it must always be accompanied by a proportional amount of frequency filtering. The principal advantage of filtering in time is that it corrects timing jitter directly, rather than indirectly as the frequency filtering does. Thus (when combined with filtering in the frequency domain), it can offer error-free transmission over an indefinitely long distance, at least in principle [73, 74].

Unfortunately, however, filtering in time shares several of the most fundamental disadvantages of electronic regeneration. For one, it is incompatible with WDM. (To do WDM, at each regenerator, the N channels must be demultiplexed, separately regenerated, and then remultiplexed, in a process that is at once cumbersome and expensive.) Second, each regenerator requires a quantity of complex, active hardware, including clock recovery, adjustable delay lines, and modulator drive, in addition to the modulator itself. (Compare with the extremely simple, inexpensive, and strictly passive etalon filters of the pure frequency filtering.) There are serious technical difficulties as well, such as the fact that nonchirping intensity modulators, whose insertion loss is polarization independent, simply do not exist at present. Thus, filtering in time would not seem to be economically or technically competitive.

Chapter 4

Soliton Interactions

In previous chapters, we have discussed only single solitons in the absence of noise fields (radiation) or other solitons. In a typical communication system, there will be many solitons in each channel as well as many different channels, and there will be noise, so it is important to understand their interactions. In this chapter, we discuss the theory of soliton collisions in different WDM channels, that is, of solitons of significantly different frequencies. Also, we apply some results of the inverse scattering transform (IST) to the interaction of small linear noise fields with solitons, and to the interaction of solitons in the same channel, that is, with the same central frequency, that get close together in time. For the highest possible bit rates, one wants the solitons in each channel as close together in time as possible, and the channels as close together in frequency as possible.

4.1. Soliton–soliton Collisions in WDM

In WDM, solitons of different channels gradually overtake and pass through each other (see Fig. 4.1). Because the solitons interact with each other, the time of overlap is known as a "collision." An important parameter here is the collision length, L_{coll}, or the distance the solitons must travel down the fiber together in the act of passing through each other. If L_{coll} is defined to begin and end with overlap at the half-power points, then transparently

$$L_{coll} = \frac{2\tau}{D\Delta\lambda}, \tag{4.1}$$

where $\Delta\lambda = \lambda_1 - \lambda_2$. For example, for $\tau = 20$ ps, $D = 0.5$ ps/nm-km, and $\Delta\lambda = 0.6$ nm, $L_{coll} = 133$ km.

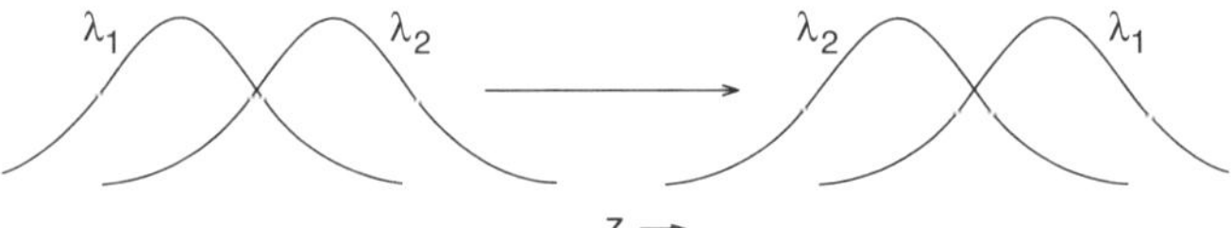

FIGURE 4.1 Two stages of a soliton–soliton collision. Because of the anomalous group velocity dispersion, the shorter wavelength soliton (λ_1) gradually overtakes and passes through the longer wavelength one (λ_2).

The interaction stems, of course, from the nonlinear term in the NLS equation. In a single-channel transmission, the only significant effect of that term is the *self*-phase modulation resulting from the self-induced index change at each pulse. During collisions, however, each pulse experiences an additional nonlinear index change as induced by the other pulse or pulses; as will be detailed shortly, the resultant *cross-phase modulation* tends to produce transient shifts in the mean frequency, or group velocity, of each pulse. Finally, the nonlinear term enables the colliding pulses to produce fields at the frequencies $\omega_S = 2\omega_1 - \omega_2$ and $\omega_A = 2\omega_2 - \omega_1$ (the "Stokes" and "anti-Stokes" frequencies, respectively) in the process known as *four-wave mixing*. In general, these effects have the potential to bring about a significant exchange of energy and momentum between the pulses, and hence to create serious timing and amplitude jitter. Under the right conditions, however, with solitons the nonlinear effects are only transient, i.e., the solitons emerge from a collision with pulse shapes, widths, energies, and momenta completely unchanged. It is this potential for nearly perfect transparency to one another that makes solitons so well adapted for WDM.

4.1.1. Soliton Collisions in Lossless and Constant-dispersion Fiber

It is useful first to consider the ideal case of lossless and constant-dispersion fiber. It is the simplest example of perfect transparency, and it is the easiest to analyze. It is of more than academic interest, however, since it can also serve as the paradigm for a number of practically realizable situations with real fibers and lumped amplifiers. In particular, it is the exact mathematical equivalent of the (at least approximately) realizable case of dispersion-tapered fiber spans.

Let us look at the collision of two solitons of equal amplitude in different channels. Except when they are actually in collision, the two solitons can be written in the more general form

$$\begin{aligned} u_1 &= \operatorname{sech}(t - t_{10} + \Omega_1 z)\exp[-i\Omega_1 t + i(1 - \Omega_1^2)z/2 + i\phi_{10}], \\ u_2 &= \operatorname{sech}(t - t_{20} + \Omega_2 z)\exp[-i\Omega_2 t + i(1 - \Omega_2^2)z/2 + i\phi_{20}]. \end{aligned} \tag{4.2}$$

We are free to choose the central frequency. For convenience then, let us set $\Omega_2 = -\Omega_1 = \Omega$; note that this makes the solitons move with equal but opposite velocities in the retarded time frame. Also note that this makes $L_{coll} = 1.763/\Omega$ ($= 1.763 \times z_c/\Omega t_c$ in ordinary units). Before and after the collision the amplitudes and frequencies of the two solitons are unchanged, but there are changes, as we shall see, in the quantities t_{10}, t_{20}, ϕ_{10}, and ϕ_{20}. These changes in the mean time and phase of the two solitons are due to the fact that the mean frequencies of the two solitons are perturbed during the collision by the cross-phase modulation. Here we discuss the case where the changes are small perturbations. For the unperturbed solitons we can set t_{10}, t_{20}, ϕ_{10}, and ϕ_{20} to zero for simplicity, since the changes in these quantities do not depend on their initial values. Now the collision occurs at $z = 0$.

Now insert $u = u_1 + u_2$ into the NLS equation for lossless fiber, expand, and group the terms according to their frequency dependencies. The channel spacing is 2Ω. The half-intensity bandwidth of each soliton is $\Delta\omega = 2 \times 1.763/\pi = 1.1222$. Thus, so long as $\Omega > 2$, the soliton spectra do not overlap appreciably. We get four equations: one for each soliton at $\pm\Omega$, and one each for the terms at $\pm 3\Omega$. The latter correspond to the previously mentioned four-wave mixing components. Since, for the special case under consideration here, these components are weak and disappear completely after the collision, they will be neglected for now. Assuming, for the moment, that the pulses are co-polarized, the equation for u_1 is

$$\frac{\partial u_1}{\partial z} = i\frac{1}{2}\frac{\partial^2 u_1}{\partial t^2} + i|u_1|^2 u_1 + 2i|u_2|^2 u_1. \tag{4.3}$$

The first two terms on the right in Eq. (4.3) correspond to the NLS equation for the isolated pulse, u_1. The last term in Eq. (4.3) corresponds to the cross-phase modulation and is zero except when the pulses overlap. It produces a shift in the phase ϕ_1 of the soliton u_1 at the rate

$$\frac{\partial \phi_1(z,t)}{\partial z} = 2|u_2(z,t)|^2. \tag{4.4}$$

This phase shift is not uniform over the pulse. Since, however, we are looking for the shift in ϕ_{10} caused by the cross-phase modulation, we need only its mean value. To calculate the mean of Eq. (4.4), we use the weighting factor $|u_1(z,t)|^2/2$ [since the weighting factor must integrate to unity, and $\int_{-\infty}^{\infty} \operatorname{sech}^2(t)\,dt = 2$], and use the unperturbed solitons in the integral. We thus obtain

$$\frac{d}{dz}\delta\phi_{10}(z) = \int_{-\infty}^{\infty} \operatorname{sech}^2(t - \Omega z)\operatorname{sech}^2(t + \Omega z)\,dt. \tag{4.5}$$

This expression can be easily integrated over z, since then $t - \Omega z$ and $t + \Omega z$ can be taken as independent variables. The resulting phase shift from this perturbation calculation is $\delta\phi_{10} = 2/\Omega$. Interchanging subscripts in Eq. (4.5) shows that the phase shift $\delta\phi_{20}$ of soliton u_2 is equal to $\delta\phi_{10}$ of soliton u_1.

We now consider the frequency shifts during the collision and the resulting shifts in the mean time of the solitons. It can be shown rigorously from the NLS equation that the inverse group velocity of a pulse is given by minus one times its mean frequency. Hence the mean time delay is the integral over z of minus the mean frequency. We will now see how the mean frequency of a soliton pulse varies during a collision.

The "instantaneous" frequency within a pulse is minus the time derivative of its phase. Let $\Omega_1 + \omega_1(t,z)$ be the instantaneous frequency of soliton u_1. The frequency shift $\omega_1(t,z)$ exists only during the collision. The phase shift of Eq. (4.4) creates the frequency shift ω_1 at the rate

$$\frac{\partial \omega_1}{\partial z} = -\frac{\partial}{\partial z}\frac{\partial \phi_1}{\partial t} = -\frac{\partial}{\partial t}\frac{\partial \phi_1}{\partial z} = -\frac{\partial}{\partial t}\left(2|u_2|^2\right). \tag{4.6}$$

Again, this frequency shift is not uniform across the pulse, but all we need is its mean value. Using the same weighting factor $|u_1(z,t)|^2/2$ as before, and labeling $\langle \omega_1 \rangle$ as $\delta\Omega_1$, we get

$$\frac{d}{dz}\delta\Omega_1 = -\int_{-\infty}^{\infty} |u_1|^2 \frac{\partial}{\partial t}|u_2|^2 \, dt = -\frac{1}{2}\int_{-\infty}^{\infty}\left(|u_1|^2\frac{\partial}{\partial t}|u_2|^2 - |u_2|^2\frac{\partial}{\partial t}|u_1|^2\right) dt. \tag{4.7}$$

The equivalence of the two expressions on the right of Eq. (4.7) can be shown by partial integration. Interchanging subscripts in Eq. (4.7) shows that the mean frequency shift $\delta\Omega_2$ of soliton u_2 is minus the frequency shift $\delta\Omega_1$ of soliton u_1. Using the unperturbed solitons in the integral, we get from the second one

$$\frac{d}{dz}\delta\Omega_2 = -\frac{d}{dz}\delta\Omega_1 = \frac{1}{2\Omega}\frac{d}{dz}\int_{-\infty}^{\infty} \operatorname{sech}^2(t-\Omega z)\operatorname{sech}^2(t+\Omega z)\, dt, \tag{4.8}$$

whence

$$\begin{aligned}\delta\Omega_2 = -\delta\Omega_1 &= \frac{1}{2\Omega}\int_{-\infty}^{\infty} \operatorname{sech}^2(t-\Omega z)\operatorname{sech}^2(t+\Omega z)\, dt \\ &= \frac{4\Omega z\cosh(2\Omega z) - 2\sinh(2\Omega z)}{\Omega \sinh^3(2\Omega z)}.\end{aligned} \tag{4.9}$$

The peak frequency shift, at $z=0$, is $\delta\Omega_2 = -\delta\Omega_1 = 2/(3\Omega)$. Finally, Eq. (4.9) can be integrated to yield the net time displacements:

$$\delta t_{10} = -\delta t_{20} = 1/\Omega^2. \tag{4.10}$$

Equation (4.8) (multiplied by -1) represents the "acceleration," i.e., the rate of change of inverse group velocity with distance into the collision.

Later in this chapter we obtain the exact expressions for the collision-induced phase and time shifts, correct for any channel spacing. For the case studied here, with solitons of equal amplitude, the exact results are

$$\delta\phi_{10} = \delta\phi_{20} = 2\tan^{-1}(1/\Omega),$$

$$\delta t_{10} = -\delta t_{20} = \ln(1 + 1/\Omega^2).$$

These results are derived from a purely mathematical analysis starting from an exact solution of the two-soliton problem and, therefore, do not shed much light on the processes involved in a collision.

The preceding expressions for the acceleration and the velocity shift [Eqs. (4.8) and (4.9)] may not be particularly transparent. When numerically evaluated and graphed, however, they are seen to be simply behaved (see Fig. 4.2). The importance of including the acceleration curve in Fig. 4.2 will become apparent in the next chapter. Note, either from the graph or from the pertinent equations, that the pulses attract each other, while their frequencies repel one another. Also note that, as advertised, the completed collision leaves the soliton intact, with the same frequency and amplitude it had before the collision. Thus, the only changes are the time shifts δt_{10} and δt_{20} and the associated phase shifts $\delta\phi_{10}$ and $\delta\phi_{20}$. As will be shown later, guiding filters tend to remove these unwanted time shifts.

The discussion in this section has thus far been almost entirely in terms of soliton units. For convenient future reference, however, we now list formulas for the principal quantities in practical units. First, in terms of the full-channel

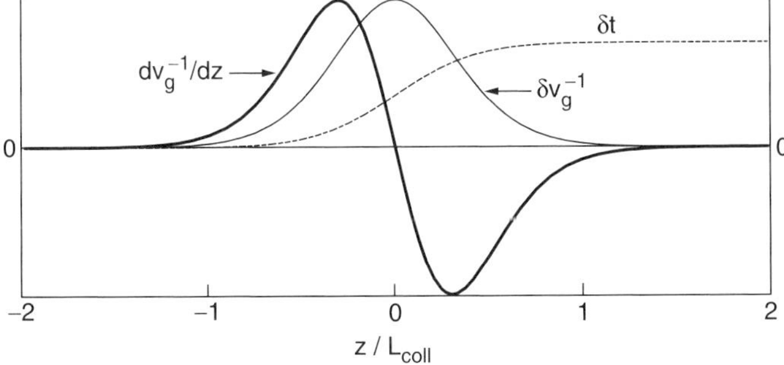

FIGURE 4.2 Acceleration (dv_g^{-1}/dz), velocity shift ($\delta v_g^{-1} = -\delta\Omega$), and time shift of the slower pulse, during a soliton–soliton collision in a lossless fiber. [For the faster (higher frequency) pulse, turn this graph upside down.] The maximum inverse group velocity shift $(\delta v_g^{-1})_{max} = 2/(3\Omega)$, and $\delta t_{max} = \Omega^{-2}$.

separation $\Delta\nu$ and the pulse width τ, the half-channel separation in soliton units is

$$\Omega = \frac{\pi}{1.763}\tau\Delta\nu = 1.782\tau\Delta\nu. \tag{4.11a}$$

The maximum frequency shift during the collision, $\delta\nu$, and the net time displacement, δt, when expressed in practical units, are, respectively,

$$\delta\nu = \pm\left(\frac{1.763}{\pi}\right)^2 \frac{1}{3\tau^2\Delta\nu} = \pm\frac{0.105}{\tau^2\Delta\nu} \tag{4.11b}$$

and

$$\delta t = \pm\frac{1.763}{\pi^2\tau\Delta\nu^2} = \pm\frac{0.1786}{\tau\Delta\nu^2}. \tag{4.11c}$$

(The number 1.763 represents $2\cosh^{-1}(\sqrt{2})$, the dimensionless ratio τ/t_c.) For example, consider a collision between 20-ps solitons in channels separated by 75 GHz (0.6 nm). Then Eqs. (4.11a), (4.11b), and (4.11c) yield, respectively, $\Omega = 2.67$ (more than big enough for effective separation of the soliton spectra), $\delta\nu = \pm 3.5$ GHz, and $\delta t = \pm 1.59$ ps. Note that in this case Ω is not much larger than one, but the general features of the collision remain valid.

4.1.2. Four-wave Mixing

We now take up the two temporarily neglected equations from the preceding expansion of the NLS equation and the four-wave mixing products that result. During a collision between solitons in channels centered at frequencies Ω_2 and Ω_1, the nonlinear term in the propagation equation produces anti-Stokes and Stokes sidebands centered, respectively, at frequencies

$$\Omega_A = 2\Omega_2 - \Omega_1 = \Omega_2 + \Omega_{21} \qquad \text{and} \qquad \Omega_S = 2\Omega_1 - \Omega_2 = \Omega_1 - \Omega_{21},$$

where $\Omega_{21} = \Omega_2 - \Omega_1$, by the four-wave mixing process (see Fig. 4.3). At the "anti-Stokes" frequency Ω_A, the propagation equation is

$$-i\frac{\partial u_A}{\partial z} = \frac{1}{2}\frac{\partial^2 u_A}{\partial t^2} + u_2^2 u_1^*, \tag{4.12}$$

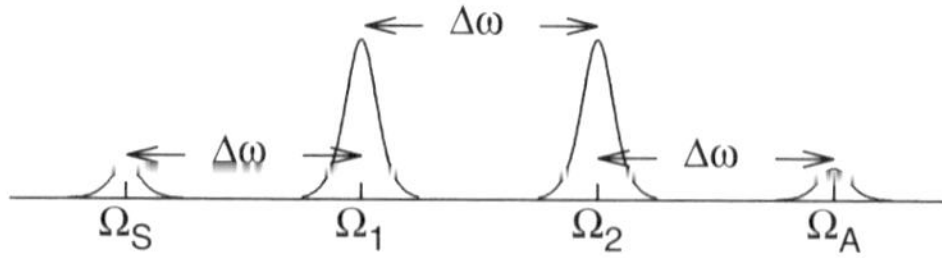

FIGURE 4.3 The spectra of pulses from different channels and the sidebands generated by their interaction through four-wave mixing.

where u_2 represents the soliton at frequency Ω_2, and u_1 represents the soliton at frequency Ω_1. The propagation equation for the Stokes sideband is obtained by substituting u_S for u_A and interchanging u_1 and u_2 in Eq. (4.12).

An important feature of the production of these sidebands is phase matching, or wavenumber matching. If the two solitons are as described in Eq. (4.2), i.e., both have amplitude $A=1$, the soliton at frequency Ω_1 has the wavenumber $k_1 = (1-\Omega_1^2)/2$ and the soliton at frequency Ω_2 has the wavenumber $k_2=(1-\Omega_2^2)/2$. Thus the driving term $u_2^2 u_1^*$ in Eq. (4.12) has the wavenumber $2k_2-k_1=(1-2\Omega_2^2+\Omega_1^2)/2$. On the other hand, small fields at frequency Ω_A have the linear dispersion relation $k_A=-\Omega_A^2/2$. As a result, there is a wavenumber mismatch Δk given by

$$\Delta k = 2k_2-k_1-k_A=(1-2\Omega_2^2+\Omega_1^2+\Omega_A^2)/2=1/2+(\Omega_2-\Omega_1)^2. \tag{4.13}$$

This is analogous to the frequency mismatch of an oscillator driven off resonance. Note that the wavenumber of a soliton with amplitude unity and central frequency Ω is the sum of (1/2), which is produced by the nonlinearity, and $-\Omega^2/2$, which is the linear dispersion relation. Later in this book there is a discussion of the effects of higher order linear dispersion, particularly the third-order term $d^3k/(d\Omega)^3$. By expanding the first expression for Δk in Eq. (4.13) as a power series in Ω around Ω_2, one sees that Δk depends only on even orders in the linear dispersion relation and so does not depend on the third-order term. That is,

$$\Delta k = 1/2+(\Omega_2-\Omega_1)^2\frac{\partial^2 k}{(\partial\Omega)^2}+O\left(\frac{\partial^4 k}{(\partial\Omega)^4}\right), \tag{4.14}$$

where the derivatives are evaluated at the frequency Ω_2. Again, results for the Stokes sideband are obtained by replacing u_A by u_S and interchanging u_1 and u_2.

Four-wave Mixing with Continuous Waves

As a preamble to dealing with the soliton problem, consider what would happen in Eq. (4.12) if u_2 and u_1 were continuous waves at the respective frequencies $\omega_2=\Omega_2$ and $\omega_1=\Omega_1$. Thus, if

$$u_1=\eta_1\exp(-i\omega_1 t+ik_1 z) \qquad \text{and} \qquad u_2=\eta_2\exp(-i\omega_2 t+ik_2 z), \tag{4.15}$$

where η_1 and η_2 are small real amplitudes, then

$$u_2^2u_1^*=\eta_2^2\eta_1\exp(-i\omega_A t+i(2k_2-k_1)z),$$

where $\omega_A=2\omega_2-\omega_1$. With this driving term, the general solution to Eq. (4.12) is

$$u_A=\frac{\eta_2^2\eta_1}{\Delta k}\exp(-i\omega_A t)\{\exp[i(2k_2-k_1)z]+a\exp(ik_A z)\}, \tag{4.16}$$

where $k_A = -\omega_A^2/2$. In this exercise, $\Delta k = 2k_2 - k_1 - k_A = (\omega_2 - \omega_1)^2$ is the linear part of the wavenumber mismatch [see Eq. (4.13)], and a is a numerical constant that depends on the assumed initial conditions. For example, if $u_A = 0$ at $z = 0$, then $a = -1$. The first term in the numerator of Eq. (4.16) represents the particular solution of Eq. (4.12), while the second term represents the homogeneous solution (at the same frequency but with the driving term absent). If $a \neq 0$, the amplitude of u_A is oscillatory with a period Δz of $2\pi/\Delta k$ due to the interference of the two terms in Eq. (4.16). For example, if a is real, then from Eqs. (4.16) and (4.15), we have

$$|u_A|^2 = \frac{|u_2|^4 |u_1|^2}{\Delta k^2} \left(1 + a^2 + 2a\cos(\Delta k\, z)\right). \tag{4.17}$$

For use in later chapters, it is helpful to transform Eq. (4.17) to ordinary units. The quantity $|u|^2$ represents the power (P) in the various channels. According to the usual procedure, in standard units, P becomes P/P_c and Δk becomes $\Delta k\, z_c$. Since $P_c z_c = 1/k_{NL}$, we get the result

$$P_A = \left(\frac{k_{NL} P_2}{\Delta k}\right)^2 P_1 \left(1 + a^2 + 2a\cos(\Delta k\, z)\right).$$

Then if we take $a = -1$ (i.e., assume that u_1 and u_2 are injected at the beginning of the fiber), we get

$$P_A = 2\left(\frac{k_{NL} P_2}{\Delta k}\right)^2 P_1 (1 - \cos(\Delta k\, z)) = 4\left(\frac{k_{NL} P_2}{\Delta k}\right)^2 P_1 \sin^2(\Delta k\, z/2), \tag{4.18}$$

where $k_{NL} = (2\pi\, n_2)/(\lambda A_{eff})$ [Eq. (1.11)], where

$$\Delta k = \frac{\lambda^2 D}{2\pi c}(\omega_2 - \omega_1)^2 = -2\pi c D \left(\frac{\delta\lambda}{\lambda}\right)^2, \tag{4.19}$$

and $\delta\lambda$ corresponds to $\omega_2 - \omega_1$. As before, the corresponding formula for P_S can be obtained by interchanging subscripts 1 and 2 in Eq. (4.18).

If the two cw waves are large enough that the nonlinear effects cannot be ignored, then the wavenumbers k_1 and k_2 are modified to read

$$k_1 = -\omega_1^2/2 + \eta_1^2 \qquad \text{and} \qquad k_2 = -\omega_2^2/2 + \eta_2^2,$$

so the wavenumber mismatch $\Delta k = 2k_2 - k_1 - k_A$ becomes $\Delta k = (\omega_2 - \omega_1)^2 + 2\eta_2^2 - \eta_1^2$. In ordinary units, this becomes

$$\Delta k = -\frac{cD}{2\pi}\left(\frac{\delta\lambda}{\lambda}\right)^2 + k_{NL}(2P_2 - P_1). \tag{4.19a}$$

Note that from the linear dispersion relation, $\Delta k = (\omega_2 - \omega_1)^2$. If we take $\omega_2 = \Omega$ and $\omega_1 = -\Omega$, then $\Delta k = 4\Omega^2$, and the period Δz is $\pi/(2\Omega^2)$, which is typically shorter than the collision length $L_{coll} = 1.763/\Omega$ for the two solitons. For example, if $\Omega = 2.5$, then $\Delta z = 0.251$, while $L_{coll} = 0.705$.

Now consider the actual collision of two solitons in different channels. As before, let us take $\Omega_2 = -\Omega_1 = \Omega$, so that $\Omega_A = 3\Omega$. The propagation equation for the anti-Stokes wave is

$$-i\frac{\partial u_A}{\partial z} = \frac{1}{2}\frac{\partial^2 u_A}{\partial t^2} + \mathrm{sech}^2(t+\Omega z)\,\mathrm{sech}(t-\Omega z)\exp[-i3\Omega t + i(1-\Omega^2)z/2]. \tag{4.20}$$

On the assumption that $4\Omega^2 \gg 1$, a first approximation is that u_A follows the driving term adiabatically. Another way of saying this is that the rate of change with z of the soliton envelopes is slow enough that derivatives of the sech functions can be neglected. The adiabatic solution is [see Eq. (4.16)]

$$u_A(z,t) = \frac{1}{4\Omega^2}\,\mathrm{sech}^2(t+\Omega z)\,\mathrm{sech}(t-\Omega z)\exp\left[-i3\Omega t + i\left(1-2\Omega^2\right)z/2\right]. \tag{4.21}$$

Except for a phase shift proportional to $1/\Omega$, this is the same as the solution given in the next section, derived there from the inverse scattering transform.

The growth and decay of the four-wave mixing (FWM) power during a two-soliton collision, as inferred from the absolute value of Eq. (4.21) squared, is shown in Fig. 4.4. Note that even for the smallest frequency separations that might be encountered in practice ($\Omega_2 - \Omega_1 \approx 5$ or $\Omega \approx 2.5$), the peak FWM sideband power ($P_c/16\Omega^4$) tends to be a very small fraction (0.16% for $\Omega = 2.5$) of the soliton power P_c. As functions of time at the center of the collision, the FWM sidebands are roughly sech^2 in shape, but have only about 53% of the width of the

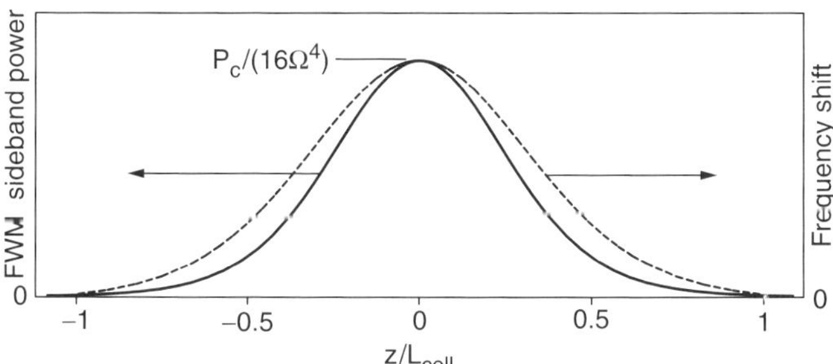

FIGURE 4.4 Heavy, solid line: Peak power of either of the two FWM sidebands produced in a collision between two solitons separated in frequency by 2Ω, plotted as a function of z/L_{coll}; the maximum of this curve is $P_c/16\Omega^4$. Dashed line: The corresponding frequency shift, shown for comparison.

solitons. Thus, the energy in each FWM sideband is smaller still, less than 0.1% of W_{sol}. Of course, it is of equal importance that the FWM sidebands disappear altogether after the collision is complete.

Despite this disarming behavior of the FWM sidebands in lossless fiber, however, in real systems with gain, loss, and other perturbations, the behavior can be quite different. The issue here is the potential for phase matching. That is, if there is a perturbation whose period closely matches $\Delta z = 2\pi/\Delta k$ [see the discussion of Eq. (4.16)], the two halves of the oscillatory cycle do not quite cancel, and, as discussed in some detail in Chapter 5, Section 5.5, the four-wave mixing components can build up considerably. This is a condition that can and must be avoided. In the case of the two-soliton collision, Eq. (4.21) corresponds only to the driven solution. If the solitons are perturbed at some location z_1, the field u_A is no longer the driven solution for the perturbed solitons. It must, therefore, be resolved into the new driven solution plus some of the homogeneous solution, which propagates with the wavenumber k_A. If the same perturbation happens again at a location z_2, a similar piece of the homogeneous solution is created. Now if $z_2 - z_1 = \Delta z$, the two homogeneous components are in phase and add constructively. Thus a perturbation with wavenumber $2\pi/\Delta k$ can cause a significant buildup of the anti-Stokes field u_A and a similar buildup of the Stokes wave u_S.

4.2. Applications of the Inverse Scattering Transform

The inverse scattering transform (IST) is a kind of nonlinear Fourier transform. It has been shown via the IST that under the conditions of nonlinearity and dispersion that allow solitons, a general field consists of solitons and dispersive radiation. In the absence of radiation, it provides exact analytical solutions for the case of many solitons in the form of sets of algebraic equations, N equations for N solitons. While these are available, they rapidly become unwieldy, so for N>2 they are not very useful. In this section we will use some results gleaned from the IST to look into the case of one soliton interacting with a small dispersive field and at the general interaction of two solitons. We reproduce the results obtained in the preceding sections of this chapter and, in addition, discuss the interaction of neighboring solitons in the same channel.

When a small linear noise-like field interacts with a soliton, its Fourier components suffer phase shifts while traversing the soliton, depending on the frequency difference between the linear field component and the central frequency of the soliton. This results in a small shift of the mean time of any small linear wave packet that passes the soliton.

Soliton interactions occur when two or more solitons come close together in time. If two solitons of the same frequency, and thus of the same group velocity, are too close, they either attract or repel each other, depending on their phase relationship. If the solitons are in phase, their fields constructively interfere in the overlap region between them, resulting in an attractive force. If they are out of phase, their fields destructively interfere in the overlap region, resulting in a repulsive force. Solitons of different central frequencies have different group velocities, so they can pass one another. While two solitons are in collision, the nonlinear term $|u|^2u$ creates sideband fields at multiples of their frequency separation via four-wave mixing. Being solitons (no scattering), we expect that these sideband fields must disappear at the end of the collision, and so they do for solitons of the basic NLS equation. The solitons themselves, while they suffer no changes in amplitude or frequency as a result of a collision, do suffer transient frequency shifts during the collision due to cross-phase modulation, which result in small time and phase shifts similar to those experienced by small linear fields. In a communication system, with losses and gains, these sidebands and frequency excursions cause problems, and must be considered carefully.

There are a few situations in which the IST can cast some light on the behavior of solitons of the basic NLS equation. There is the case of a small amplitude field when a soliton is present. There is the case of two solitons. (For those interested in pursuing the IST, we refer to the literature.)

4.2.1. *One Soliton and Noise*

Suppose that we have the situation of a small amplitude field such as noise in the presence of a soliton. Amplifiers inject noise into a transmission fiber, and the injected noise that overlaps a soliton makes small random alterations in the parameters of the soliton (mean time, phase, amplitude, and frequency). Other small amplitude fields that are orthogonal to the soliton, in the sense that they do not alter the soliton parameters, are called radiation. We look here into how the soliton influences the radiation fields. Let $u_s(z,t)$ represent the initial soliton and $u_p(z,t)$ represent the radiation. If we put $u = u_s + u_p$ in the NLS equation, and keep only terms linear in u_p, we get an equation of the form

$$-i\frac{\partial u_p}{\partial z} = \frac{1}{2}\frac{\partial^2 u_p}{\partial t^2} + 2|u_s|^2 u_p + u_s^2 u_p^*. \tag{4.22}$$

Solutions to this equation yield perturbations of the radiative field caused by the soliton. Suppose that the soliton u_s is the fundamental soliton.

$$u_s = \text{sech}(t)\exp(iz/2).$$

If the small radiative field is temporally separate from the soliton, it simply satisfies the linear dispersive equation. If the radiative field consists primarily of frequency components in the neighborhood of some mean frequency ω, note that the term $u_s^2 u_p^*$ in Eq. (4.22) creates a sideband at the image frequency $-\omega$ if the radiation encounters the soliton.

By linearization of the radiative field in the IST, the solution to Eq. (4.22) comes out as follows [75]. Let $f(z,t)$ be any solution of the linear dispersive equation,

$$-i\frac{\partial f}{\partial z}=\frac{1}{2}\frac{\partial^2 f}{\partial t^2}, \tag{4.23}$$

i.e., the NLS equation without the nonlinear term. For every such function $f(z,t)$ that satisfies the linear equation, there is a corresponding function u_p, as given by

$$u_p=-\frac{\partial^2 f}{\partial t^2}+2\tanh(t)\frac{\partial f}{\partial t}-\tanh^2(t)f+u_s^2 f^*, \tag{4.24}$$

which is a solution of Eq. (4.22). This may be verified (with some difficulty) by inserting Eq. (4.24) into Eq. (4.22).

Consider now a spectral component of f at frequency Ω, such that at location z, $f(z,t)=\tilde{f}(z,\Omega)\exp(-i\Omega t)$. Then we find

$$u_p(z,t)=(\Omega-i\tanh(t))^2\tilde{f}(z,\Omega)\exp(-i\Omega t)+u_s^2\tilde{f}^*(z,\Omega)\exp(i\Omega t). \tag{4.25}$$

Notice that u_p suffers a frequency-dependent phase shift while traversing the soliton, and that a sideband field at the mirror image frequency $-\Omega$ (remember that the central frequency of the soliton is zero) appears at the location of the soliton. The "instantaneous" frequency of the part of u_p proportional to $\exp(-i\omega t)$, given by $\omega=-\Im(u^{-1}\,\partial u/\partial t)$ is

$$\omega=\Omega\left(1+\frac{\operatorname{sech}^2(t)}{\Omega^2+\tanh^2(t)}\right)=\frac{\Omega(\Omega^2+1)}{\Omega^2+\tanh^2(t)}.$$

The instantaneous frequency of this radiative field component returns to its original value once it passes the soliton. The frequency excursion of the radiative field as it passes the soliton is always away from the soliton frequency. It can get very large if the frequency of a radiative field component approaches that of the soliton.

The frequency-dependent phase shifts encountered by the spectral components of u_p give rise to a shift in the mean time of any wave packet that passes the soliton. Suppose that a small radiative field in the form of a wave packet traverses the soliton. As a function of z and ω, spectral components of the wave packet $f(z,t)=\frac{1}{\sqrt{2\pi}}\int\tilde{f}(\omega,z)\exp(-i\omega t)$ vary as

$$\tilde{f}(z,\omega)=\tilde{f}(0,\omega)\exp(-i\omega^2 z/2).$$

Suppose that the spectrum of the wave packet consists only of positive frequencies (higher than the soliton), and that the wave packet starts well behind the soliton, at positive t, so that we can take $\tanh(t)=1$ in Eq. (4.25). As it propagates, the wave packet will pass the soliton and later will appear at negative t, where we can take $\tanh(t)=-1$. Before its encounter with the soliton, the spectral components of u_p vary as

$$\begin{aligned}\tilde{u}_p &= (\omega-i)^2\tilde{f}(0,\omega)\exp\left(-i\omega^2 z/2\right) \\ &= \left(\omega^2+1\right)\tilde{f}(0,\omega)\exp\left[-i\omega^2 z/2 - i2\tan^{-1}(1/\omega)\right].\end{aligned} \tag{4.26}$$

The mean time of any wave packet is given by

$$\langle t\rangle = W^{-1}\int_{-\infty}^{\infty}\tilde{u}^*(-i\,\partial\tilde{u}/\partial\omega)\,d\omega,$$

where $W=\int_{-\infty}^{\infty}\tilde{u}^*\tilde{u}\,d\omega$. Applied to Eq. (4.26), we get the result

$$\langle t\rangle = \langle t\rangle_{z=0} - \langle\omega\rangle z + \left\langle\frac{2}{\omega^2+1}\right\rangle, \tag{4.27}$$

where we have used $d\tan^{-1}(1/\omega)/d\omega = -1/(1+\omega^2)$. Similarly, after the encounter with the soliton, the spectral components of u_p are

$$\tilde{u}_p = \left(\omega^2+1\right)\tilde{f}(0,\omega)\exp\left[-i\omega^2 z/2 + i2\tan^{-1}(1/\omega)\right], \tag{4.28}$$

and the corresponding mean time is

$$\langle t\rangle = \langle t\rangle_{z=0} - \langle\omega\rangle z - \left\langle\frac{2}{\omega^2+1}\right\rangle. \tag{4.29}$$

The total time shift of the wave packet is the mean value of $-4/(\omega^2+1)$. We see that this wave packet, at frequencies higher than the soliton, is shifted ahead in time. Similarly, a wave packet at frequencies lower than the soliton is shifted back in time by the same amount in its encounter with the soliton. This behavior is reflected in the collisions of two solitons, as we have seen previously, and which we now consider from the vantage point of the IST.

4.2.2. *Two Solitons*

Solutions of the NLS equation corresponding to an integral number N of solitons have been given in the form of the solutions of a set of N algebraic equations. For completeness, we give here the form of those equations, even though we will consider only pairs of solitons.

Let A_j be the amplitude, and Ω_j the frequency, of the jth soliton in a set of N solitons of various amplitudes and frequencies, locations, and phases. At various locations, solitons of different frequencies will collide, but since there is no scattering in these collisions, the set of amplitudes and frequencies remains the same before and after any collisions. The parameters A_j and Ω_j are therefore basic constants of the motion of the solitons. Whenever the jth soliton is apart from the others, it will always have the amplitude A_j and the frequency Ω_j. On the other hand, as we have already seen, collisions occasion transient frequency shifts of the colliding solitons, resulting in position and phase shifts, as well as transient sidebands at multiples of their frequency separation. The multisoliton function accounts for all of these effects.

The general N soliton function is given by [76]

$$u(z,t)=\sum_{k=1}^{N}u_k(z,t). \tag{4.30}$$

The $u_k(z,t)$ satisfy the set of algebraic relations

$$\sum_{k=1}^{N}M_{jk}\left(\gamma_j^{-1}+\gamma_k^*\right)u_k=1, \quad j=1\ldots N, \tag{4.31}$$

where

$$M_{jk}=(A_j+A_k-i(\Omega_j-\Omega_k))^{-1} \tag{4.32}$$

and

$$\gamma_j=\exp\{A_j(t-t_{j0}+\Omega_j z)+i[-\Omega_j z+\frac{1}{2}(A_j^2-\Omega_j^2)z+\phi_{j0}]\}. \tag{4.33}$$

In the case of one soliton ($N=1$), there is only one equation, which is easily solved. The result (omitting the j subscript) is

$$u=A\,\mathrm{sech}[A(t-t_0+\Omega z)]\exp\{i[-\Omega z+\frac{1}{2}(A^2-\Omega^2)z+\phi_0]\}, \tag{4.34}$$

which reproduces the general form of the single soliton. In the general case of N solitons, the jth soliton peaks near where

$$t=t_{j0}-\Omega_j z.$$

However, it is shifted both in time and phase in order to account for the interactions that occur when the solitons collide. This multisoliton function is symmetrical with respect to the soliton collisions. That is, it takes each collision into account by ascribing oppositely directed time and phase shifts before and after the collision. One can show that these shifts are additive. It does not matter whether a set of solitons collide successively or all at once, the total shifts are the same.

The case of two solitons is already somewhat cumbersome, but it is instructive. To simplify the result without losing insight, we take $A_1 = 1 + a$ and $A_2 = 1 - a$, where $|a| \leq 1$, so that the sum of their amplitudes is 2, and also $\Omega_1 = \Omega(1-a)$, $\Omega_2 = -\Omega(1+a)$, so that they have equal and opposite "momenta" $A\Omega$. Also we take $t_{10} = (1-a)t_0$ and $t_{20} = -(1+a)t_0$, so that the solitons collide when $\Omega z = t_0$, and finally we take $\phi_{10} = -\phi_{20} = \phi_0$. In what follows we will always assume that $\Omega \geq 0$, so that soliton number one is always traveling with or faster than soliton number two, and if we are to witness a collision, soliton number one must start behind soliton number two, that is, at positive time. If $a > 0$, soliton number one is larger than soliton number two.

Using the above parameters, we have

$$M_{11}^{-1} = 2(1+a) \quad M_{22}^{-1} = 2(1-a) \quad M_{12}^{-1} = 2(1-i\Omega) \quad M_{21}^{-1} = 2(1+i\Omega),$$
$$\gamma_1 = \exp(t + y + i(\theta + \phi)),$$
$$\gamma_2 = \exp(t - y + i(\theta - \phi)),$$

where

$$y = at + (1-a^2)(\Omega z - t_0),$$
$$\theta = a\Omega t + \left(\frac{1}{2}\right)(1+a^2)(1-\Omega^2)z,$$
$$\phi = -\Omega t + a(1+\Omega^2)z + \phi_0.$$

Note that y is a surrogate for $\Omega z - t_0$. The first soliton is to be found near $t = -(1-a)(\Omega z - t_0)$, and in this region, $y \approx (1-a)(\Omega z - t_0)$. The second soliton is to be found near $t = (1+a)(\Omega z - t_0)$, and in this region, $y \approx (1+a)(\Omega z - t_0)$.

Solution of the two equations for the two-soliton function yields

$$u(z,t) = \frac{\exp(i\theta)(f_{11} + f_{12} + f_{21} + f_{22})}{(a^2+\Omega^2)\cosh(2t) + (1+\Omega^2)\cosh(2y) - (1-a^2)\cos(2\phi)}, \tag{4.35}$$

where

$$f_{11} = (1+a)(1+i\Omega)(a-i\Omega)\exp(t - y + i\phi),$$
$$f_{12} = (1+a)(1-i\Omega)(a+i\Omega)\exp(-t + y + i\phi),$$
$$f_{21} = (1-a)(1+i\Omega)(-a-i\Omega)\exp(-t - y - i\phi),$$
$$f_{22} = (1-a)(1-i\Omega)(-a+i\Omega)\exp(t + y - i\phi).$$

This function describes the two solitons at all locations and all times. We will look into some of its many guises. In the special case $\Omega = 0$, the two solitons have the same frequency and never completely separate. Otherwise, if $\Omega > 0$, the two solitons pass through each other. Assume, for the moment, that $t_0 = 0$, so that the soliton collision occurs at $z=0$. Well before the collision, at large negative z, y is large and negative near the solitons. The first (faster) soliton appears when t is large and positive. Hence, in Eq. (4.35) the exponential $\exp(t-y)$ in f_{11} is much larger than the others, so that we can neglect f_{12}, f_{21}, and f_{22}. Likewise, the second (slower) soliton appears where t is large and negative, and here the exponential in f_{21} is dominant. After the collision, at large positive z, y is large and positive. The first soliton appears when t is negative, so f_{12} is the dominant term, and the second soliton appears when t is positive, so f_{22} is the dominant term.

Consider the first soliton. Before the collision, at large negative y and positive t, we need keep only the term f_{11} in the numerator, and only the exponentials $\exp(2t)$ and $\exp(-2y)$ in the denominator of Eq. (4.35). If we label u here as u_{11}, we get

$$u_{11} = \frac{2(1+a)(1+i\Omega)(a-i\Omega)e^{i(\theta+\phi)}}{(a^2+\Omega^2)e^{t+y}+(1+\Omega^2)e^{-t-y}}$$

$$=(1+a)\operatorname{sech}(t+y-\eta/2)\exp[i(\theta+\phi-\delta\phi_1/2)], \tag{4.36}$$

where

$$\eta=\ln\left[(1+\Omega^2)/(a^2+\Omega^2)\right] \qquad \text{and} \qquad \delta\phi_1=2\left(\tan^{-1}(1/\Omega)-\tan^{-1}(a/\Omega)\right).$$

We can write this result, using the original parameters of the first soliton, as

$$u_{11}=A_1\operatorname{sech}[A_1(t+\Omega_1 z)-\eta/2]\exp[i(-\Omega_1 t+\left(\frac{1}{2}\right)(A_1^2-\Omega_1^2)z-\delta\phi_1/2)]. \tag{4.37}$$

This describes the first soliton before having suffered a phase shift and a time displacement due to the collision with the second soliton. The displacements that occur here are purely kinematic. They are predictive of the soliton collision rather than of its result.

After the collision, at large positive Ωz, the first soliton is expected at negative time, where $\exp(-2t)$ and $\exp(2y)$ are very big. Thus f_{12} dominates the numerator of Eq. (4.35), and we get

$$u_{12}=\frac{2(1+a)(1-i\Omega)(a+i\Omega)e^{i(\theta+\phi)}}{(a^2+\Omega^2)e^{-t-y}+(1+\Omega^2)e^{t+y}}$$

$$=A_1\operatorname{sech}[A_1(t+\Omega_1 z)+\eta/2]\exp\left[i(-\Omega_1 t+\left(\frac{1}{2}\right)(A_1^2-\Omega_1^2)z+\delta\phi_1/2)\right]. \tag{4.38}$$

In a similar fashion, we find that well before the collision, at large negative Ωz, the second soliton is found at negative time, so f_{21} dominates the numerator of

Eq. (4.35). In this region we find

$$u_{21} = A_2 \operatorname{sech}[A_2(t+\Omega_2 z)+\eta/2] \exp\left[i\left(-\Omega_2 t+\left(\frac{1}{2}\right)\left(A_2^2-\Omega_2^2\right)z-\delta\phi_2/2\right)\right], \tag{4.39}$$

where

$$\delta\phi_2 = 2\big(\tan^{-1}(1/\Omega)+\tan^{-1}(a/\Omega)\big).$$

After the collision, at large positive Ωz, the second soliton is found at positive time. Here, f_{22} dominates the numerator of Eq. (4.35), and we get

$$u_{22} = A_2 \operatorname{sech}[A_2(t+\Omega_2 z)-\eta/2] \exp\left[i\left(-\Omega_2 t+\left(\frac{1}{2}\right)\left(A_2^2-\Omega_2^2\right)z+\delta\phi_2/2\right)\right]. \tag{4.40}$$

(As a sanity check, note that if $a=1$, so that the second soliton is missing, both the phase shift and the time displacement of the first soliton are reduced to zero.)

Solitons in Different Channels

Of particular pertinence to this book is the case where $a=0$ ($A_1=A_2=1$) and $\Omega^2 \gg 1$, corresponding to solitons in different channels. In this case the total phase shift is $\approx 2/|\Omega$ and the total time shift is $\approx -1/\Omega^2$. The time shifts are created by frequency shifts during the collision, and these are important to know. A way to look at the collision is to note that each soliton passes though a potential well caused by the other. If we take $a=0$ and $\Omega^2 \gg 1$, the part of u proportional to $f_{11}+f_{12}$, which is primarily the first soliton, is

$$u_1 = \frac{2e^{i(\theta+\phi)}\left[\cosh(t-\Omega z)-i\frac{1}{\Omega}\sinh(t-\Omega z)\right]}{\cosh(2t)+\left(1+\frac{1}{\Omega^2}\right)\cosh(2\Omega z)-\frac{1}{\Omega^2}\cos(2\phi)}. \tag{4.41}$$

Since $|\Omega|^2 \gg 1$, in order to find the frequency variation of the first soliton during the collision, we can ignore $1/\Omega^2$ factors in Eq. (4.41). Then, using the formula $\cosh(2t)+\cosh(2\Omega z) = 2\cosh(t+\Omega z)\cosh(t-\Omega z)$ in the denominator of Eq. (4.41), it can be reduced to

$$u_1 \approx \operatorname{sech}(t+\Omega z)\exp\left[i\left(\theta+\phi-\frac{1}{\Omega}\tanh(t-\Omega z)\right)\right]. \tag{4.42}$$

Note that Eq. (4.42) reproduces the collision's phase shift, but not the time shift, because the latter is proportional to $1/\Omega^2$. The mean frequency of the first soliton during the collision is given by

$$\langle\omega\rangle = \frac{i}{2}\int_{-\infty}^{\infty} dt\, u_1^* \frac{\partial u_1}{\partial t}. \tag{4.43}$$

Only the phase of u_1 contributes to this integral. The result is

$$\langle\omega\rangle = \Omega + \frac{1}{2\Omega}\int_{-\infty}^{\infty} dt\,\mathrm{sech}^2(t-\Omega z)\,\mathrm{sech}^2(t+\Omega z), \tag{4.44}$$

in agreement with the earlier result in Eq. (4.9).

The net time displacement due to the collision should be minus the integral over z of the transient frequency change during the collision. Integration of Eq. (4.44) over z yields the result $\delta t = -1/\Omega^2$, in agreement with our earlier result. The frequency shift, and time displacement for the second soliton, at frequency $-\Omega$, are the negatives of those for the first soliton.

Four-wave Mixing

In the previous paragraphs we ignored the effect of the $\cos(2\phi)$ term in the denominator of Eq. (4.41). For the solitons in different channels, as discussed earlier, this term produces small sidebands (four-wave mixing products). The first soliton, at frequency Ω, is accompanied during the collision by sidebands at frequencies 3Ω and $-\Omega$. Likewise, the second soliton, at frequency $-\Omega$, is accompanied during the collision by sidebands at frequencies -3Ω and Ω. In this chapter we are looking at the behavior of solitons of the basic NLS equation, and in this case the sidebands are small and disappear after the collision. Later on, when we discuss systems with losses and gains, the sidebands can persist or grow, and must be taken into account. Since $\Omega^2 \gg 1$, we can expand Eq. (4.41) in a power series in $\cos(2\phi)$ and keep only the first term. We still ignore other terms in Eq. (4.41) of order Ω^{-2}. The anti-Stokes sideband u_a at 3Ω is

$$\begin{aligned} u_a = {} & \frac{1}{4\Omega^2}\,\mathrm{sech}^2(t+\Omega z)\,\mathrm{sech}(t-\Omega z) \\ & \times \exp\left[i\left(-3\Omega t + \frac{1}{2}(1-\Omega^2)z - \frac{1}{\Omega}\tanh(t-\Omega z)\right)\right], \end{aligned} \tag{4.45}$$

where we have replaced ϕ by its value $-\Omega t$ and θ by $\left(\frac{1}{2}\right)(1-\Omega^2)z$. The other sideband originating from the first soliton coincides in frequency with the second soliton, but since it is of order Ω^{-2}, its effects are mostly ignorable. A similar argument applied to the second soliton finds a Stokes sideband at frequency -3Ω and a sideband at frequency Ω that adds to soliton number one.

Solitons in the Same Channel

To examine the interactions of solitons in the same frequency channel, we need to examine the case where both solitons have nearly the same size ($a^2 \ll 1$) and nearly the same frequency ($\Omega^2 \ll 1$). In this case, the $\cos(2\phi)$ term of Eq. (4.35), which creates the four-wave mixing products when the two solitons are in different channels, modulates the behavior of solitons in the same channel. Let $\rho = a - i\Omega$. If we neglect terms of order $|\rho|^2$ with respect to unity, the two-soliton function, Eq. (4.35), reduces to

$$u(z,t) = \frac{\exp(i\theta)(f_{11}+f_{12}+f_{21}+f_{22})}{|\rho|^2\cosh(2t)+\cosh(2y)-\cos(2\phi)}, \tag{4.46}$$

where

$$f_{11} = \rho\exp(t - y + i\phi + \rho^*),$$

$$f_{12} = \rho^*\exp(-t + y + i\phi + \rho),$$

$$f_{21} = -\rho^*\exp(-t - y - i\phi - \rho),$$

$$f_{22} = -\rho\exp(t + y - i\phi - \rho^*),$$

and

$$y + i\phi = \rho t + i\rho z - t_0 + i\phi_0.$$

The problem is now to find the solitons represented by this function. In the numerator, we add the parts that pertain to positive time, f_{11} and f_{22}, and the parts that pertain to negative time, f_{12} and f_{21}. In the denominator, we use the relation $\cosh(2y) - cos(2\phi) = 2\sinh(y+i\phi)\sinh(y-i\phi)$ and, in $\cosh(2t)$, keep only the positive exponential at positive time and the negative exponential at negative time. The latter approximation means that the result is not valid in a small interval around $t = 0$, or alternately that it is valid only when the solitons are pretty well resolved. We get the suggestive result

$$u(z,t) = u_+(z,t) + u_-(z,t), \tag{4.47}$$

where u_+ is the part of u at positive time,

$$u_+(z,t) = e^{i\theta}\frac{-4\rho\sinh(y - i\phi - \rho^*)}{|\rho|^2\exp(t) + 4\sinh(y+i\phi)\sinh(y-i\phi)\exp(-t)},$$

and u_- is the part of u at negative time,

$$u_-(z,t) = e^{i\theta}\frac{4\rho^*\sinh(y + i\phi + \rho)}{|\rho|^2\exp(-t) + 4\sinh(y+i\phi)\sinh(y-i\phi)\exp(t)}.$$

Next, we rewrite $y + i\phi$ to separate its time-dependent factor. Since ρ is small, the motion of the solitons is slow, and we will be looking in regions where ρt is small, while ρz may not be small. Define

$$\zeta = i\rho z - t_0 + i\phi_0 = \Omega z - t_0 + i(az + \phi_0) \tag{4.48}$$

so that $y+i\phi=\rho t+\zeta$, while $y+i\phi+\rho=\rho(t+1)+\zeta$, and $y-i\phi-\rho^*=\rho^*(t-1)+\zeta^*$. Now expand the sinh terms using $\sinh(a+b)=\cosh(a)\sinh(b)+\sinh(a)\cosh(b)$. Using the presumed smallness of ρt, we get

$$\sinh(\rho t+\zeta)\approx\sinh(\zeta)+\rho t\cosh(\zeta)\approx\sinh(\zeta)\exp(\rho t\coth(\zeta)). \tag{4.49}$$

The crucial step in our quest for the solitons is the transformation

$$\sinh(\zeta)=i\frac{1}{2}\rho\exp(q+i\psi), \tag{4.50}$$

which converts Eq. (4.47) to exponential form. The initial factor i on the right side of Eq. (4.50) is not necessary, as it could be incorporated in $\exp(i\psi)$, but it is convenient in what follows. As we shall see, the variable q is approximately half the interval between the solitons. It is always positive when the solitons are resolved. Taking the z derivative of Eq. (4.50), we get

$$\cosh(\zeta)=\frac{1}{2}(q_z+i\psi_z)\exp(q+i\psi), \tag{4.51}$$

where the subscript z represents differentiation with respect to z. Dividing these last two equations yields

$$\rho\coth(\zeta)=\psi_z-iq_z. \tag{4.52}$$

Putting these relations into Eq. (4.49) yields

$$\sinh(\rho t+\zeta)=i\frac{1}{2}\rho\exp(q+\psi_z t+i(\psi-q_z t)). \tag{4.53}$$

Similarly,

$$\sinh(\rho(t+1)+\zeta)=i\frac{1}{2}\rho\exp(q+\psi_z t+i(\psi-q_z t)+\psi_z-iq_z) \tag{4.54}$$

and

$$\sinh(\rho^*(t-1)+\zeta^*)=i\frac{1}{2}\rho\exp(q+\psi_z t-i(\psi-q_z t)-\psi_z-iq_z). \tag{4.55}$$

The solitons appear when we insert these relations into Eq. (4.47). The result is

$$\begin{aligned}u=&\exp[i(\theta-q_z)]\{(1-\psi_z)\operatorname{sech}[(1-\psi_z)t-q]\exp[-i(\psi-q_z t)]\\&+(1+\psi_z)\operatorname{sech}[(1+\psi_z)t+q]\exp[i(\psi-q_z t)]\}.\end{aligned} \tag{4.56}$$

In this expression one of the solitons has amplitude $1+\psi_z$ and frequency q_z while the other has amplitude $1-\psi_z$ and frequency $-q_z$. The temporal separation between the two solitons is $2q$ (to first order in ψ_z). These values change while the solitons interact. Equation (4.56) covers a multitude of behaviors of the two-soliton system. It is generally difficult to find the basic parameters of solitons from knowledge of some initial state of the field. In the present case this is possible. From a field in the form of Eq. (4.56), the field variables $[q, \psi, q_z, \psi_z]$ at any location can be used to find the basic constants of the two-soliton state $[a, \Omega, t_0, \phi_0]$. Particularly important is the frequency parameter Ω, as it tells whether the solitons separate and how fast. Suppose that we create a two-soliton state in the form of Eq. (4.56), say at $z=0$. We can recover the basic parameters of the solitons from Eqs. (4.50) and (4.51). Using these two equations and applying the formula $\cosh^2(\zeta)-\sinh^2(\zeta)=1$, we obtain the important relation

$$(a-i\Omega)^2=\rho^2=(\psi_z-iq_z)^2+4\exp(-2q-i2\psi). \tag{4.57}$$

Apart from an ambiguity in sign, Eq. (4.57) gives the values of a and Ω. The sign ambiguity is resolved by the convention that Ω is always positive, so that if and when the two solitons are completely separate, soliton number one has positive frequency and amplitude $1+a$, while soliton number two has negative frequency and amplitude $1-a$. If, in fact, the solitons are far apart, so that $q\gg 1$, we find from Eq. (4.57) that

$$a-i\Omega=\pm(\psi_z-iq_z).$$

If q_z is positive, then $\Omega=q_z$ and $a=\psi_z$, while if q_z is negative, then $\Omega=-q_z$ and $a=-\psi_z$. In the former case, we see from Eq. (4.56) that soliton number one is ahead of soliton number two and that the temporal distance between them is increasing. In the latter case, soliton number one is behind soliton number two, and the temporal distance between them is decreasing. Having found the values of a and Ω, the values of t_0 and ϕ_0 can also be derived from Eqs. (4.50) and (4.51), as we will illustrate in a few cases.

The equations of motion of the two solitons can be gleaned by differentiating Eq. (4.57) with respect to z. Since ρ is a constant, we get the relation

$$\psi_{zz}-iq_{zz}=i4\exp(-2q-i2\psi).$$

Separating the real and imaginary parts of this relation gives

$$\psi_{zz}=4\exp(-2q)\sin(2\psi) \qquad \text{and} \qquad q_{zz}=-4\exp(-2q)\cos(2\psi), \tag{4.58}$$

which shows that the soliton motion can be described by distance-dependent forces that vary with the solitons' relative phase angle. If $\psi=0$, the two solitons are approaching one another, while if $\psi=\pi/2$, they are moving apart. Note that $\psi=0$ is an unstable situation, while $\psi=\pi/2$ is stable. The two-soliton function is a solution of these two equations subject to the appropriate initial conditions. Another look at the soliton motion comes from Eq. (4.50). Multiplying Eq. (4.50) by its conjugate, and using $2\sinh(\zeta)\sinh(\zeta^*)=\cosh(\zeta+\zeta^*)-\cosh(\zeta-\zeta^*)$ and Eq. (4.48), gives

$$|\rho^2|\exp(2q)=2[\cosh(2\Omega z-2t_0)-\cos(2az+2\phi_0)]. \tag{4.59}$$

Equation (4.59) shows that unless Ω is zero, the two solitons eventually separate. It also shows that the cosine term modulates the temporal location of the solitons only when they are relatively close to each other, because, of course, the modulation occurs due to their interaction, which depends on their separation and their relative phase.

Some initial conditions deserve comment. Assume that the two solitons are initially injected at $z=0$. The subscript 0 indicates an initial value. Suppose that the two solitons are injected with equal amplitudes and frequencies, but are not necessarily in phase, so that the initial state has the form

$$u=\operatorname{sech}(t-q_0)\exp(-i\psi_0)+\operatorname{sech}(t+q_0)\exp(i\psi_0). \tag{4.60}$$

In the general Eq. (4.56) this result is obtained by setting both ψ_{z0} and q_{z0} equal to zero. The common phase θ is not important, but we could set it to zero also. For this case, Eq. (4.57) becomes

$$\rho=a-i\Omega=\pm 2\exp(-q_0-i\psi_0).$$

The plus sign applies when $0\le\psi_0<\pi$ and the minus sign applies otherwise, since Ω is nonnegative. The values of the soliton parameters a and Ω thus depend on the relative phase of the two injected solitons. This is a reminder that a and Ω are constant parameters that represent the amplitudes and frequencies of the solitons when and if they are far apart; they do not represent the actual amplitudes and frequencies of the solitons while they are interacting. Only in the case $\psi_0=0$ or π, when the two solitons are initially in phase, do they not eventually separate. Assume now that $0\le\psi_0<\pi$ so that the plus sign applies. Equation (4.48) gives $\zeta_0=-t_0+i\phi_0$. From Eqs. (4.50) and (4.51) we get $\sinh(\zeta_0)=i$ and $\cosh(\zeta_0)=0$, so that $t_0=0$ and $\phi_0=\pi/2$. This completes the discovery of the soliton parameters from the initial condition. Using these values, Eq. (4.59) yields the result

$$\exp(2q)=\frac{1}{2}[\cosh(2\Omega z)+\cos(2az)]\exp(2q_0). \tag{4.61}$$

If $\psi_0=0$, so that the two solitons are initially in phase, we see that $\Omega=0$ and $a=2\exp(-q_0)$. In this case, $\exp(2q)=\frac{1}{2}[1+\cos(2az)]\exp(2q_0)$. Since we require q to be a positive quantity, taking the square root of this last expression gives

$$\exp(q)=\exp(q_0)|\cos(az)|.$$

The solitons are seen to oscillate about each other. Our approximation describes the two-soliton state only while the solitons are resolved, so it does not describe the collision in detail. However, the exact function is available if one wants to see the details of the collision. One might be tempted to think of this last situation as a bound state of the two solitons. However, we have seen that the attractive force is phase dependent. Thus, any small perturbation that alters the phase relation of the two solitons will give Ω a finite value, so that the two solitons will eventually separate.

Another case that illustrates the behavior of the two-soliton system occurs when the solitons are injected with the same frequency and the same phase, but not necessarily with the same amplitude. This case is reproduced by taking $q_{z0}=\psi_0=0$ in Eq. (4.56). This time we find from Eq. (4.57) that $\rho=a-i\Omega$ is real, with

$$a^2=\psi_{z0}^2+4\exp(-2q_0).$$

From Eqs. (4.48), (4.50), and (4.51), the initial conditions yield

$$\sinh(-t_0+i\phi_0)=ia\exp(q_0)/2 \qquad \text{and} \qquad \cosh(-t_0+i\phi_0)=i\psi_{z0}\exp(q_0)/2.$$

Expanding the cosh and sinh terms into real and imaginary parts shows that $\phi_0=\pm\pi/2$ and $\cosh(t_0)=(a/2)\exp(q_0)$. With these results Eq. (4.59) yields

$$\begin{aligned}\exp(2q)&=\frac{2}{a^2}[\cosh(2t_0)+\cos(2az)]=\frac{4}{a^2}[\cosh^2(t_0)-\sin^2(az)]\\&=\exp(2q_0)-4\sin^2(az)/a^2.\end{aligned} \tag{4.62}$$

We see in this case that the two solitons oscillate about fixed positions, in general not passing each other. This happens because the relative phase of the two solitons varies with time and so they alternately attract and repel one another. This case and the previous one have a common situation, the latter when the solitons have the same initial phase, and here when the two solitons have the same initial amplitude. If $\psi_{z0}=0$, then $a^2=4\exp(-2q_0)$, and we get from Eq. (4.62) that $\exp(2q)=\exp(2q_0)\cos^2(az)$, in agreement with the previous result.

A third simple case is the insertion of two solitons with equal amplitudes and phases, but with slightly different frequencies. Here we find from Eq. (4.57) that $(a-i\Omega)^2=4\exp(-2q_0)-q_{z0}^2$. The two solitons remain together ($\Omega=0$) so long as the initial frequency difference is small enough. The two solitons stay together only in special cases. It is clear from Eq. (4.57) that small changes of the initial soliton parameters in these special cases can give a finite value to Ω, indicating that the solitons will eventually separate. Thus the two solitons are never bound in the usual sense.

Chapter 5

Wavelength Division Multiplexing with Ordinary Solitons

5.1. Introduction

In the previous chapter, we have seen that in lossless and constant-dispersion fiber, collisions between ordinary solitons of distinctly different frequencies are perfectly elastic. That is, after the collision is completed, there is no net exchange of energy or momentum of either soliton. This perfect transparency would seem to make ordinary solitons ideal for use in dense WDM. Unfortunately, however, the loss and/or varying dispersion of real fibers tend to destroy the symmetries necessary for such transparency, so that the emerging solitons suffer significant frequency shifts (which dispersion then converts into timing shifts) and loss of energy. In principle, these defects can be perfectly corrected through the use of fibers whose dispersion profile, $D(z)$, tracks the loss/gain-induced intensity profile $I(z)$ (for example, fibers with exponentially tapered D used between lumped amplifiers.) Indeed, as will be reported on later in this chapter, step-wise approximations to such exponentially dispersion-tapered spans have allowed for successful experimental demonstration of six-channel WDM over distances of 10,000 km or more. Further expansion of the channel count, however, is severely limited by the very narrow range of wavelengths over which single fiber types can provide the correct sub-picosecond/nn-km dispersion values necessary for ordinary solitons, and the cost of production of fibers with custom-shaped dispersion profiles is seen as prohibitively expensive. Thus, in the commercial world, dense WDM with solitons is now concentrated almost entirely on the much more versatile and practical dispersion-managed solitons, as will be described in the next chapter.

In the meantime, the material of this chapter is worthy of the reader's careful study, as a necessary background for a complete understanding and appreciation of dense WDM with dispersion-managed solitons, if for nothing else. For one thing, as we shall see later, the more complex collisions of dispersion-managed solitons really consist of many ordinary soliton-like "mini-collisons," so the present study is a most useful foundation for understanding of those more complex collisions. Second, comparison of the different scaling properties of collisions of ordinary solitons with those of dispersion-managed solitons is illuminating. Third, in the experiments to be described here, sliding-frequency guiding filters provided a very strong and useful control over the WDM with ordinary solitons in several different ways. It is interesting to consider the possibility of extending such control to dense WDM with dispersion-managed solitons.

5.2. Effects of Periodic Loss and Variable Dispersion

The periodic intensity fluctuations in a system with real fiber and lumped amplifiers can serve to destroy the perfect asymmetry of the acceleration curve of Fig. 4.2, and hence to result in a net residual velocity shift and associated timing displacement [77]. For the purposes of illustration, Fig. 5.1 shows an extreme such case, where

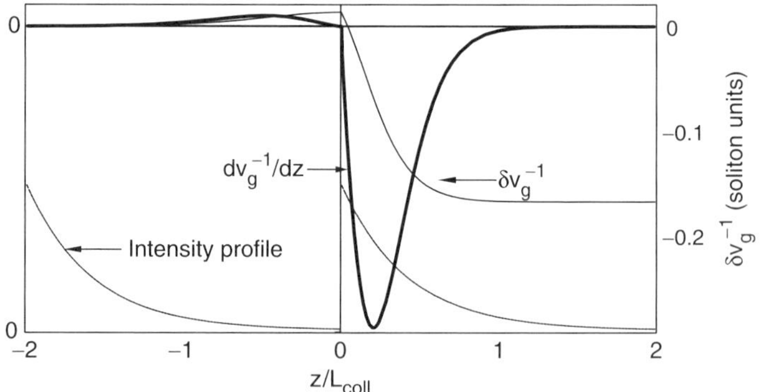

FIGURE 5.1 (Heavy curve) acceleration and (thin curve) inverse velocity shift ($\propto -\delta f$) for a collision centered at an amplifier, and where $L_{coll} = L_{amp}/2$. The sawtooth curve at the bottom of the figure shows the relative intensities of the spans when $L_{amp} = 100$ km. A set of parameters yielding $L_{coll} = 50$ km is $\tau = 20$ ps, $\bar{D} = 0.5$ ps/nm-km, and $\Delta\lambda = 1.6$ nm ($\Delta f = 200$ GHz). Note the severe asymmetry of the acceleration curve, and the resultant large residual velocity shift (−0.165 in soliton units, corresponding to $\delta f = 4.63$ GHz).

the collision length is half the amplifier spacing, and where the collision is centered at an amplifier. Note that just prior to the amplifier, where the intensity is low, the acceleration curve is correspondingly attenuated, while just the opposite happens in the space immediately following the amplifier. Thus, most of the integral of the acceleration curve comes from the right half of the graph, and, as a result, there is a large residual velocity (frequency) shift. The residual frequency shift here (4.63 GHz) really is large. Note that when its wavelength equivalent is multiplied by $\bar{D} = 0.5$ ps/nm-km and by a characteristic filter damping length of say, 600 km, the resultant time shift is ≈ 11 ps. When further magnified by the typical spread of nearly zero to at least several tens of collisions in a transoceanic length, that time shift would result in a completely disastrous timing jitter.

On the other extreme, where the collision length is large relative to the amplifier spacing, one might reasonably expect the velocity curve to look much like that of Fig. 4.2. For example, Fig. 5.2 shows what happens when L_{coll} is just $2.5L_{amp}$. Although the acceleration curve in Fig. 5.2 contains large discontinuities at each amplifier, its integral looks remarkably close to the ideal velocity curve (Fig. 4.2). Most importantly, the velocity returns almost exactly to zero following the collision. In the following section, it will be shown analytically that the residual velocity is essentially zero as long as the condition

$$L_{coll} \geq 2L_{amp} \tag{5.1}$$

is satisfied. Note that this condition, combined with Eq. (4.1), puts an upper bound on the maximum allowable channel spacing:

$$\Delta\lambda_{max} = \frac{\tau}{DL_{amp}}. \tag{5.2}$$

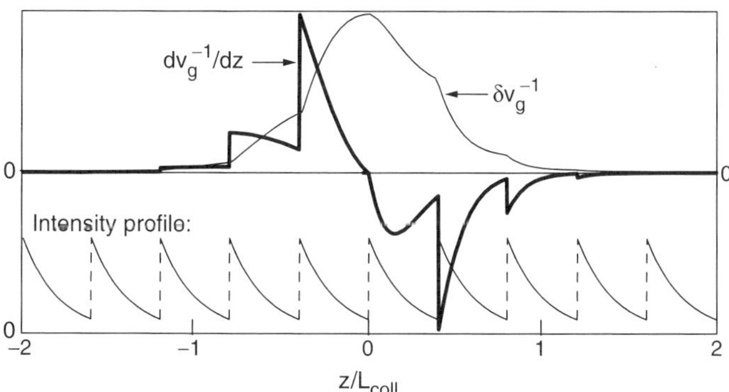

FIGURE 5.2 Acceleration and velocity shift for a collision centered at an amplifier, in constant D fiber, and where $L_{coll} = 2.5L_{amp}$. Note how the velocity curve here closely approximates the ideal of lossless fiber (Fig. 4.2).

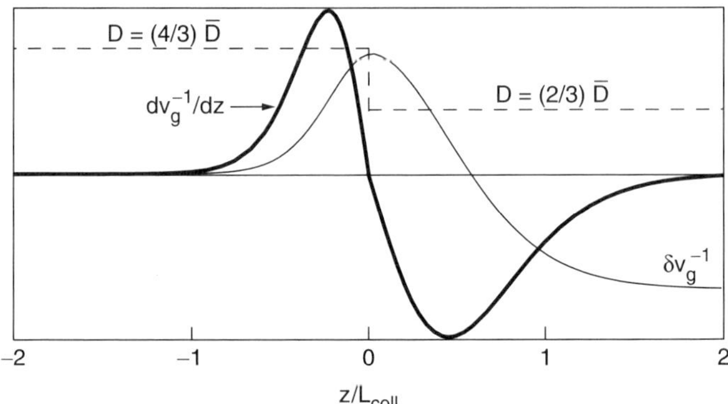

FIGURE 5.3 Acceleration and velocity curves for a collision centered at a discontinuity in D, and where, for simplicity, the fiber is lossless.

For example, let $\tau = 20$ ps, $D = 0.5$ ps/nm-km, and $L_{amp} = 33$ km. Equation (5.2) then yields $\Delta\lambda_{max} \approx 1.2$ nm; with nearest channel wavelength spacings of 0.6 or 0.4 nm, the maximum allowable number of channels is just three or four, respectively. As will be shown later, however, this limit can be expanded through the use of dispersion-tapered spans.

Variation in the fiber's chromatic dispersion can also upset the symmetry of the collision. For example, Fig. 5.3 illustrates what happens when the collision is centered at a discontinuity in D. To keep the example pure, here the fiber is lossless. Note the different length scales on either side of the discontinuity in D. These occur, or course, because the length scale for the collision, or for any part of it, is L_{coll}, which is in turn inversely proportional to D [see Eq. (4.1)].

5.3. Analytical Theory of Collisions in Perturbed Spans

We are concerned here with soliton propagation and collisions where there are variations of gain (loss) and dispersion on a distance scale much less than the characteristic dispersion length z_c [Eq. (1.18)], i.e., we assume that we have the conditions for good path-average solitons. The propagation equation for the optical field can then be written as

$$-i\left[\frac{\partial u}{\partial z} - \frac{\alpha(z)}{2}u\right] = \frac{\Delta(z)}{2}\frac{\partial^2 u}{\partial t^2} + |u|^2 u, \tag{5.3}$$

where, by assumption, in a distance much smaller than unity, the gain (loss) coefficient α averages to zero, while the normalized dispersion parameter $\Delta(z) = D(z)/\bar{D}$ averages to one.

We can change Eq. (5.3) to a more useful form by the following transformation: Let

$$u' = u/\sqrt{G} \text{ and } dz' = \Delta\, dz, \tag{5.4}$$

where the parameter $G(z)$ satisfies the equation $dG/dz = \alpha G$, and averages over z to unity. [Thus, note that $G(z)$ can be thought of as a normalized power gain factor $= P(z)/\bar{P}$.] To specify z' completely as a function of z, we can set $z' = 0$ when $z = 0$. The effect of this transformation is to remove the amplitude and dispersion variations from u. Note that because Δ averages to one in a short distance, z and z' never get very far apart. If the variation of Δ is periodic, for example, then z' and z are equal again after each period. The modified equation has the form

$$-i\frac{\partial u'}{\partial z'} = \frac{1}{2}\frac{\partial^2 u'}{\partial t^2} + g|u'|^2 u', \tag{5.5}$$

where $g(z') = G(z)/\Delta(z)$. Note that $g(z')dz' = G(z)dz'$, so that $g(z')$ averages over z' to unity. The original variations in dispersion and gain loss are thus equivalent to a variation of the nonlinear coefficient.

Even before any further analysis, it is useful to take stock of what has already been accomplished. First, it is of the greatest importance that G and Δ are no longer needed separately, but can be folded into a single parameter (g, equal to their ratio). Second, note that when $g(z) = 1$, Eq. (5.5) becomes just the standard NLS equation. This happens whenever $G(z)$ and $\Delta(z)$ track each other, as, for example, when $D(z)$ falls off between lumped amplifiers at the same exponential rate as does the signal power. Thus we come to the powerful conclusion that all of the perfect transparencies of Chapter 4 for lossless and constant-dispersion fiber apply equally under the more general condition that one simply has $g(z') = 1$. We shall make use of this result in the following section.

We can now take up the case of colliding solitons with nonoverlapping frequency spectra. The analysis proceeds exactly as it did in Chapter 4, if [in Eqs. (4.7) and (4.8)] we merely put primes on all us and on z, and multiply all nonlinear terms by $g(z')$. Thus, Eq. (4.8) becomes

$$\frac{d\Omega}{dz} = \frac{g}{2\Omega}\frac{d}{dz'}\int \operatorname{sech}^2(t + \Omega z')\operatorname{sech}^2(t - \Omega z')\,dt, \tag{5.6}$$

which, of course, reduces to Eq. (4.8) when $g = 1$. There is a net frequency shift from a completed collision when g depends on z', however. Integrating Eq. (5.6) over z', and doing a partial integration on the right-hand side, yields for this frequency shift

$$\delta\Omega = -\frac{1}{2\Omega}\int dz' \frac{dg}{dz'} \int \operatorname{sech}^2(t+\Omega z')\operatorname{sech}^2(t-\Omega z')\,dt. \qquad (5.7)$$

Finally, Eq. (5.7) is reduced to a simpler form by doing a spatial Fourier resolution of g. Thus, let $g(z') = \int dk\,\tilde{g}(k)\exp(ikz)$. After inserting this transform in Eq. (5.7), doing the z' derivative, and changing integration variables from z' and t to $\Omega z' \pm t$, Eq. (5.7) reduces to

$$\delta\Omega = \frac{-i}{4\Omega^2}\int dk\,k\,\tilde{g}(k)\left[\int ds \exp\left(i\frac{ks}{2\Omega}\right)\operatorname{sech}^2(s)\right]^2. \qquad (5.8)$$

The integral over s in Eq. (5.8) evaluates to $2\chi/\sinh(\chi)$, where $\chi = \pi k/(4\Omega)$. Also, since $g(z')$ is real, we have $\tilde{g}(-k) = \tilde{g}^*(k)$. Thus, we can write Eq. (5.8) as

$$\delta\Omega = \operatorname{Im}\frac{32}{\pi^2}\int_0^\infty dk \frac{\tilde{g}(k)}{k}\frac{\chi^4}{\sinh^2(\chi)}. \qquad (5.9)$$

Equation (5.9) is the fundamental result of our analysis.

We are primarily interested in cases where the variations of gain and dispersion are periodic. For such cases, $\tilde{g}(z')$ reduces to a sum over spatial harmonics of the fundamental period, with wavenumbers $k = 2\pi n/L_{pert}$, and Eq. (5.9) becomes

$$\delta\Omega = \frac{16L_{pert}}{\pi^3}\sum_{n=1}^{\infty}\operatorname{Im}\tilde{g}_n \frac{n^3\chi^4}{\sinh^2(n\chi)}, \qquad (5.10)$$

where $\chi = \pi^2/(2\Omega L_{pert})$, and $\tilde{g}_n$ is the average of $g(z)\exp(-i2\pi nz'/L_{pert})$, that is,

$$\tilde{g}_n = \frac{1}{L_{pert}}\int_0^{L_{pert}} (G/\Delta)\exp(-i2\pi nz'/L_{pert})\,dz'. \qquad (5.11)$$

Note that since the (z') collision length $L_{coll} = 1.7627/\Omega$, we have $\chi = 2.7995\ L_{coll}/L_{pert}$. Figure 5.4 plots the factor $n^3\chi^4/\sinh(n\chi)$ of Eq. (5.10) vs. χ and L_{coll}/L_{pert}. The plot reveals first, for $L_{coll}/L_{pert} > 1$, that only the fundamental makes significant contribution to $\delta\Omega$, and second, and of greatest importance, that $\delta\Omega$ becomes negligibly small for L_{coll}/L_{pert} not much greater than 2. Note how this agrees very well with the behavior we have seen in Figs. 5.1 and 5.2.

In the preceding development, the collisions are centered at $z = z' = 0$. To extend the results to collisions centered at some general z'_{coll}, note that moving the collision forward a distance z' along the fiber is equivalent to moving the fiber (amplifiers and spatial variations in D included) backward by the same distance, which has the result of multiplying $\tilde{g}_n$ by the factor $\exp(i2n\pi\, z'_{coll}/L_{pert})$. To express this result in z requires replacement of z'_{coll} by $\int_0^{z_{coll}} \Delta\, dz$.

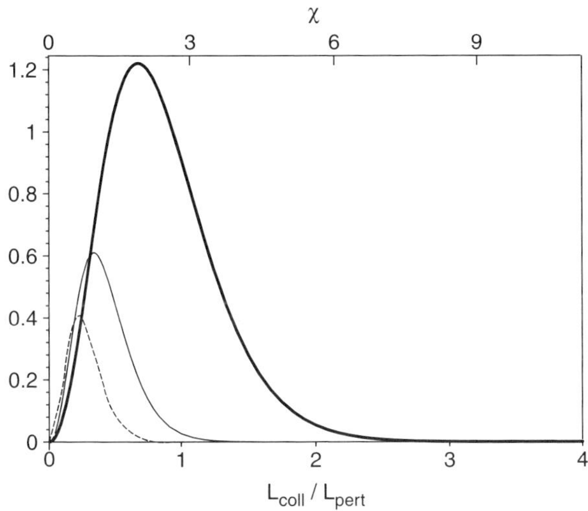

FIGURE 5.4 The function $n^3\chi^4/\sinh(n\chi)$ vs. χ or $L_{coll}/L_{pert} = \chi/2.7995$ for $n=1$ (the fundamental) (heavy solid line), $n=2$ (the second harmonic) (thin solid line), $n=3$ (the third harmonic) (dashed line).

It may be helpful to look at a few simple cases. For example, for lumped amplifiers between spans of constant-dispersion fiber, we have $\Delta=1$ and $z'=z$ everywhere, and $L_{pert}=L_{amp}$. From Eq. (5.11) one can easily compute

$$\tilde{g}_n = \frac{a\exp(in\phi_c)}{a+i2\pi n}, \tag{5.12}$$

where $\phi_c = 2\pi\, z_{coll}/L_{amp}$ and $a=\alpha L_{amp}$. For the case $L_{coll}/L_{amp} > 1$, where only the fundamental is important, the imaginary part of Eq. (5.12) can be written as

$$\mathrm{Im}\,\tilde{g} = \frac{1}{\sqrt{1+(2\pi/a)^2}}\,\sin(\phi_c - \arctan(2\pi/a)). \tag{5.13}$$

Figure 5.5 is a plot of Eq. (5.13) for $a=1.92$ (corresponding to $L_{amp}=40$ km and $\alpha=0.048$/km). Thus, it could apply to the situation of Fig. 5.2, or to the same but with greater channel separation for shorter L_{coll} (and hence much greater residual frequency shift than the very small one of Fig. 5.2). Note the sinusoidal variation (as would be expected when only the fundamental is important) and the fact that the extremes of frequency shift occur when the collisions are centered just ahead of the amplifiers (but not exactly at them), and just ahead of the mid-span position.

For cases like that of Fig. 5.3 (lossless fiber with dispersion variation), let the dispersion change from Δ_1 to Δ_2 at a (z) distance $L_1 < L_{pert}$. Since Δ must

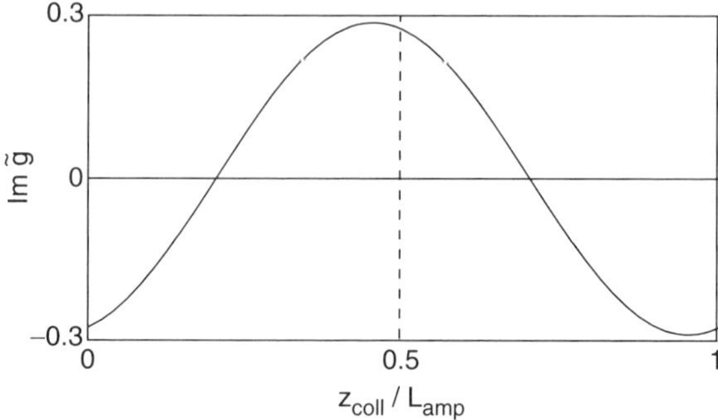

FIGURE 5.5 Im $\tilde{g}$ vs. z_{coll}/L_{amp} where $L_{amp}=40$ km ($a=1.92$). (Amplifiers are located at 0 and 1.) Note that the extrema correspond to collisions centered just ahead of the amplifiers and just ahead of mid-span.

average to unity, we must have $\Delta_1 L_1 + \Delta_2(L_{pert} - L_1) = L_{pert}$. In this case we can use $dz' = \Delta dz$ to express Eq. (5.11) as an integral in z, with the result

$$\tilde{g}_n = \frac{\Delta_1 - \Delta_2}{\pi n \Delta_1 \Delta_2} \sin(n\phi_1/2)\exp(in(\phi_c - \phi_1/2)), \tag{5.14}$$

where $\phi_1 = 2\pi \Delta_1 L_1/L_{pert}$ and (essentially as before), $\phi_c = 2\pi z'_{coll}/L_{pert}$. Equation (5.14) can be used to generalize on results like that of Fig. 5.3.

5.4. Dispersion-tapered Fiber Spans

In the previous section, we have seen proof that when $D(z)$ perfectly tracks the signal intensity profile factor [$G(z)$], the propagation equation [Eq. (5.5)] reverts to the unperturbed NLS equation, with the result that collisions have the perfect transparency described in Chapter 3. Unfortunately, however, fiber spans having the ideal exponential taper in D (for use between lumped amplifiers) are not available commercially (and almost certainly never will be). Thus, at present it is necessary to use a step-wise approximation. Figure 5.6 illustrates the optimum three-step approximation to the ideal taper. There, the length of each step is inversely proportional to the D value of the step. Note that that makes the steps all have equal lengths, as measured in soliton units.

Using such an N-step approximation to the ideal dispersion taper increases the limit on maximum channel spacing imposed by Eq. (5.2) by a factor of N. That is, one now has

$$\Delta\lambda_{max} = \frac{N\tau}{DL_{amp}}. \tag{5.2a}$$

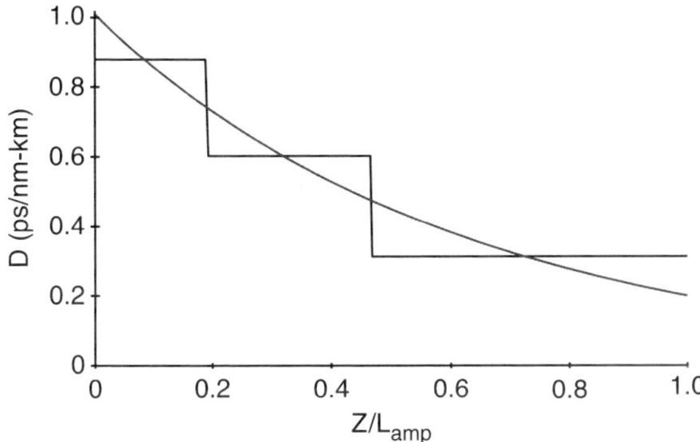

FIGURE 5.6 Ideal exponential taper of D and the best, three-step approximation to it for a fiber span with $L_{amp} = 33.3$ km and loss rate of 0.21 dB/km.

In practice, once N is large enough, the intensity variation over each step becomes so small that the perturbations become acceptably small in any event. In that case, the limit of Eq. (5.2a) is no longer significant.

5.5. Pseudo Phase Matching of Four-wave Mixing in WDM

In Chapter 4, we found that the maximum FWM energy produced during a soliton–soliton collision in lossless fiber is small, typically less than 0.1% of W_{sol}, and that it disappears entirely after the collision is over. In real systems involving periodic gain and loss or other perturbations, however, the situation can be very different. In particular, if the transmission line has periodic perturbations with k_{pert} in resonance with the phase mismatch of the FWM, i.e., when

$$Nk_{pert} = \Delta k \;\; (N = 1, 2, 3, \ldots), \tag{5.15}$$

then one has pseudo phase matching, and the FWM product can grow steadily [78]. The perturbations can correspond to the gain/loss cycle whose period is the amplifier spacing, L_{amp}, and/or to periodic variations of the fiber parameters (dispersion, mode area). For the case of lumped amplifiers, $k_{pert} = 2\pi/L_{amp}$ and the pseudo phase-matching conditions are met when

$$L_{amp} = NL_{res} \equiv 2\pi N/\Delta k. \tag{5.16}$$

Figure 5.7 illustrates such pseudo phase matching in the development of FWM between two cw waves as they traverse four 40-km spans of fiber with loss factor

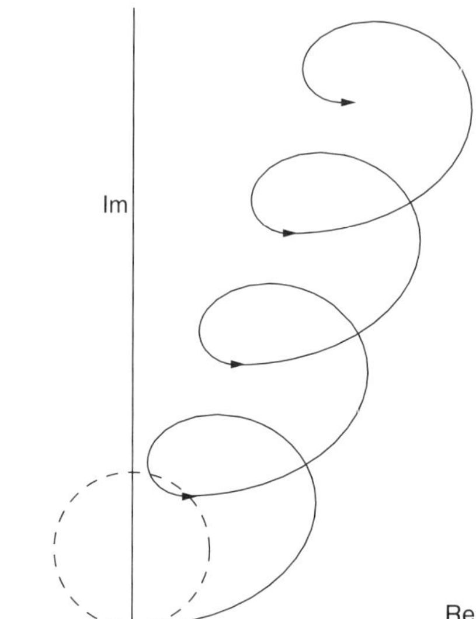

FIGURE 5.7 Solid, spiral curve: Growth of complex amplitude of FWM resulting from interaction of two cw waves as they traverse four, 40-km spans of fiber with loss factor 0.048/km and when $\Delta k = (2\pi/40)\,\text{km}^{-1}$. Note how the resonant pseudo phase matching ($N = 1$) creates a spiral that increases in net length with passage through successive amplifier spans. Dashed circle: The same, but in lossless fiber and with the same path-average signal powers. In this case, the circle simply closes on itself after each 40-km fiber span.

$\alpha = 0.048$/km and separated by lumped amplifiers. Keep in mind that the intensities of the FWM products scale as the square of the amplitudes shown there, and compare the ever-growing spiral of the pseudo phase-matched case with the closed circular path in lossless fiber. From those considerations, the efficiency of pseudo phase matching in promoting rapid FWM growth should be well evident.

Although FWM generation during a soliton–soliton collision is more complicated than with continuous waves, the basic features remain the same. Figure 5.8 shows the numerically simulated growth in energy of the FWM products at $\omega_{A,S}$ during a single collision of two solitons. [The particular parameters represented there are those of certain experiments [28, 29], viz., $\tau = 20$ ps, adjacent channel separation $\Delta f = 75$ GHz ($\Delta\lambda = 0.6$ nm at $\lambda = 1556$ nm), and where the path-average dispersion $\bar{D} = 0.5$ ps/nm-km.] For these parameters, $L_{res} = 44.4$ km. Note that for the case of lossless fiber of constant dispersion, and for the case of real fiber with exponentially tapered dispersion, the FWM energy disappears completely following the collision. Also note that due to the fact that the solitons have finite temporal and spectral envelopes, and due to the effect of cross-phase

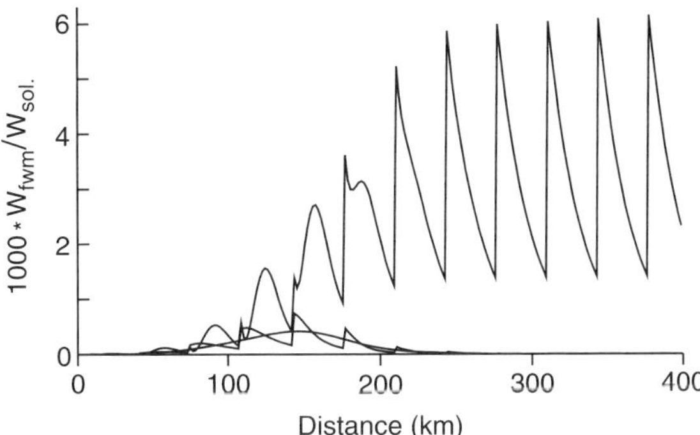

FIGURE 5.8 Growth of FWM energy during a single soliton–soliton collision, for three different conditions: lossless fiber with constant dispersion (small, smooth curve), real fiber with lumped amplifiers spaced 33.3 km apart and exponentially tapered dispersion (small, jagged curve), and real fiber with lumped amplifiers spaced 33.3 km apart and constant dispersion (large, jagged curve). The FWM energy is for a single sideband and is normalized to the soliton pulse energy. Note that for the first two cases, the FWM energy disappears completely following the collision, while for the third case, where there is effective pseudo phase matching, the FWM energy builds to a large residual value.

modulation (which shifts the pulse carrier frequencies during the collision), the oscillations of the FWM energy with the period of L_{res} are almost completely washed out. Finally, for the case of real fiber with constant dispersion, the collision produces a residual FWM energy several times larger than the (temporary) peak obtained with lossless fiber.

Although the residual energy from the pseudo phase-matched collision of Fig. 5.8 may seem small, the fields from a succession of such collisions can easily build to a dangerously high value. Such uncontrolled growth of the FWM imposes penalties on the transmission by two different mechanisms. First, since the energy represented by the FWM fields is not reabsorbed by the solitons, the solitons tend to lose energy with each collision. Since the net energy loss of a given soliton depends on the number of collisions it has suffered, and upon the addition of four-wave mixing fields with essentially random phases, it directly creates amplitude jitter. The energy loss leads to timing jitter as well, both through the intimate coupling between amplitude and frequency inherent in filtered systems, and through its tendency to asymmetrize the collision, and hence to induce net velocity shifts. Finally, certain noise fields in the same band with the FWM products can influence the solitons' frequencies, in a kind of extended Gordon–Haus effect. Thus, even in a two-channel WDM, there can be serious penalties (see Fig. 5.9). Moreover, if the wavelengths of the FWM products coincide with the wavelengths of other

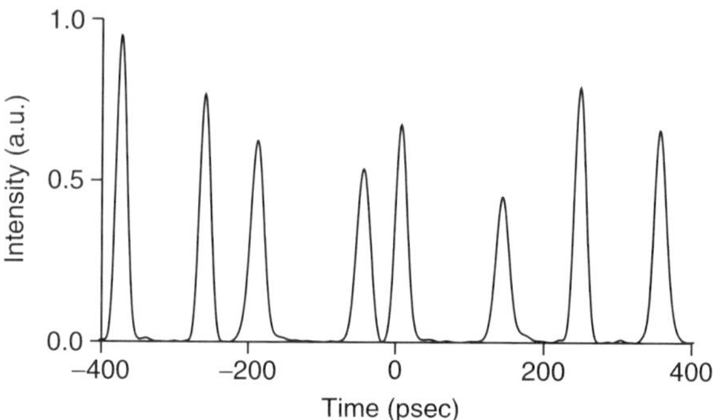

FIGURE 5.9 Pulses that have traversed a 10-Mm transmission line with $L_{amp} = 33$ km and constant $D = 0.5$ ps/nm-km, and that have undergone collisions with an adjacent channel, 0.6 nm away, containing all ones. In this numerical simulation, there were no guiding filters. A small seed of noise was added, but only in the FWM sidebands; thus, the large amplitude and timing jitter seen here are from uncontrolled growth of FWM alone.

WDM channels (possible only when there exist three or more channels), the runaway FWM becomes an additional source of noise fields to act on those channels. In that way, the well-known amplitude and timing jitter effects of spontaneous emission are enhanced.

The growth of FWM can often be controlled adequately with the use of one or another of the N-step approximations to the ideal exponential dispersion taper discussed in Section 5.4. Figure 5.10 plots the residual FWM intensity resulting from a single collision, as a function of L_{amp}, for various numbers of steps in D per L_{amp}, for the channel separation of 0.6 nm, and for the $\tau = 20$-ps solitons and $\bar{D} = 0.5$ ps/nm-km of Fig. 5.8. Figure 5.11 does the same, but for twice the channel separation (1.2 nm). First, note that the intensity scale in Fig. 5.10 is approximately a factor of $2^5 = 32$ times that of Fig. 5.11, just as implied by Eqs. (5.12) and (5.14), and by the fact that L_{coll} scales inversely as the channel spacing. This scaling is easily generalized; for channels spaced n times the adjacent channel spacing, the FWM intensity should scale as n^{-5}. This apparently rapid falloff in FWM effect is tempered somewhat by the fact that the number of collisions tends to increase as n increases, and that it is really the vector addition of residual field quantities from at least several successive collisions that is to be feared here. Also note that the number of steps required for total suppression of the FWM intensity increases with increasing channel spacing. For example, in Fig. 5.10, just two steps are required for L_{amp} in the neighborhood of 30 km, while four steps are required for the same in Fig. 5.11. Finally, note that because of the finite nature of the pulse widths and collision lengths, the resonances in Figs. 5.10 and 5.11 are fairly broad.

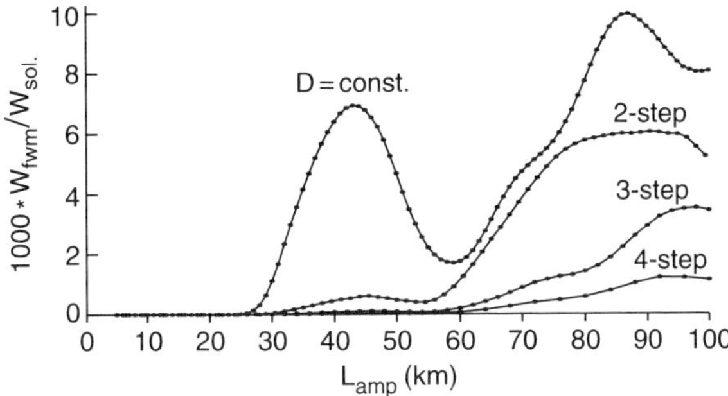

FIGURE 5.10 Residual FWM energy following a single collision of 20-ps solitons in channels spaced 0.6 nm apart in a chain of fiber spans with $\bar{D} = 0.5$ ps/nm-km, as a function of the amplifier spacing, for constant D, and for the optimal 2-, 3-, and 4-step approximations to the ideal exponential taper. The FWM energy is for a single sideband and is normalized to the soliton pulse energy. No noise seed was used in these simulations.

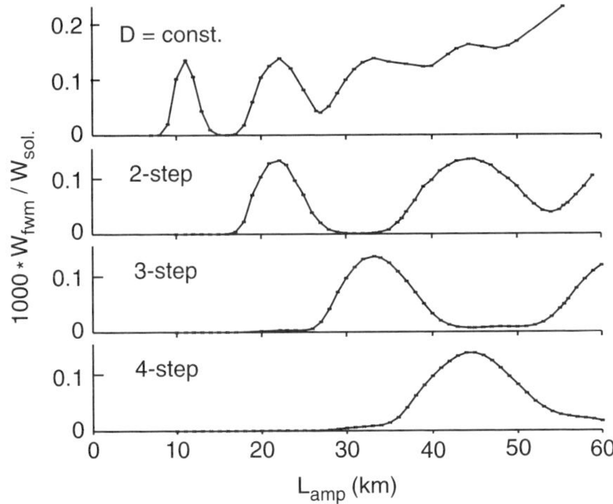

FIGURE 5.11 The same as Fig. 5.10, except that here the channel spacing is twice as great, i.e., it is 1.2 nm.

As a further potential danger due to pseudo phase matching, we note that four-wave mixing in its presence can cause large ($\approx$10%) changes in the soliton's energies and equally significant changes in their frequencies in the course of a single collision of three solitons, one in each of three adjacent equally spaced channels [79].

5.6. Control of Collision-induced Timing Displacements

With the use of dispersion-tapered fiber spans, the only major penalty in WDM comes from the collision-induced timing displacements given by Eq. (4.10) or (4.11c). Since some pulses undergo dozens of collisions, while others nearly none, in the course of a transoceanic transmission, the resultant timing jitter can be substantial, even when the time displacement from a single collision is no more than a picosecond or two. It so happens, however, that the frequency-guiding filters nearly eliminate that jitter as well [80]. The argument can be made as follows: First, to simplify the notation, let $\delta v_g^{-1} \rightarrow v$ and $\partial v_g^{-1}/\partial z \rightarrow a$. Without filtering, let the colliding solitons accelerate each other by $a_0(z)$, whose first z integral is $v_0(z)$, and whose second z integral is $t_0(z)$. We require only that the completed collision must leave no residual velocity shift (see Fig. 5.2, for example). Thus,

$$v_0(\infty) = \int_{-\infty}^{\infty} a_0(z)\,dz = 0. \tag{5.17}$$

For simplicity, we assume that the filters are continuously distributed, and they provide a damping (acceleration) $a_d = -\gamma v = -v/\Delta$, where γ and Δ are the damping constant and characteristic damping length, respectively [see Eq. (3.52)]. The equation of motion then becomes:

$$\frac{dv}{dz} = a_0(z) - v(z)/\Delta. \tag{5.18}$$

Equation (5.18) can be rewritten as

$$v(z) = \left[a_0(z) - \frac{dv}{dz} \right] \Delta. \tag{5.18a}$$

To get t, we simply integrate Eq. (5.18a):

$$t(z) = \int_{-\infty}^{z} v(x)\,dx = \Delta \times \int_{-\infty}^{z} \left[a_0(x) - \frac{dv}{dx} \right] dx. \tag{5.19}$$

We are primarily interested in $t(\infty)$:

$$t(\infty) = \Delta \times \int_{-\infty}^{\infty} a_0(x)\,dx - \Delta \times \int_{-\infty}^{\infty} dv. \tag{5.19a}$$

The first term on the right $= 0$ by virtue of Eq. (5.16). The second term is just $v(\infty)$, which, for a filtered system with no excitation beyond a certain point, must $= 0$. Figure 5.12 shows the quantities δf corresponding to $-v(z)$ and $t(z)$, numerically simulated for the case of lumped filters. One can see from the figure how the timing displacement is nulled: The filters reduce the maximum frequency (velocity) shift,

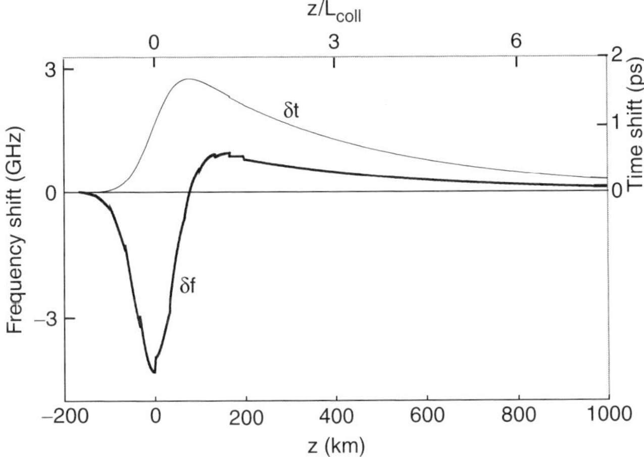

FIGURE 5.12 The frequency and time shifts, as a function of distance, resulting from a collision in a transmission line with lumped amplifiers and lumped filters spaced by 33.3 km; the other parameters are $D = 0.5$ ps/nm-km; $\tau = 20$ ps; channel spacing, $\Delta\lambda = 0.6$ nm; damping length, $\Delta = 400$ km.

so that the acceleration in the second half of the collision causes an overshoot in frequency; the area under the long tail of the frequency curve thus produced just cancels the area under the main peak.

Real filters, such as etalons, do not always perform exactly as in the simplified model given here. In the first place, with real filters, the damping force is not always strictly proportional to $-v$, or equivalently, to δf. Second, the time delay through the filters exhibits a certain dispersion as the signal frequency moves off the filter peak. Nevertheless, numerical simulation shows that the etalons used in the theoretical examples and in the experiments cited here tend to cancel out at least 80% of the timing displacements. That large improvement has turned out to be sufficient for most purposes.

5.7. Effects of Polarization

Thus far in the discussion, we have assumed that the colliding pulses are co-polarized. It is important to note, however, that all of the effects discussed so far (XPM and FWM) are significantly affected by polarization, and all in the same way. Although we defer a full-blown discussion of polarization until the following section, the state of affairs can be summarized as follows: First, just as for the path-averaged solitons, the residual birefringence of even the highest quality transmission fibers available at present is large enough that over the (tens or hundreds of kilometers long) path of a single collision, the Stokes vectors representing the polarization states of the individual pulses tend to rotate, more or less at random, many times over and around the Poincaré sphere. Thus, in that way, the polarization tends to be well averaged over a representative sample of all possible states during a collision. On the other hand, the relative polarization, i.e., the angle between the Stokes vectors, of the colliding pulses tends to be only mildly affected. (The relative polarization is affected by two things: the dispersion in the fiber's linear birefringence and, as will be detailed in Chapter 7, a nonlinear birefringence induced by the collision itself.) Thus, even in the worst case, to first order at least, one can treat the relative polarization during a collision as a constant, so polarization does not significantly affect the symmetries of the collision. Nevertheless, under those conditions (of thorough averaging over absolute polarization states, while relative polarization states are preserved), the frequency shifts induced by XPM and the intensities of FWM products are both just half as great for orthogonally polarized pulses as they are for co-polarized pulses. This fact is of obvious practical importance.

5.8. Gain Equalization with Guiding Filters

The inevitable variation of amplifier gain with wavelength presents a problem for WDM, one that becomes ever more serious with increasing system length. In linear transmission (such as NRZ), where no self-stabilization of the pulse energies is possible, custom-designed, wavelength-dependent loss elements must be inserted periodically along the line to try to compensate for the variable amplifier gain. Even then, however, in practice it has proved difficult to maintain even roughly equal signal levels among the various channels. For soliton transmission using guiding filters, however, the guiding filters themselves provide a powerful, built-in feedback mechanism for controlling the relative strengths of the various signal channels in the face of variable amplifier gain. As should be evident from the discussion of guiding filters already presented (Chapter 3), the control stems from the fact that the guiding filters provide a loss that increases monotonically with the soliton bandwidth, and hence with the soliton energy. Thus, for a channel seeing excess amplifier gain, a modest increase in soliton energy quickly creates a compensating loss, and the signal growth is halted.

Since the soliton bandwidth scales as $\tau^{-1} \propto W/D$, where W is the soliton pulse energy [see Eq. (1.20a)], the soliton's energy loss from the guiding filters can be written as a monotonically increasing function of (W/D). For Gaussian filters, the relation $f(W/D)$ is quadratic; for the shallow etalon filters used in practice, however, $f(W/D)$ is more nearly linear. The energy evolution of N WDM channels in a soliton transmission line with sliding filters can then be described by the following system of coupled nonlinear equations [81]:

$$\frac{1}{W_i}\frac{dW_i}{dz} = \frac{\alpha_i}{1+mR(W_1+W_2+\cdots+W_N)/P_{sat}} - \alpha_{Li} - f(W_i/D_i). \tag{5.20}$$

Here the subscript $i=1,\ldots,N$ identifies that the particular channel, $W_i(z)$ is its energy, α_i is its small-signal gain coefficient, α_{Li} is its linear loss rate, m is the mark-to-space ratio (usually 1/2), R is the per-channel bit rate, P_{sat} is the saturation power of the amplifiers, and $D_i(z)$ is the dispersion at the ith channel wavelength (the dispersion could change with distance z due to the combined action of the frequency sliding and third-order dispersion). Note that the Eq. (5.20) fix the equilibrium values of W_i/D_i according to the various α_i. Thus, when the α_i are all nearly the same, the various channel energies will scale in direct proportion to $D(\lambda)$. In the usual situation where the third-order dispersion is essentially a constant, then the channel energies will be in direct proportion to their separation in wavelength from λ_0, the wavelength of zero D. It should also be noted that the Eqs. (5.20) are essentially the same as Eq. (3.49a), with the frequency offset term omitted. [Recall that in soliton units, one has $W=A$ (not $W=A^2$), and recognize that the quantity α in Eq. (3.49a) is just a compact way of writing the sum of the first two terms on the right-hand side of the Eq. (5.20).] Finally, for Gaussian filters, $f(W/D)=\text{const.}\times W^2$. Thus, just as a linearized version of Eq. (3.49a) was shown to be a damping equation, the linearized version of the Eq. (5.20) is essentially the same damping equation, with $W df/dW$ as the damping constant. Thus, one has

$$W\frac{df}{dW} = \frac{1}{\Delta}, \tag{5.21}$$

where, at least for Gaussian filters, Δ is the same characteristic damping length as discussed in Section 4.3.

Figure 5.13 shows the solution to the Eq. (5.20) for the case of three channels with significantly different small-signal gain rates, both with and without filters. Note how the filters quickly lock the signal energies to tightly clustered equilibrium values. By contrast, the large divergence in channel energies that occur when no filters are used will clearly lead to large penalties and disastrous error rates. Thus, the ability of the guiding filters to regulate the relative channel energies constitutes a major and very important advantage for solitons over all other possible modes of WDM.

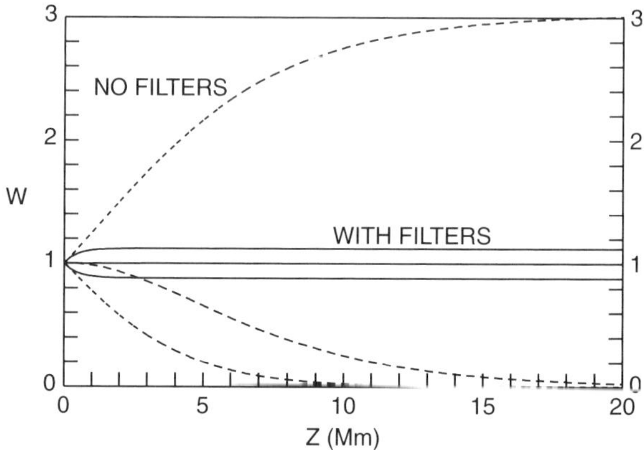

FIGURE 5.13 Growth with distance of signal pulse energies in WDM channels having relative gain rates of ± 1 and 0 dB/Mm, respectively, for a system using sliding-frequency filters (damping length $\Delta = 400$ km) and for no filters at all.

5.9. Experimental Confirmation

Some years ago, the ideas in this chapter were tested in transmissions of up to eight channels at per-channel rates of 5 and 10 Gbit/s (with the major emphasis on the greater rate) [82, 83]. While many aspects of the transmissions were monitored, the ultimate criterion of success was to achieve measured BER rates of 1×10^{-9} or less on all channels, over at least the trans-Pacific distance of 9 Mm.

Figure 5.14 is an overall schematic of the signal source. The soliton pulse shaper, based on a $LiNbO_3$ Mach–Zehnder type modulator, both carves out the pulses and provides them with a controlled and desired chirp [84]. (A full description of the pulse carver is provided in Chapter 8, Section 8.1.1.) After a second modulator imposes the data (a 2^{15}-bit, random pattern), the 4-km length of standard fiber ($D \approx 17$ ps/nm-km at 1557 nm) compresses the pulses to $\approx$20 ps; it also separates the bits of adjacent channels by 40 ps, and this separation prevents the occurrence of half-collisions at the input to the transmission line for all but the most widely spaced channels. Finally, the 3.7-m length of polarization-maintaining (PM) fiber, used as a multiple half-wave plate, enables adjacent channels to be launched with orthogonal polarizations.

The orthogonal polarizations provide two significant benefits: First, they reduce the interchannel interaction by a factor of two over that obtained with co-polarization states. Second, at least where the number of channels is even, the net optical power in the transmission line is essentially unpolarized, so the amplifiers exhibit no significant polarization-dependent gain from polarization

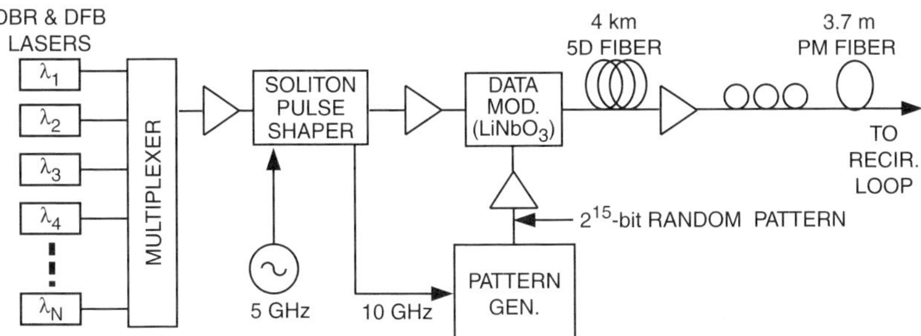

FIGURE 5.14 Source for soliton WDM experiments at 10 Gbit/s per channel.

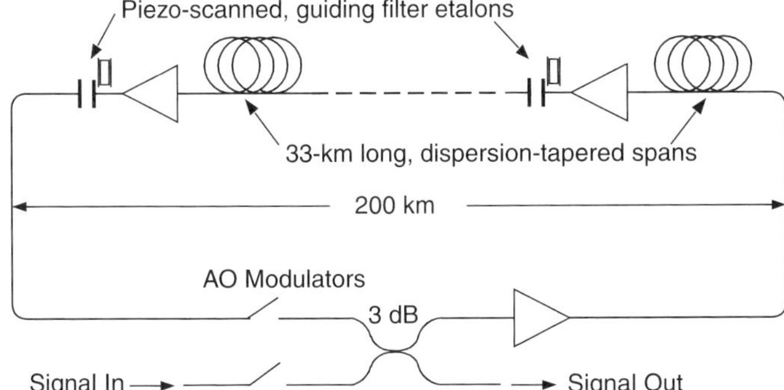

FIGURE 5.15 Recirculating loop with sliding-frequency filters and dispersion-tapered fiber spans. The acousto-optic (AO) modulators act as optical switches with very large on–off ratios to control the sequencing of each transmission. That is, initially, the lower switch is held closed, and the upper switch is open long enough for the source to fill the recirculating loop with data and to bring the amplifier chain to equilibrium. The conditions of the two switches are then simultaneously reversed, so that the loop is closed on itself, and there is no longer an external source. During each such transmission, a linear ramp voltage is applied to the piezo-driven filters to produce the desired sliding of the filter frequencies. Samples of the signal, corresponding to successive round trips, emerge more or less continuously from the signal-out port.

hole burning. Thus, here there is no need for polarization scrambling of the individual channels. There is more here than just the avoidance of unnecessary and expensive hardware, however, since the fiber's birefringence tends to convert polarization scrambling into timing jitter. It should also be noted that the order of the pulse shaper and the data modulator is of no consequence.

The recirculating loop (see Fig. 5.15) contained six spans of 33.3 km each between (Er C band) amplifiers, each span dispersion-tapered typically in three or four steps, with span path-average value $\bar{D}=0.5\pm0.05$ ps/nm-km at 1557 nm. Two out of every three loop amplifiers are immediately followed by piezo-driven, Fabry–Perot etalon filters, each having a 75-GHz (0.6 nm at 1557 nm) free spectral range and mirror reflectivities of 9%; this combination provides the optimum path-average filter strength of $\eta\approx 0.4$ at 1557 nm.

At the receiver, the desired 10-Gbit/s channel is first selected by a wavelength filter, and is then time-division demultiplexed to 2.5 Gbit/s by a polarization-insensitive electro-optic modulator having a 3-dB bandwidth of 14 GHz and driven by a locally recovered clock (see Fig. 5.16). The demultiplexer provides a nearly square acceptance window in time, one bit period wide.

Figure 5.17 shows a typical set of BER data. Note the tight clustering of the BER performance for all channels. Figure 5.18 plots the measured error-free distances vs. the number, N, of 10-Gbit/s channels. For each of these points, the BER was better than 1×10^{-9} on all N channels. Note that most of the data points correspond to loop amplifiers pumped at 1480 nm, with corresponding high noise figure ($\approx$6 dB) and narrow gain-bandwidth. The last point corresponds to pumping at 980 nm, however, where the noise figure is much closer to 3 dB and the gain bandwidth is improved, at least toward shorter wavelengths. The error-free distances represented in Fig. 5.18 tend to be determined by a low-level error floor, which is only very weakly dependent on distance. This dependence is especially noticeable for the largest values of N. Thus, even small future improvements should enable both the error-free distances and the maximum number of allowable channels to be increased.

An experiment was also performed at 5 Gbit/s per channel (six channels), with the same apparatus, simply by programming the pattern generator to eliminate every second pulse of the otherwise 10-Gbit/s data. As the only other substantial change, consistent with the lower bit rate, the time-acceptance window of the

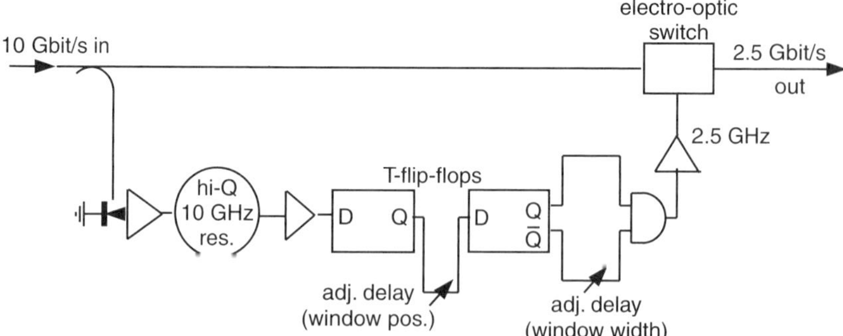

FIGURE 5.16 Time division demultiplexer (T-flip-flops = toggle flip-flop chips).

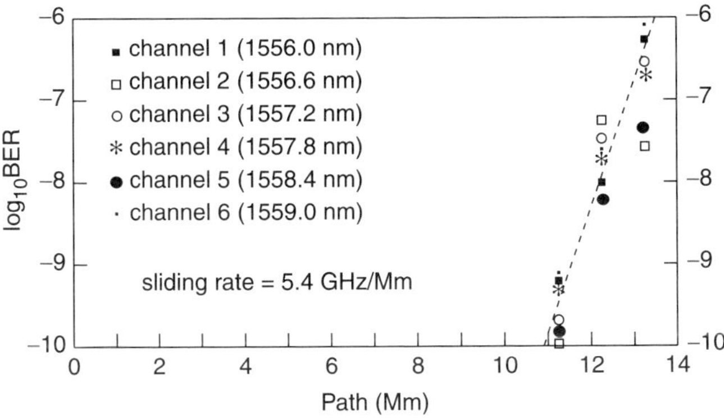

FIGURE 5.17 Measured BER vs. distance for a 6×10 Gbit/s WDM transmission.

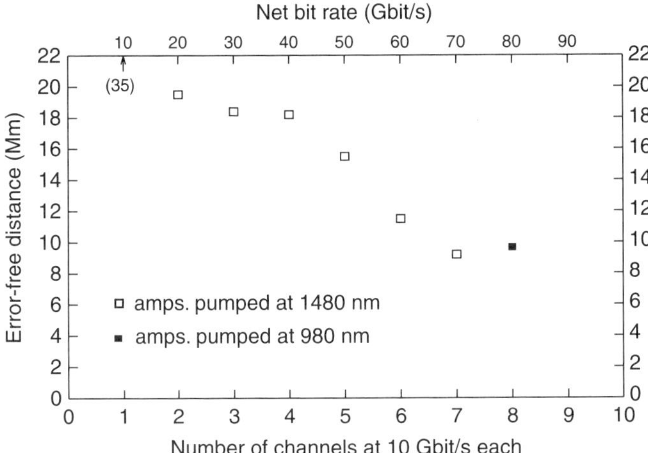

FIGURE 5.18 Measured error-free distances vs. number, N, of channels at 10 Gbit/s per channel.

demultiplexer (Fig. 5.16) was opened up by a factor of two. For that experiment, the error-free distance was greater than 40 Mm on all channels. The great increase in error-free distance was due primarily to two factors: First, at half the bit rate, the rate of collisions was decreased by half. Second, of course, the doubled time-acceptance window greatly increased the tolerance to timing jitter.

Finally, Fig. 5.19 shows an example of the spectrum of the WDM transmission. Although the example corresponds to a particular distance (10 Mm), the spectrum looks just the same at any but the very shortest distances. In all of the experiments, regardless of the number of channels, the spectra all had the same feature,

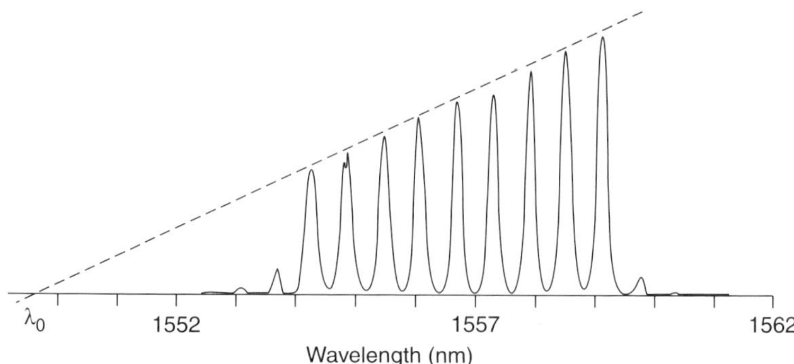

FIGURE 5.19 Optical spectrum of 9 × 10-Gbit/s WDM transmission, as measured at 10 Mm. Note that the dashed line connecting the spectral peaks of the individual channels passes accurately through the wavelength of zero dispersion, λ_0. This behavior results from the strong regulation of the soliton pulse energies provided by the sliding-frequency filters. Following initial adjustment, it becomes independent of distance.

viz., that the spectral peaks could all be joined by a straight line that passed through the zero intensity axis at λ_0, the wavelength of zero dispersion. All of this occurred in the face of considerable amplifier gain variation over the total wavelength span. Thus, in these spectra we have direct and complete confirmation of the ideas presented in Section 5.7, where the guiding filters were shown to provide a very tight control over the relative signal strengths of the various channels.

Chapter 6

Wavelength Division Multiplexing with Dispersion-managed Solitons

6.1. Soliton–soliton Collisions

In Chapter 5, we have just seen how the perfect transparency of ordinary solitons to one another in lossless fiber begins to break down in real fiber, and thus to cause certain problems and limitations for dense WDM. In Chapter 2, however, we have already seen how the phase mismatch created by the large values of local dispersion tends to make four-wave mixing during soliton–soliton collisions negligibly small in dispersion-managed systems. Thus, in contrast to ordinary solitons, the only nonlinear penalty for dense WDM with dispersion management stems from cross-phase modulation during the collisions, and the timing jitter that results from it. As will soon be evident, however, even in this regard there are great differences, one of the most fundamental of these being the fact that with dispersion-managed solitons, the collision length is essentially independent of channel spacing, again with generally positive consequences. Finally, we shall see how a recently invented trick of dispersion management tends to render the jitter from collisions so small that the performance of a dense WDM system becomes nearly indistinguishable from that of a single, isolated channel.

6.1.1. General Description of Collisions and the Collision Length

As we have just seen in the previous two chapters, in WDM with ordinary solitons, pulses in a shorter wavelength channel steadily overtake and pass through the pulses of a longer wavelength channel. With dispersion management, however,

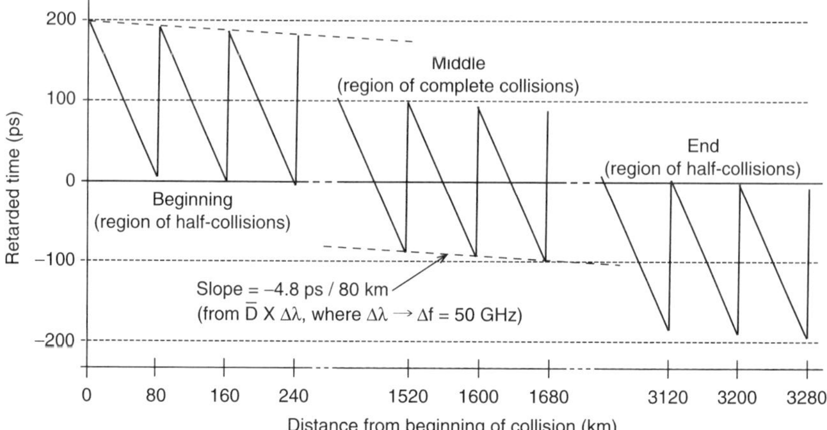

FIGURE 6.1 Relative motion in retarded time of a pair of soliton pulses from adjacent channels separated by 50 GHz; for convenience, the lower frequency pulse (horizontal line at $st = 0$) is fixed in retarded time—thus, the higher frequency pulse displays the entire relative motion (the sawtooth path). The dispersion map consists of 80-km spans of $D = 6$-ps/nm-km fiber, compensated by a coil of DCF to yield $\bar{D} = 0.15$ ps/nm-km. Note the two breaks in the distance scale, necessitated by the extremely great length of the overall collision.

the situation is more complicated: In response to the large and rapidly alternating D values, solitons from different channels race back and forth with respect to each other in retarded time. Thus, collisions between pairs of solitons tend to consist of fast, repeated, "mini-collisions," [35,85] that individually tend to produce only small displacements of the pulses in frequency and time. But when the ratio of local-to-path-average dispersion is very high (as it usually is), then the colliding pair tends to undergo a very large number of such mini-collisions before the solitons cease to cross each other's paths. Thus, the net length for an overall collision tends to be long—typically several thousands of kilometers.

Figure 6.1 illustrates a typical case of the relative motion of a pair of colliding, dispersion-managed solitons in adjacent channels. The sawtooth represents the motion of the higher frequency pulse, while the lower frequency pulse (for convenience, made stationary in retarded time) is represented by the horizonal line at $t = 0$. Note that the time displacement during passage through the (D_+) main transmission span (the more gently sloping line of the sawtooth) is almost, but not quite, reset after transmission through the coil of DCF (the very steep segment of the sawtooth). Thus, the sawtooth pattern descends only very slowly, with slope $\bar{D}\Delta\lambda$, where $\Delta\lambda$ is the channel separation.

From the figure one can also see that the overall collision tends to consist of three distinct phases: First, it begins where the pulses tend to achieve maximum

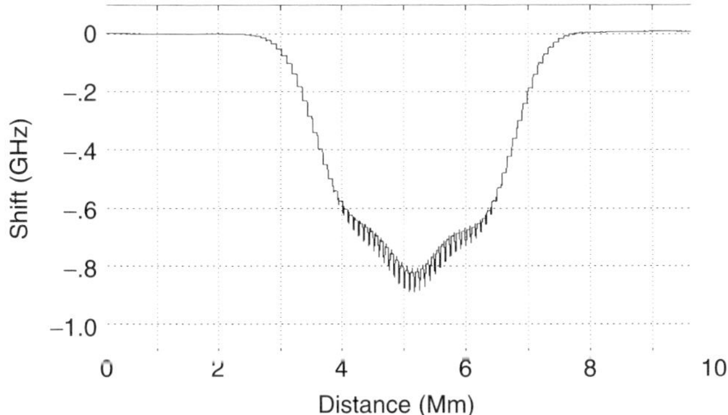

FIGURE 6.2 Frequency shift, as a function of distance, of the lower frequency pulse of the colliding pair of dispersion-managed solitons of Fig. 6.1, as determined by exact numerical solution of the NLS equation. (An equal but opposite frequency shift is induced into the higher frequency pulse.) The colliding pulses are orthogonally polarized. Mid-span Raman pumping minimizes signal intensity variation along the main span, in order to make the frequency-shift curve symmetric.

overlap at the junction between the +D and −D fibers; note that each of these half-collisions tends to produce a net frequency shift that is approximately twice as great (when $|D_+| \approx |D_-|$) as the peak shift of a collision completed in just one kind of fiber. Thus, the net effect of these half-collisions is to produce a steep wall of rise (or fall) in frequency shift (see Fig. 6.2). The middle part consists of complete collisions that tend to produce only small net effects, especially when they take place in a region of small intensity gradient. (The curve of Fig. 6.2 corresponds to the use of mid-span backward Raman pumping to produce a relatively small and symmetrical intensity profile, as, for example, is shown in Fig. 2.9.) Finally, at the end, once again we have half-collisions, but this time at the other junction of the +D and −D fibers; these produce a steep decline (or increase) of frequency, back to zero net shift (again, see Fig. 6.2).

If we define L_{coll} as beginning and ending when the pulses completely overlap during the half-collisions, a study of Fig. 6.1 reveals that over the course of an entire collision, the sawtooth motion must descend by its peak-to-peak amplitude, equal to $(D_+ - \bar{D})L_+ \Delta\lambda$, where D_+ and L_+ refer to the +D segment of the dispersion map. It is natural to think of that time displacement as an effective width, τ_{eff}, of the colliding pulse. As should be evident from the figure, τ_{eff} must be equal to L_{coll} times the slope (cited above) of the sawtooth pattern. Thus we have

$$L_{coll} = \frac{\tau_{eff}}{\bar{D}\Delta\lambda} = \frac{(D_+ - \bar{D})L_+}{\bar{D}} \cong \frac{D_+ L_+}{\bar{D}}. \tag{6.1}$$

Note that when the specific parameters of the map of Fig. 6.1 are entered into it, Eq. (6.1) yields $L_{coll} = 3200$ km, in excellent agreement with the figure. Also, note that $\tau_{eff} = 192$ ps, in accord with the peak-to-peak amplitude of the sawtooth in Fig. 6.1. Most important of all, however, note that L_{coll} as defined by Eq. (6.1) is *completely independent of the channel separation*, in stark contrast to the case of collisions between ordinary solitons. This fact is of great importance for the practicality of dense WDM with dispersion-managed solitons, as it obviously avoids one of the major problems with ordinary solitons encountered in Chapter 5.

It should be noted that for a more general definition of L_{coll} that will remain accurate and meaningful when the map strength is made very small or zero, it is necessary to add the collision length for ordinary solitons $[2\tau/(\bar{D}\Delta\lambda)]$ to the right-hand side of Eq. (6.1). Nevertheless, as the simpler definition of Eq. (6.1) is perfectly adequate when the dispersion map has the usual strength ($S \approx 2$ or more), that is the one we shall use in the following sections.

6.1.2. Collision-induced Frequency and Time Shifts

From the description just given, it should be clear that an exact analytic theory of collisions between dispersion-managed solitons is probably not feasible. Nevertheless, in this section, we outline a "quasi-analytic" model, based on careful observation of many exact numerical simulations like that of Fig. 6.2 and on analytic treatment of some of the simpler constituent elements. In our model, we replace the actual shape of the frequency-shift curves of Fig. 6.2 with a perfect rectangle that yields approximately the same maximum frequency shift and, more importantly, the same time shift as the real thing. While admittedly a bit "rough and ready," our model reveals the fundamentally important scaling with L_{coll} and the channel spacing, and it yields numerical results that are at once essentially accurate and surprisingly invariant to change in certain details. Finally, it provides the basis for a satisfactory mathematical model of the all-important timing jitter that results from the collisions.

As already discussed in Chapter 4, the relative state of polarization of colliding pulses is important, since the XPM (and hence, all of the subsequent collisional effects) are just half as great when the pulses are orthogonally polarized as when they are co-polarized. Although the absolute states of polarization of the colliding solitons tend to evolve rapidly along the fiber (with major changes typically occurring every few meters), in low-polarization-mode dispersion (PMD) fiber, their relative states of polarization tend to remain fixed over long distances (many thousands of kilometers); this is especially true for pulses in adjacent channels that are initially either co-polarized or orthogonally polarized. Therefore, to handle this matter in our model, we introduce the polarization coefficient C_{pol}, which equals 1 when the interacting channels are co-polarized, and 1/2 when they are orthogonally polarized.

The frequency shifts of the half-collisions are the sum of two shifts: that created when the pulses come together in the +D fiber, and that created when they subsequently back away from each other in the −D fiber. From the known effects of XPM on colliding pulses in fiber of constant D, we can then obtain

$$\delta f_{half\text{-}coll} = C_{pol}\frac{n_2}{A_{eff}\lambda}\left[\frac{1}{D_+}+\frac{a}{|D_-|}\right]\frac{W_{sol}}{\Delta\lambda\,\tau}, \tag{6.2}$$

where A_{eff} is the effective area of the +D fiber and where a is the ratio of A_{eff} to the core area of the −D fiber. (Note that when the −D fiber is DCF, the major contribution comes from the +D fiber.) To obtain the net frequency shift of the overall collision, we must first calculate the effective number of half-collisions, i.e., the number of collisions required to advance the turn-around point of the half-collisions by 2τ, where τ is the pulse intensity FWHM. Since the time advance per collision is $L_{map}\bar{D}\Delta\lambda$, the effective number of half-collisions is

$$N_{coll} = \frac{2\tau}{L_{map}\bar{D}\Delta\lambda}. \tag{6.3}$$

The net frequency shift of the overall collision is then obtained by multiplying Eq. (6.2) by Eq. (6.3):

$$\delta f = C_{pol}\frac{n_2}{A_{eff}\lambda}\left[\frac{1}{D_+}+\frac{a}{|D_-|}\right]\frac{W_{sol}}{\Delta\lambda\,\tau}\times\frac{2\tau}{L_{map}\bar{D}\,\Delta\lambda}. \tag{6.4a}$$

For $A_{eff} = 50\,\mu\text{m}^2$, $\lambda = 1550$ nm, and expressing W_{sol} in fJ and the other quantities in the usual units of ps/nm-km, nm, and ps, respectively, Eq. (6.4a) becomes

$$\delta f(\text{GHz}) = \pm 0.335 C_{pol}\left[\frac{1}{D_+}+\frac{a}{|D_-|}\right]\frac{2W_{sol}}{L_{map}\bar{D}(\Delta\lambda)^2}. \tag{6.4b}$$

[Note that the numerical coefficient in Eq. (6.4b) is not dimensionless.] If we put the particular parameters (the data from the captions, plus $C_{pol} = 0.5$ and $a = 2.5$) for the collision of Figs. 6.1 and 6.2 into Eq. (6.4b), we get $\delta f = 0.74$ GHz, in good agreement with the mean of the "middle" of the numerical solution shown in Fig. 6.2.

The time shift associated with the collision is exactly

$$\delta t(z) = -\frac{\lambda^2}{c}\int_{-\infty}^{z}\delta f(z')D(z')\,dz', \tag{6.5}$$

where $D(z)$ step changes back and forth between D_+ and D_-. Figure 6.3 plots $\delta t(z)$ for the collision of Fig. 6.2. The large oscillations in δt seen there result, of course,

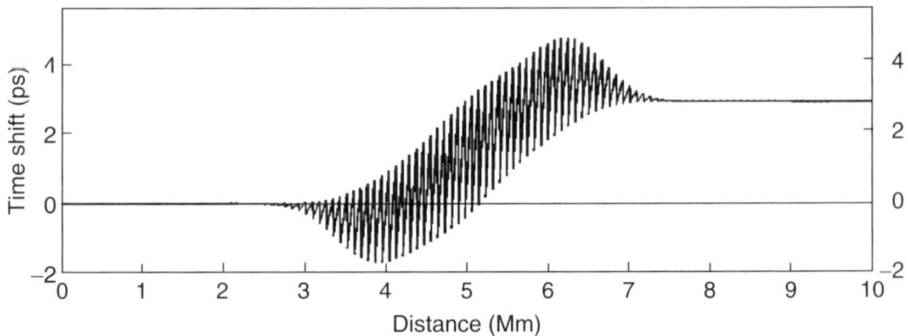

FIGURE 6.3 Time shift, as a function of distance, of the lower frequency pulse of the colliding pair of dispersion-managed solitons of Fig. 6.1, as determined by exact numerical solution of the NLS equation. (An equal but opposite time shift is induced into the higher frequency pulse.) Conditions are exactly as in Fig. 6.2.

from the fact that the δf of the moment is alternately multiplied by the large positive and negative terms D_+L_+ and D_-L_-, respectively. For our purposes in calculating the timing jitter, however, the oscillations are of little interest, since all we really need is the net time shift obtained at the very end of the collision. Note further that at the extreme ends of the overall collision, the result of each completed mini-collision is in the "wrong" direction, i.e., in the direction of negative δt. The reason is that $\delta t(z)$ grows as the sum of two terms: the area under the curve of frequency shift (which, of course, must increase monotonically with z) and a second term that oscillates between negative and positive values. Since the second term vanishes at the end of the collision [86], the net time shift of the overall collision can be estimated as

$$\delta t \equiv \delta t(\infty) = -\frac{\lambda^2}{c}\bar{D}\sum_n L_{map}\,\delta f_n, \tag{6.5a}$$

where δf_n is the height of the nth step of the mini-collisions and where the sum in Eq. (6.5a) is the area under the frequency-shift curve, such as that of Fig. 6.2. Now, since in our simple model the area under the frequency-shift curve can be written as the δf from Eq. (6.4) times L_{coll}, we finally have

$$\delta t = -\frac{\lambda^2}{c}\bar{D}\,\delta f\,L_{coll}. \tag{6.5b}$$

Inserting Eqs. (6.4b) and (6.1) into Eq. (6.5b) and regrouping terms, we obtain

$$\delta t\ (\text{ps}) = \pm 0.335\,C_{pol}\left[1 + a\frac{D_+}{|D_-|}\right]\frac{L_+}{L_{map}}\,\frac{2W_{sol}}{c\bar{D}}\,\frac{\lambda^2}{(\Delta\lambda)^2} \tag{6.5c}$$

Once again, W_{sol}, c, and all terms in D must be in fJ, km/s, and ps/nm-km, respectively, in order to make this equation valid for the particular numerical

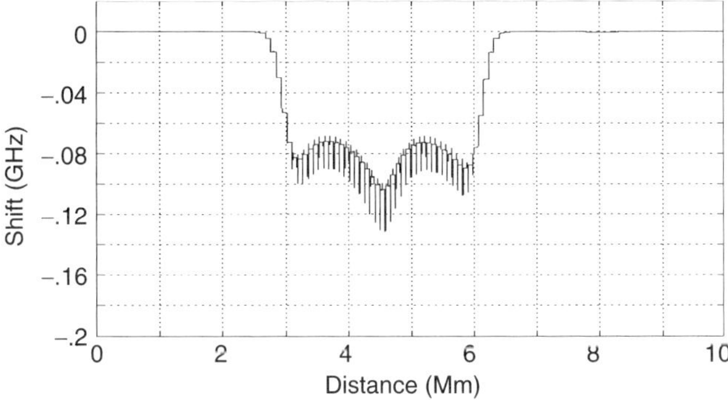

FIGURE 6.4 Frequency shift, as a function of distance, for a collision like that of Fig. 6.2, except that now the channel spacing is 150 GHz.

coefficient shown. (Note that all other factors on the right-hand side of this equation, save the numerical coefficient itself, are dimensionless.) For the parameters of the collision of Fig. 6.2, Eq. (6.5c) yields $\delta t \approx 2.9$ ps, in good agreement with the numerical result of Fig. 6.3. (Again, the result is for orthogonally polarized colliding pulses; δt is twice as great when the pulses are co-polarized.)

It is extremely important, as we shall discuss shortly, that the XPM-induced frequency and time shifts scale as the inverse *square* of the channel separation ($\Delta\lambda$). It should be noted that this scaling has been nicely confirmed in many exact numerical simulations. For example, Fig. 6.4 shows the same collision as Fig. 6.2, but for a three times greater channel separation. Note that the peak frequency shift is indeed nine times smaller. Note further, however, that except for its steeper walls, the collision has about the same shape and essentially the same width as with the smaller channel spacing.

It is also extremely important that *complete* collisions exhibit essentially zero residual frequency shifts, as evidenced by the curves of Figs. 6.2 and 6.4 and by countless other exact numerical simulations. Later we shall see evidence that this rule does not break down until L_{coll} becomes shorter than about four periods of the dispersion map. Although here, too, there is no exact analytical proof, the behavior undoubtedly has a certain similarity to that obtained with ordinary solitons when L_{coll} is long enough with respect to L_{pert}, just as we have already seen in Chapter 5.

For *incomplete* collisions, however, the story is different. It may happen that, at or near the transmitter, the colliding solitons are on top of each other in the middle of a span, so that the overall collision begins near the middle of a complete collision (see Fig. 6.5). Such partial collisions can produce even greater time shifts, since they produce a large residual frequency shift that does not disappear. In that case, the length factor in Eq. (6.5) is not L_{coll}, but nearly the entire distance of the transmission. Also note that the algebraic sign of the frequency shift (and hence

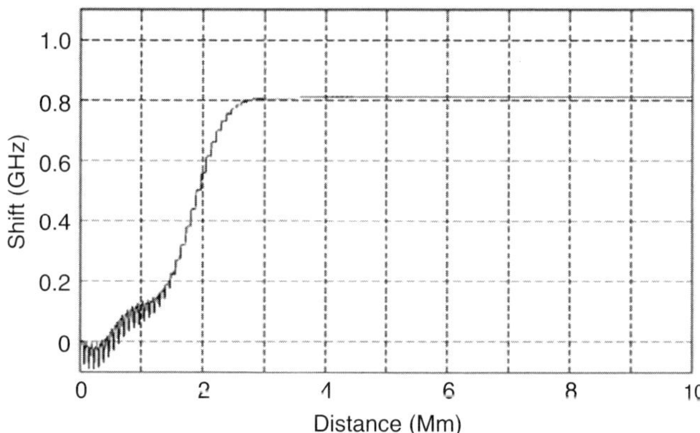

FIGURE 6.5 Frequency shift, as a function of distance, of the half-collision version of the collision of Fig. 6.2. Note that the time shift is now proportional to the area underneath the tail of the curve.

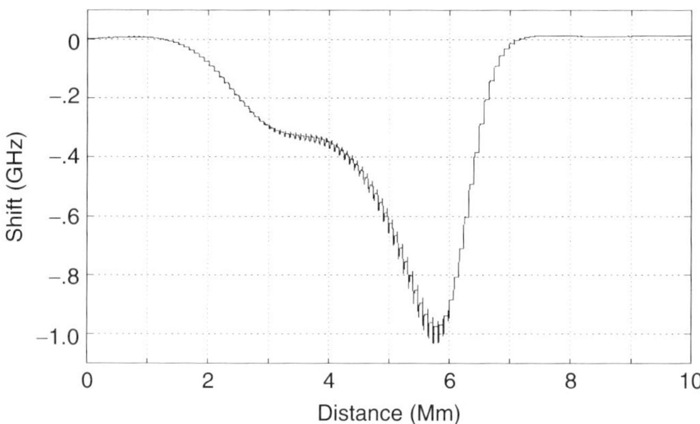

FIGURE 6.6 Frequency shift, as a function of distance, of the collision like that of Fig. 6.2, except that in this case, the lack of mid-span pumping produces a considerably greater variation of signal intensity across the main span.

that of the corresponding time shift) is opposite to those produced by the complete collision.

Finally, we note that the more usual distributions of signal intensity with distance over the span (as, for example, in Fig. 2.18) tend to introduce a significant asymmetry into the curves of collision-induced frequency shift (see Fig. 6.6). This occurs because at the beginning of the overall collision, the half-collisions take place at the low-intensity junction of the main and DCF spans, while at the end of the overall collision, the half-collisions take place at the high-intensity junction of those spans. Despite the rather dramatic change in shape, however, the area under the curve (and hence the resultant net time shift) does not seem to change, at least not significantly, as has been verified through many numerical simulations. Thus, our simple model can still yield the correct time shift.

6.1.3. Calculating the Jitter

Thus far we have discussed the collision interaction of just two solitons. In the course of a long-haul transmission, however, each pulse of a given channel experiences many collisions with pulses of all of the other channels. In this section, we calculate the timing jitter resulting from those collisions.

For the purposes of this section, it is convenient to reserve the term $\Delta\lambda$ for the nearest neighbor channel separation only, so that the general separation becomes $m\Delta\lambda$, where m is a positive integer, and to define δt_m as the time displacement resulting from collision with one pulse from a channel separated by $m\Delta\lambda$. Thus, from Eq. (6.5c), we have both

$$\delta t_m = \delta t_1/m^2 \tag{6.6}$$

and a way to calculate δt_1 from all the pertinent parameters of the system.

Let Δt_m represent the net time displacement of a pulse from collisions with all of the bits in the channel separated by $m\Delta\lambda$, as the pulse traverses the entire transmission distance Z. Thus we have $\Delta t_m = \delta t_m n_{coll}$, where n_{coll} is the particular number of collisions occurring in the transmission distance Z. Since, if all pulses experienced the same Δt_m, there would be no jitter, what we really want is the variance in Δt_m,

$$\left\langle \Delta t_m^2 \right\rangle = \delta t_m^2 \left\langle \delta n_{coll}^2 \right\rangle. \tag{6.7}$$

Clearly, our task is now to calculate $\langle \delta n_{coll}^2 \rangle$, as follows: First, the minimum distance between subsequent collisions with pulses from the mth channel, l_{coll}, is

$$l_{coll} \cong \frac{T}{\bar{D} m \Delta\lambda}, \tag{6.8}$$

where T is the bit period. Thus, the maximum possible number of collisions with pulses of the mth channel is

$$N_m = \frac{\bar{D} m \Delta\lambda Z}{T}. \tag{6.9}$$

Now, we assume random data where, in any given bit period, the presence of a pulse or its absence has probabilities $p = q = 1/2$, respectively. The probability distribution of n_{coll} is then the binomial distribution

$$P(n_{coll}; N_m, p) = \frac{N_m(N_m-1)\ldots(N_m-n_{coll}+1)}{(n_{coll})!} p^{(n_{coll})} q^{(N_m-n_{coll})}, \tag{6.10}$$

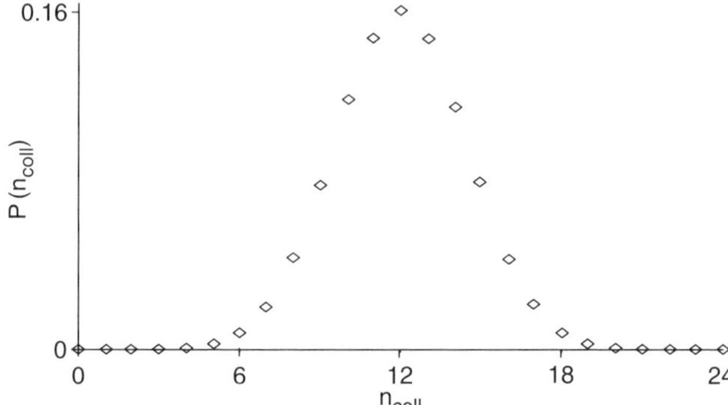

FIGURE 6.7 The binomial probability distribution for $N_m = 24$. As N_m becomes larger, its width becomes a smaller fraction of N_m, and the peak probability becomes smaller.

whose variance is simply $N_m pq = N_m/4$. (For those unfamiliar with it, the envelope of this discrete distribution looks approximately like a Gaussian; see Fig. 6.7.) Substituting $1/4\times$ the right-hand side of Eq. (6.9) for $\langle \delta n_{coll}^2 \rangle$, and $\delta t_1/m^2$ for δt_m [from Eq. (6.6)] in Eq. (6.7), and regrouping terms, we obtain

$$\left\langle \Delta t_m^2 \right\rangle = \delta t_1^2 \frac{\bar{D}\Delta\lambda}{4T} \frac{Z}{m^3} \tag{6.11}$$

for the variance in timing jitter induced by interaction with the mth channel.

Since the Δt_m are statistically independent, the variance of the net jitter of a given channel is just the sum of the $\langle \Delta t_m^2 \rangle$ over all other channels. There are two more significant variables to be considered here: (1) whether adjacent channels are orthogonally polarized or co-polarized and (2) whether the channel in question is at one far edge of the WDM band or somewhere in its middle. To facilitate matters, once again it will be convenient to make a slight change in notation. That is, we now remove C_{pol} from the δt_m (and hence from the Δt_m) in order to display it explicitly. Let the total number of channels be $M+1$, with M_1 on one side of the affected channel and M_2 on the other. (Thus $M_1 + M_2 = M$.) The variance of the net jitter can then be written as

$$\left\langle \Delta t^2 \right\rangle = \sum_{m=1}^{M-1} \left\langle \delta t_m^2 \right\rangle = \delta t_1^2 \frac{\bar{D}\Delta\lambda}{4T} Z \left[\sum_{m=1}^{M_1} (C_{pol})_m^2 m^{-3} + \sum_{m=1}^{M_2} (C_{pol})_m^2 m^{-3} \right], \tag{6.12}$$

where $(C_{pol})_m^2 = 1$ when all channels are co-polarized and $(C_{pol})_m^2 = (5/8 + 3/8(-1)^m)$ when adjacent channels are orthogonally polarized. (Note that for the latter case, C_{pol}^2 alternates between 1/4 and 1 as m goes from odd to even values.) The standard deviation for each of these cases is given, of course, by the square root

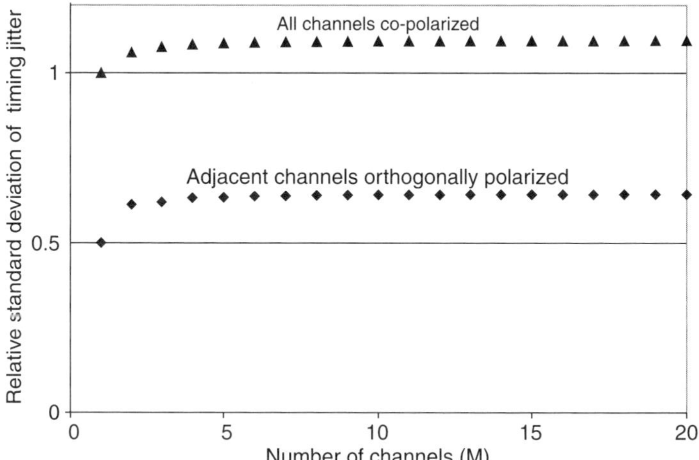

FIGURE 6.8 The square root of the final sum in Eq. (6.12) for the cases where the affected channel is at one far end of the WDM band, plotted as a function of the total number of interacting channels. Note that these quantities scarcely change for $M > 3$. The asymptotic values here are approximately 1.0964 and 0.6428 for the co-polarized and orthogonally polarized cases, respectively (ratio $\cong 1.70$). When the affected channel is at least several channels in from either edge of the WDM band, multiply each of these values by $\sqrt{2}$.

of the respective variance. Figure 6.8 plots the square root of the final sums (the quantities in square brackets) in Eq. (6.12), for the cases where $M_1 = M, M_2 = 0$ (i.e., where the affected channel is at one far edge of the WDM band). When the affected channel is at least three or more channels in from an edge, the corresponding variances are essentially doubled, so the standard deviations are increased by $\sqrt{2}$. Notice how quickly the sums converge to an asymptotic value, such that the net effect of all channels removed more than about $3\Delta\lambda$ is almost negligible. One important consequence of this fact is that numerical simulations need employ no more than about six to eight channels to yield realistic results.

In order to have some idea of the magnitude of jitter created by Eq. (6.12) and its effects, let us apply it to a system with 100-km, DCF-compensated spans, and orthogonally polarized adjacent channels, whose experimental behavior will be reported on later in this chapter. First, from Eq. (6.5a), we insert $D_+ = 7$ ps/nm-km, $|D_-| = 100$ ps/nm-km, $\bar{D} = 0.15$ ps/nm-km, $a = 2.5$, $L_+ = 100$ km, $L_{map} \cong 107$ km, $\lambda = 1570$ nm, $\Delta\lambda = 0.4$ nm, and $W_{sol} = 25$ fJ, and obtain $\delta t_1 = 6.297$ ps. We then insert δt_1, $T = 100$ ps, and for the sum, the quantity $2 \times (0.6427)^2$ (appropriate for a channel in the middle of a group with orthogonal polarization of adjacent channels) into Eq. (6.12) and take the square root. The result is plotted in Fig. 6.9. It is also realistic for this particular system to assume that the bit error rate is completely dominated by the

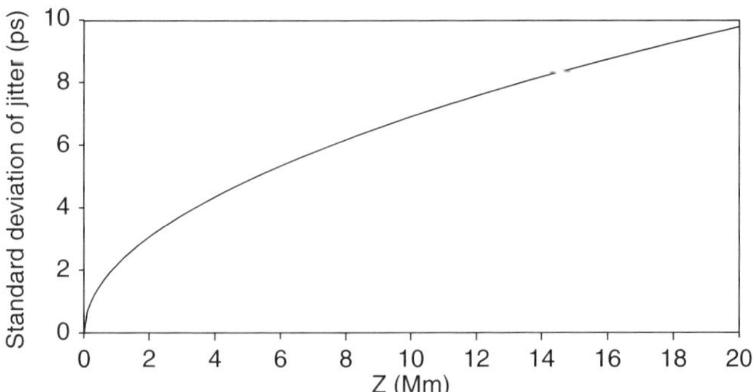

FIGURE 6.9 The standard deviation of collision-induced timing jitter as derived from the square root of Eq. (6.12), plotted as a function of transmission distance Z, for a channel anywhere in the middle of a large number of channels with spacing $\Delta\lambda = 0.4$ nm and orthogonal polarization of adjacent channels. Major system parameters: 100-km spans with $D = 7$ ps/nm-km, compensated by DCF such that $\bar{D} = 0.15$ ps/nm-km; $W_{sol} = 25$ fJ. For other details, see text.

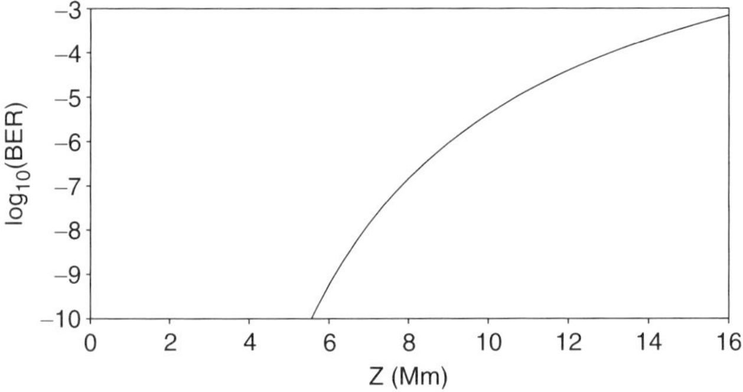

FIGURE 6.10 Computed bit error rate as a function of transmission distance Z when the errors are due entirely to timing jitter having the standard deviation given in Fig. 6.9. The assumed acceptance window at the detector is 70 ps wide.

collision-induced jitter. The computed bit error rate, based on the jitter curve of Fig. 6.9 and an assumed acceptance window at the detector of 70 ps, is plotted in Fig. 6.10. We shall soon see that the BER of Fig. 6.10 is in excellent accord with experiments.

At the beginning of this chapter, we asserted that the only significant nonlinear defect in WDM with dispersion-managed solitons was the timing jitter we have just calculated. Figure 6.11 offers direct proof of that assertion from an exact numerical simulation: the eye diagram of one in the middle of eight channels after 8000-km transmission through the system referred to in Figs. 6.9 and 6.10.

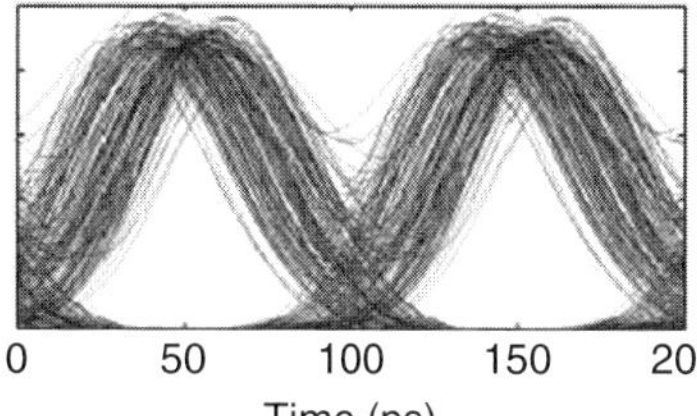

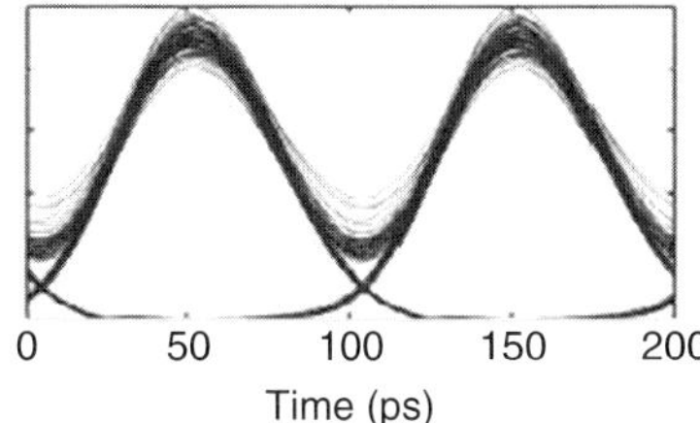

FIGURE 6.11 Numerically simulated electrical eye diagrams after 8000-km transmission through the system referred to in Figs. 6.9 and 6.10, for a middle channel out of eight channels. ASE is not included; the only effect is that of the collisions. A different, 2^7-bit random pattern was used for each channel. Left: The unaltered eye. Right: The eye with timing jitter artificially removed.

Note from Fig. 6.11 that artificial removal of the timing jitter almost completely opens the eye (there remains only a very small residual closure from modest amplitude jitter). Thus, we can conclude that the rather severely penalizing closure of the unaltered eye does indeed stem almost entirely from the collision-induced timing jitter, and little else.

6.1.4. Large Reduction of Jitter through Use of Periodic-group-delay Dispersion Compensation

From Fig. 6.10, it should be evident that collision-induced timing jitter is the limiting factor for low BER transmission distance. (For Raman-amplified systems, at least, ASE noise typically does not cause significant error until the transmission distance exceeds 10,000 km or more.) Thus, it would be very desirable to have a way to significantly reduce the collision-induced jitter. In particular, if only a way could be found to reduce L_{coll}, the time shift per collision (the δt_m) and hence the jitter would be reduced as well. At first it would appear, however, that there is no way to improve the situation, since the factors involved in L_{coll} [Eq. (6.1)] seem to be fixed by the requirements for well-behaved dispersion-managed solitons with the correct energy. But there happens to be a way to create a new and independent *interchannel* path-average dispersion, $\bar{D}_{inter}$, to determine L_{coll}, without in any way disturbing the $\bar{D}$ effective within a given channel (which can now be designated the *intrachannel* dispersion, or $\bar{D}_{intra}$), required for proper maintenance of the solitons and their energies. With introduction of the new $\bar{D}_{inter}$, Eq. (6.1) becomes

$$L_{coll} = \frac{(D_+L_+)}{\bar{D}_{inter}}. \tag{6.1a}$$

The way to accomplish this change is to use a "periodic-group-delay dispersion compensation module," or PGD-DCM, whose group-delay period is equal to the

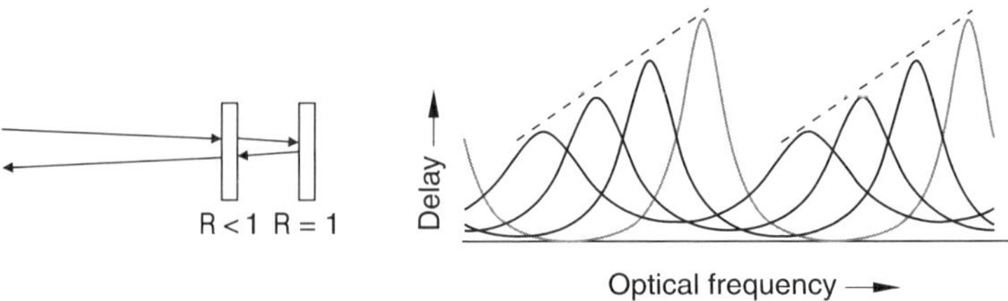

FIGURE 6.12 Left: Basic scheme of a Gires–Tournois etalon, used to obtain frequency-dependent delay times. Right: Showing how cascaded etalons of different resonance frequencies and different peak delay times can create net delay linearly dependent on frequency, as indicated by the dashed lines.

WDM channel spacing, to do at least part of the dispersion compensation of each span [87].

While several different types of PGD-DCMs have been developed [88–90], perhaps the most successful are those [91] based on a set of Gires–Tournois etalons (see Fig. 6.12). Gires–Tournois etalons are used in reflection, and the back mirror has 100% reflectivity, so that the net reflectivity of the device is nominally independent of frequency. Time delay is a different matter, however: Light in resonance bounces back and forth several times, while that off resonance tends to be returned immediately, thereby yielding a frequency-dependent delay. By cascading a number of such etalons where each is tuned to a different frequency and each has a different peak delay (again, see Fig. 6.12), it is possible to create an amazingly smooth curve of net delay with linear slope across a good fraction of the free spectral range. For our purpose here, however, the fundamentally important fact is that *the mean group delay does not change from channel to channel* (see Fig. 6.13 for typical behavior).

To understand how the new $\bar{D}_{inter}$ is created by the use of a PGD device, we need to look at the relative motion in retarded time of a pair of colliding pulses, similar to that we have already made in Fig. 6.1. Now, however, we must consider the more general case where a fraction f of the span dispersion is compensated by a PGD device, while the remainder is compensated by fiber (see Fig. 6.14).

Note from Fig. 6.14 that of the reset time formerly supplied by the DCF, now the fraction f is missing, so the sawtooth descends with a new, steeper slope, which we write as $\bar{D}_{inter}\,\Delta\lambda$. Since the slope of the sawtooth is also $(f|D_-|L_-/L_{map}+\bar{D})\Delta\lambda = fD_+L_+/L_{map}+(1-f)\bar{D}$, we have

$$D_{inter} = fD_+L_+/L_{map}+(1-f)\bar{D} \approx fD_+, \tag{6.13}$$

where the approximation is a good one for all but very small f. Note that $\bar{D}_{inter}$ ranges from the very small value $\bar{D}$ at $f=0$ to a many times greater value

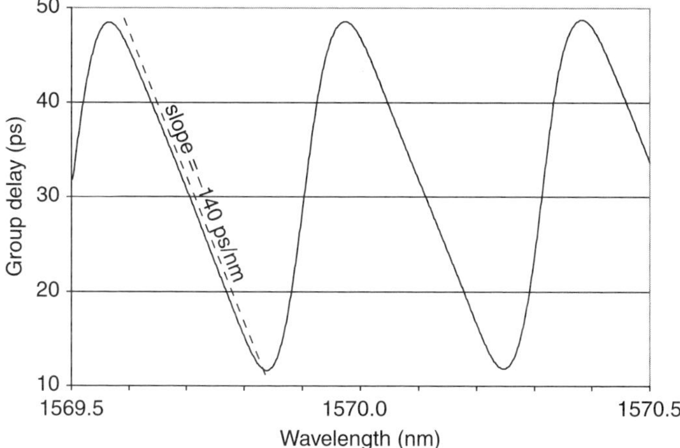

FIGURE 6.13 Measured group delay of a Gires–Tournois etalon-based DCM, with channel spacing of 50 GHz (0.4 nm). Note that the mean group delays for each channel are the same. Also note the lack of detectable delay ripple in the usable frequency regions. Insertion loss = 2.6 dB. This extremely stable and robust device was made by Avanex, Inc., of Freemont, CA.

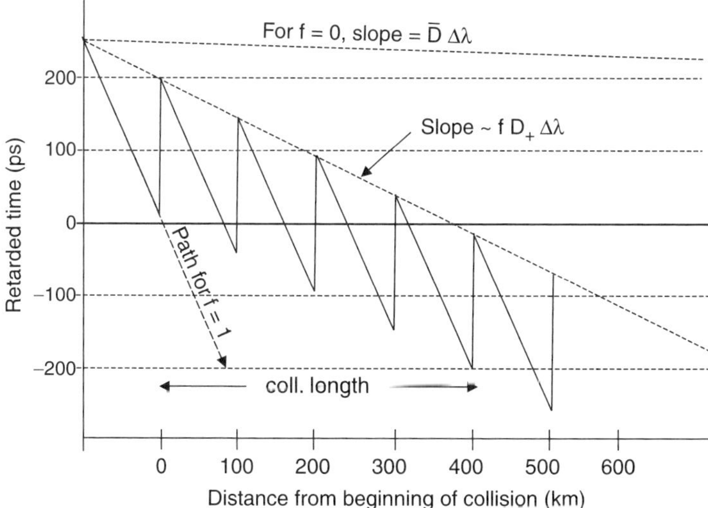

FIGURE 6.14 Relative motion in retarded time of a pair of colliding pulses from neighboring channels separated by 50 GHz, where 100-km spans of $D = 6$-ps/nm-km fiber, save for a small $\bar{D}$, are fractionally compensated f by a PGD device, and $(1-f)$ by a DCF module. (For the particular behavior shown here, $f = 0.2$.)

$\approx D_+$ at $f=1$. Note also that for the particular value $f=0.2$ used in Fig. 6.14, $L_{coll} \cong 400$ km, approximately a $10\times$ reduction from the case where $f=0$. Thus, L_{coll} decreases very rapidly as f increases.

Note that in our theory of the collisions and collision-induced jitter, in all equations based solely on the rate that the pulses move through each other [viz., Eqs. (6.1), (6.3), (6.4), (6.8), and (6.9)], $\bar{D}$ must be replaced by $\bar{D}_{inter}$. On the other hand, while the $\bar{D}$ explicitly displayed in Eq. (6.5), as the scale factor for converting wavelength shifts to time shifts, must retain the intrachannel value $\bar{D}$, the other two factors in Eq. (6.5) each contain a $\bar{D}$ in the denominator that must be converted to $\bar{D}_{inter}$. Thus, the factor $\bar{D}^{-1}$ in Eq. (6.5c) must be replaced by $\bar{D}/\bar{D}^2_{inter}$, so that Eq. (6.5c) becomes

$$\delta t_1\ (\mathrm{ps}) = \pm 0.335 C_{pol} \left[1 + a\frac{D_+}{|D_-|}\right] \frac{L_+}{L_{map}} \frac{2W_{sol}}{c} \frac{\lambda^2}{(\Delta\lambda)^2} \frac{\bar{D}}{\bar{D}^2_{inter}}. \tag{6.5d}$$

Finally, since the $\bar{D}$ explicitly displayed in Eqs. (6.11) and (6.12) is a $\bar{D}_{inter}$, the net scaling of the variances of the timing jitter is as $\bar{D}^{-3}_{inter}$. With those changes in place, it is reasonable to assume that our theory of the collision effects will continue to work as $\bar{D}_{inter}$ increases, or, equivalently, as L_{coll} decreases. Thus, we should also expect to see a great decline in the collision-induced jitter as L_{coll} decreases. Exact numerical simulations show that this assumption is essentially correct, at least until L_{coll} becomes too short.

Indeed, the rapid decline in L_{coll} as f is increased from zero [Eq. (6.13)] has profound effects, as can be clearly seen from the curves of collision-induced frequency shifts shown in Fig. 6.15. First, note from Fig. 6.15 that for $f=0$ and $f=0.2$, the frequency shifts ultimately return to zero. The most striking feature of these "frequency-conserving" collisions is the great shrinkage in all measures of the collision size as f goes from 0 to 0.2: First, as already seen in Fig. 6.15, L_{coll} is reduced by a factor of nearly 10 times, and the peak frequency shift is reduced by a similar factor. From the same numerical simulations, we find that the corresponding time displacements are 4.62 and 0.07 ps, respectively, for a 66-fold reduction. More generally, exact numerical simulations show that the time shifts scale as $L^{1.85}_{coll}$ as f increases (and L_{coll} decreases) throughout the region of frequency-conserving collisions (see Fig. 6.16). As already noted, however, Eq. (6.5) of our theoretical model predicts that δt_1 scales as $\bar{D}^{-2}_{inter}$, thus as L^2_{coll}. At this point, there is no ready explanation for this difference in power law between the exact simulations and our simple theory. On the other hand, because the difference is small, the potential reduction in jitter is still very great.

We also find that as f becomes significantly greater than about 0.2 (so that L_{coll} ceases to cover at least several span lengths), the collisions are in general no longer frequency conserving, as illustrated in Fig. 6.15 for the particular case $f=0.8$.

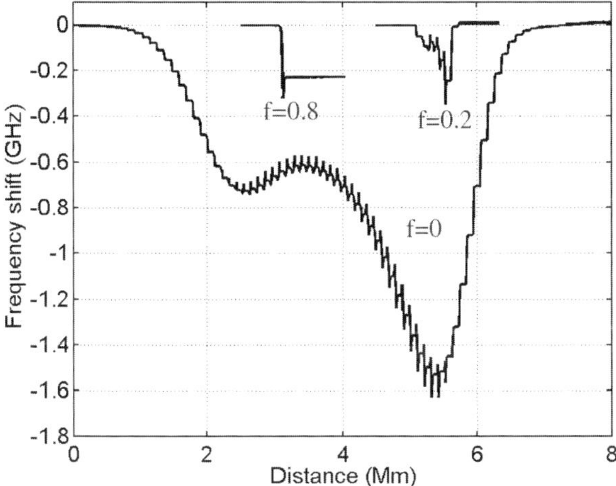

FIGURE 6.15 Frequency shifts of the lower frequency of two colliding pulses from adjacent channels in the system of Fig. 6.2, for the indicated values of f; in all cases the channels are co-polarized, the effective core area $= 50\ \mu m^2$, $\bar{D} = 0.15$ ps/nm-km, and the spans are backward Raman pumped only.

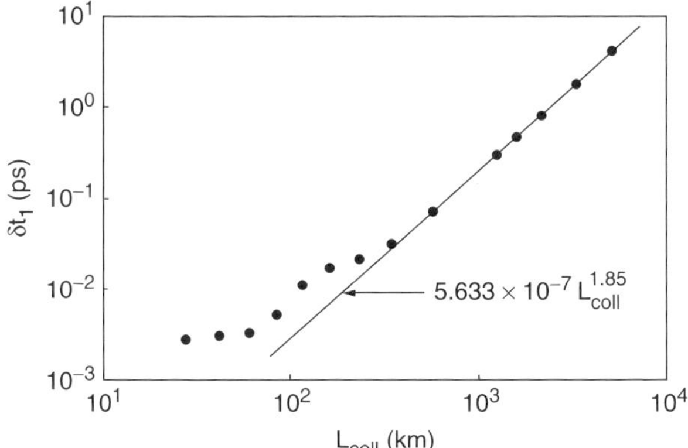

FIGURE 6.16 Collision-induced time shift δt_1 as a function of L_{coll}, determined from exact numerical simulations for the dispersion map of Fig. 6.13; $\Delta\lambda = 0.4$ nm ($\Delta f = 50$ GHz).

Figure 6.17 shows the absolute value of the maximum residual frequency shift as a function of L_{coll}; note that the values tend to become very small for L_{coll} greater than about four span lengths. The situation is similar to that discussed in Chapter 3 for dense WDM with ordinary solitons, and is principally associated with the tendency of intensity gradients to destroy the symmetry of the collisions. And again, as with ordinary solitons, the effect is to be avoided, since otherwise the residual

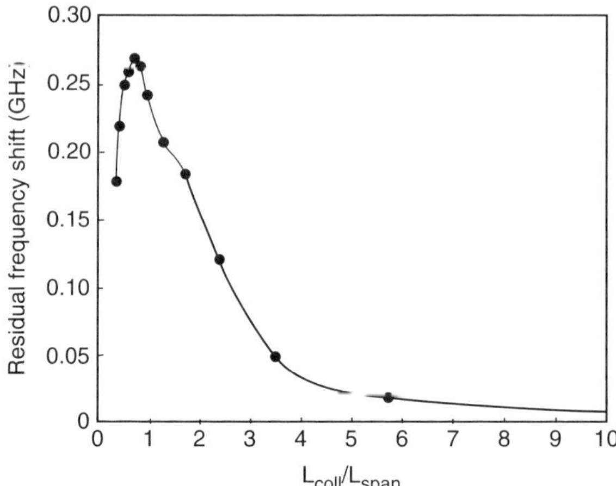

FIGURE 6.17 Absolute values of maximum residual frequency shifts for collisions in the map of Fig. 6.14, as a function of L_{coll}/L_{span}; again, $\Delta\lambda = 0.4$ nm ($\Delta f = 50$ GHz). To gain this data, for a given L_{coll}, the simulation was repeated over and over again with different starting points for the collision, until an approximate maximum residual shift was found.

frequency shifts, when compounded with the dispersion remaining to the end of the transmission, can once again produce large time shifts.

Exact numerical simulations [87] also provided the first indication of just how very much smaller the jitter would become with reduction of L_{coll}. The simulations were for 100-km spans of fiber with $D = +6$ ps/nm-km, dispersion-compensated with various combinations of DCF and PGD-DCMs. Other details were essentially as earlier in this chapter (unchirped pulse width $\tau = 33$ ps, channel spacing 50 GHz, etc.). Figure 6.18 shows a representative set of results from those many simulations, the eye diagrams as seen after 8000 km. As is immediately obvious from the figure, the quality of the eyes changes dramatically as f is varied. Note in particular that while the eyes are more or less uniformly bad at $f = 0$ (100% compensation by DCF), independently of the Raman pumping conditions, at $f = 0.2$ they are excellent for the case of uniform intensity and still very good for 25% forward/75% backward Raman pumping. (Also here the adjacent channels are co-polarized, yielding a factor of 1.7× greater jitter than the orthogonally polarized case of Fig. 6.9.) When the residual frequency shifts are zero (the case of lossless fiber), the very small remaining jitter seen in the plots probably corresponds mainly to the effects of partial collisions at the input (which our simple model does not take into account). (In our simple model, without the partial collisions, $\langle \Delta t^2 \rangle$ is expected to scale as L_{coll}^{-3}, implying a reduction by a factor of about 30× in its standard deviation when L_{coll} is reduced by a factor of 10. Even when

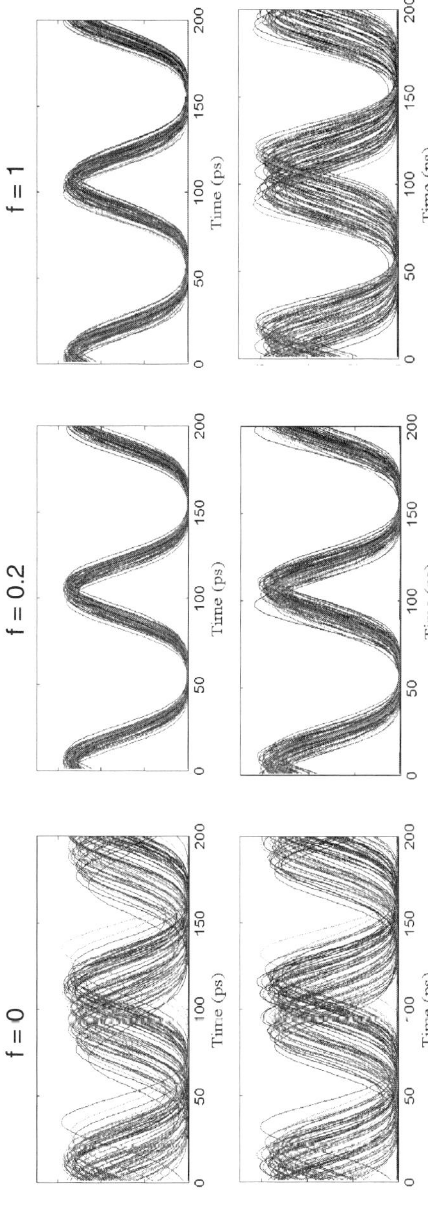

FIGURE 6.18 Optical eye diagrams at 8000 km in dense WDM with 10-Gbit/s channels spaced 50 GHz apart, with all channels co-polarized, and with no ASE, for the values of f shown. Top row: Lossless fiber. Bottom row: With 25% forward/75% backward Raman pumping. All other parameters are the same as those of the system of Fig. 6.13.

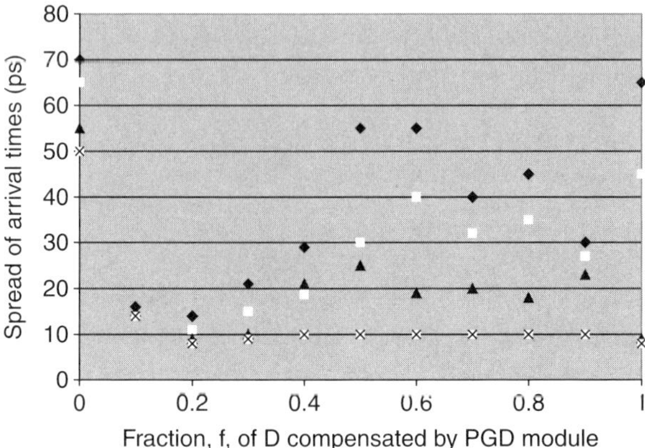

FIGURE 6.19 Total spread in pulse arrival times as measured from numerically simulated eye diagrams at 8000 km in dense WDM with 10-Gbit/s channels spaced 50 GHz apart, with all channels co-polarized, and with no ASE, plotted as a function of f. Black diamonds: With backward Raman pumping only. White squares: With 25% forward/75% backward Raman pumping. Black triangles: With mid-span backward Raman pumping. White squares with $\times$: Lossless fiber. Spans 80 km long; all other parameters are the same as those of the system of Fig. 6.13.

corrected for the empirical scaling factor shown in Fig. 6.16, the standard deviation is reduced by the very substantial factor of $\approx 20\times$. By either standard, the jitter should become almost invisible in the plots.) Finally, the extra jitter seen in the lower row and in the $f = 0.2$ and 1 columns corresponds mainly to the effects of the residual frequency shifts resulting from the intensity gradients in the spans.

A great many simulations like those of Fig. 6.18 have been carried out for the entire range $0 \le f \le 1$ and for a number of different intensity profiles. Except for the relatively minor difference that the spans in those simulations were 80 km rather than 100 km long, all of the other conditions were the same as those of Figs. 6.14 and 6.18. Figure 6.19 plots, as a function of f, the total spread in pulse arrival times (a quantity at least roughly proportional to the standard deviation) measured directly from the simulated eye diagrams. There is one plot for each of the following intensity profiles (listed here in ascending order of intensity gradient): (1) lossless fiber (zero gradient), (2) mid-span backward Raman pumping, (3) 25% forward/75% backward Raman pumping, and (4) 100% backward Raman pumping. Note that for the region $f \le 0.2$ (the region of "frequency-conserving" collisions), the points for all of those profiles tend to remain tightly clustered and rapidly descend to a minimum in the neighborhood of $f = 0.2$, while for $f > 0.2$, they begin to fan out and, for all but the lossless case, begin a rapid rise in response to the effects of growing residual frequency shifts. Thus, despite the diversity and complexity of the behavior seen in the region $f > 0.2$, there is a well-defined and nearly universal minimum with respect to f in the jitter. Already fairly broad for the case of the steepest intensity profile (100% backward Raman pumping), the minimum becomes ever broader as the intensity profiles become less severe.

To calculate the conditions for minimum jitter analytically, it would be necessary to have a proper analysis of the jitter induced by the residual frequency shifts. As suggested by the behavior seen in the right side of Fig. 6.19 (the region $f > 0.2$, where such jitter dominates), however, it is clear that such analysis would be at best complex, as, among other things, it would have to take into account details of the particular intensity profile. It would also have to deal with the "resonances" (like that seen in Fig. 6.19 at $f \approx 0.5$), where the relative time advance of colliding pulses is an integral multiple of the bit period, such that successive collisions tend to begin at the same place in the intensity profile. In passing, perhaps it should also be noted that the power law of dependence of the jitter on Z will be like that of the Gordon–Haus effect (i.e., $\langle \delta t^2 \rangle \propto Z^3$), rather than the linear dependence on Z [Eq. (6.12)] we have found for the frequency-conserving collisions. That is, just as in the Gordon–Haus effect, the residual frequency shift of each collision must be multiplied by the dispersion (hence distance) remaining to the end of the transmission. Because of all that complexity, and because the final result would not tell us much of practical interest beyond what can be easily found out (as in Fig. 6.19) from a few quick numerical simulations, we shall not attempt such analysis here.

6.2. Experimental Tests

The experiments reported on here [92–94] were carried out in 2003, when, for the first time, hardware became available for efficient, large-scale tests of dense WDM at 10 Gbit/s per channel using dispersion-managed solitons. That was also a time not long after the concept of the PGD technique had been born and was beginning to be digested. Thus, it was possible to make, in a rapid succession of experiments, a thorough and very meaningful comparison between dense WDM using the conventional dispersion compensation and that using the PGD devices. As the reader will soon see, the experimental results make amazingly good fit to the theoretical models we have just explored here. Finally, part of the record-setting results are very closely related to recently active commercial practice.

6.2.1. *Experimental Setup*

For all tests, the transmitter (Fig. 6.20) contained two sets of 80 DFB lasers, the lasers of each set tuned to a grid of 100 GHz spacing, with their combined outputs

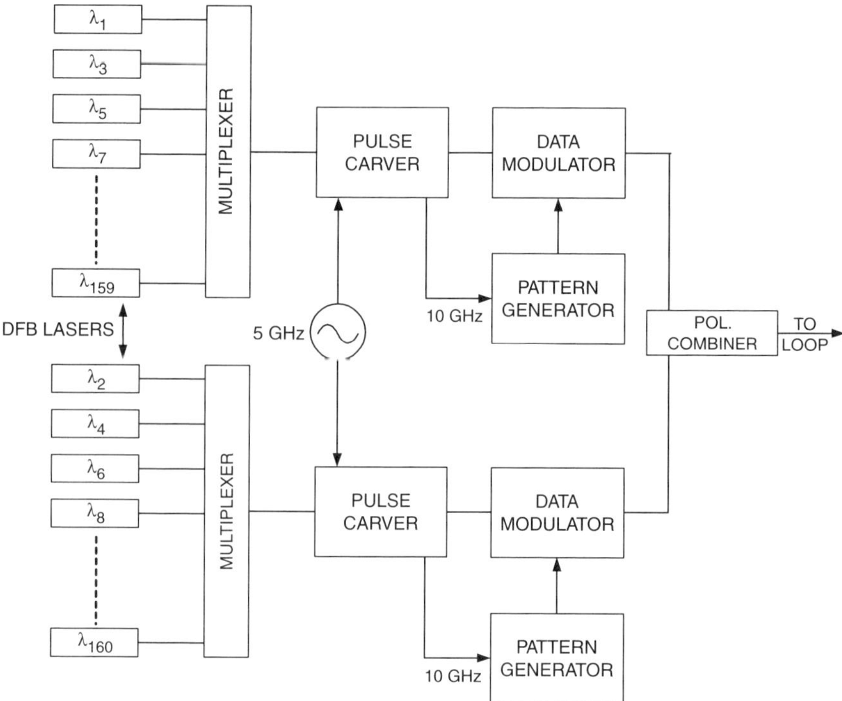

FIGURE 6.20 Schematic of the 160-channel transmitter. All leads up to the polarization combiner were polarization-maintaining.

fed through a common, $LiNbO_3$, Mach–Zehnder modulator-based pulse carver, and then through a second, similar modulator used to impose data. The pulse carvers were driven sinusoidally at 5 GHz and biased to yield two pulses per cycle in a very good approximation to Gaussian pulses of 33 ps FWHM. (For further details of the pulse carvers, see Chapter 8, Section 8.1.) Each data modulator was driven with its own independent pseudo-random pattern. (Typically, pattern lengths were 2^{15} bits long, but extensive tests with pattern lengths ranging from 2^7 to 2^{21} bits showed little or no dependence of BER performance on pattern length.) Finally, the two sets of channels, with their frequencies offset from each other by 50 GHz, were brought together with a polarization combiner to yield an array of as many as 160 channels, such that adjacent channels had orthogonal polarizations.

The recirculating loop contained six 100-km spans of Lucent "True Wave Extra Reduced Slope Fiber" ($D \sim 7$ ps/nm km at 1570 nm), which could be compensated either 100% by DCF alone, or ≈80% by DCF and most of the remainder by the PGD modules referred to in Fig. 6.13. Figure 6.21 is a schematic of the loop as compensated by the latter, and Fig. 3.11 shows exactly the same loop, but with

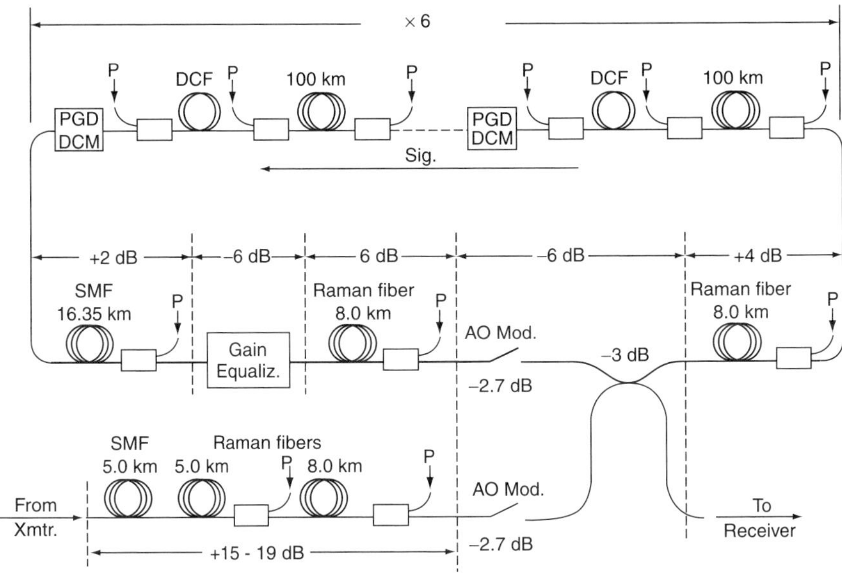

FIGURE 6.21 The recirculating loop of Fig. 3.11, but with PGD modules added to each of the six spans.

the pure DCF compensation. It was possible to switch between either of these two compensation modes in just an hour or so.

Pre- and post-dispersion compensation were carried out as discussed in Chapter 2, Section 2.1.3; for details of the pre-compensation, see Fig. 6.21. About half of the final DCF coil, plus the SMF and 8 km of Raman fiber ($D = -18$ ps/nm-km) shown in Fig. 6.21, form the major part of the post-compensation; the remainder was in one or another Raman-pumped coil of DCF just ahead of the temporal lens and receiver. Thus, the "jitter reducer" part of the post-dispersion compensation could be adjusted, but was usually at the optimum value for a distance of about 5000 km; once again, however, that value was not at all critical.

To reduce the noise penalty to the absolute minimum, the transmission spans were backward Raman pumped every 50 km, to just a few decibels less than net zero gain; the excess loss was made up by net gain in the backward Raman-pumped DCF modules. That pumping scheme yielded about a 1-dB advantage over forward/backward pumping (see Fig. 3.10), and it avoided some of the potential mode equipartition noise problems associated with forward pumping. A mild etalon guiding filter (with free spectral range = the channel spacing) was used once every 300 km to overcome the effects of a weak adjacent-pulse interaction.

The temporal lens (see Chapter 8, Section 8.2) was used just ahead of the detector, to yield the widest possible square acceptance window in time (the effective width was about 70 ps). Finally, a polarization scrambler was inserted into the recirculating loop, to more realistically simulate the random polarization-mode dispersion (see Chapter 7) and polarization-dependent loss (PDL) effects of a real system. [The scrambler consisted of nothing more than two piezoelectric transducers, each driven at its respective (and somewhat different) resonance frequency in the neighborhood of 1 MHz, and squeezing on a fiber to form oscillating wave plates.] The measured PMD parameter for the entire loop averaged 0.05 ps/$\sqrt{\text{km}}$ (typical for high-quality transmission fibers), and the measured PDL was a relatively benign 0.5 dB/600 km.

6.2.2. *Experimental Results*

Figure 6.22 shows the principal experimental result, the measured BER vs. distance of a typical WDM channel, for WDM with the two different forms of dispersion compensation mentioned previously, and for a single, isolated channel (i.e., with all other WDM channels turned off). The most remarkable thing here is the very large displacement toward greater transmission distances of the PGD-WDM curve from the conventionally compensated WDM curve, and its very close proximity to the single-channel results. Since there is no possibility of collision-induced jitter in the single-channel experiment, while the WDM with conventional dispersion compensation is clearly dominated by such jitter, the experimental curves of Fig. 6.22 make it unambiguously clear that the PGD dispersion compensation has reduced the collisional jitter to almost negligible proportions, just as predicted.

The thin dashed line next to the single-channel result is a theoretical curve based on amplitude errors from ASE only; the derivation of this curve will be given in the following section. The thin solid line next to the curve of WDM with conventional compensation is based on our theoretical model for the collisional timing jitter, so it is essentially that of Fig. 6.10, but as adjusted to account for the mild damping effect of the guiding filters. Note the good fit for both of these curves, except in the region of very low bit error rates. The differences here are almost certainly due to the effects of polarization-mode dispersion (discussed in greater detail in Chapter 7). That is, with the polarization scrambler turned off, a fixed polarization controller in the loop can always be adjusted to change the polarization state of the observed channel to the orthogonal polarization upon each circuit around the loop, thereby (at least temporarily) virtually nulling out the PMD. As observed many times, whenever that is done, the BER plunges by several orders of magnitude in the region of very low error rates. Thus, there is every reason to believe that the fits would have been better in the low error-rate

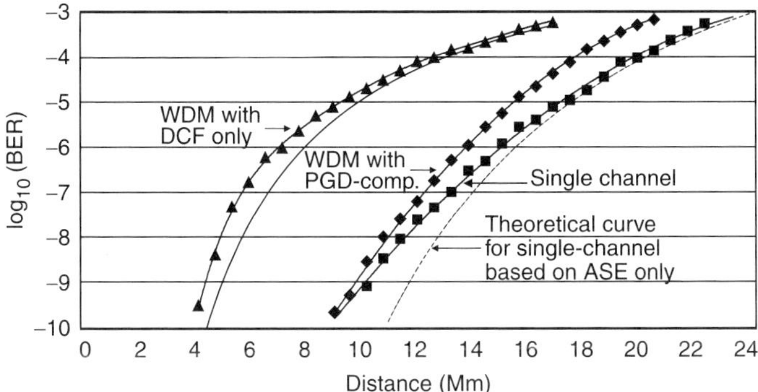

FIGURE 6.22 Experimentally measured BER vs. distance for a typical WDM channel near the middle (≈1570 nm) of a WDM band covering ≈1549–1594 nm; channel spacing 50 GHz (≅0.4 nm); adjacent channels orthogonally polarized. Triangles: WDM using conventional dispersion compensation. The nearby thin solid line is the associated theoretical curve assuming errors from collision-induced timing jitter only. Diamonds: WDM using the PGD-complemented dispersion compensation. Squares: Single-channel result (the same as diamonds but with only one channel present). The thin dashed line is a theoretical curve assuming only amplitude errors resulting from ASE. Except for the indicated changes, all other conditions were identical for all three experiments. No forward error correction was used. For other details, see text.

region if the measurements had been done in that way. Constant drift in the correct (PMD-nulling) polarization adjustment with changing temperature and the poor statistics of BER measurements made over short time intervals, however, make such measurements difficult and not entirely reliable, so they are not usually done.

It is important to understand the relative significance of the BER data at the low and high error-rate ends of the range shown in Fig. 6.21. Years ago, when fiber optic transmission was in its infancy, the standard was that raw bit error rates had to be extremely low, less than about 1×10^{-9} or perhaps even smaller. With the development of highly efficient forms of forward error correction (FEC), however, the once highly conservative telecommunications industry began to fall in line with the rest of the digital world, and much higher raw bit error rates began to be acceptable. In particular, now there exist FEC algorithms that allow for correction, with a mere 7% overhead of extra bits, of raw error rates in excess of 1×10^{-3} to $<1\times10^{-15}$. Furthermore, the high-speed chips for implementing such FEC have become relatively cheap and readily available. Thus, raw bit error rates as great as or even higher than the highest shown in Fig. 6.22 are now generally considered acceptable in the telecommunications industry.

Figure 6.23 shows the measured intrachannel path-average dispersion ($\bar{D}_{intra}$) when the PGD compensation scheme is in place. [Because $\bar{D}_{intra}$ is valid only

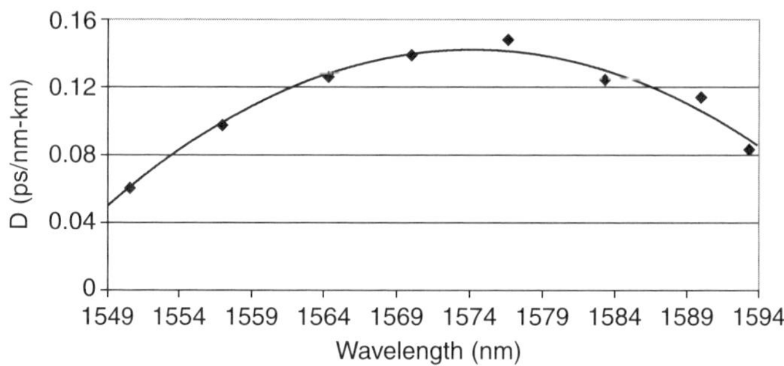

FIGURE 6.23 $\bar{D}_{intra}$ vs. wavelength for the recirculating loop of Fig. 6.21.

over frequency intervals narrower than the 50-GHz channel spacing (Fig. 6.13), the relative time of flight measurements had to be made with a special technique, described in Section 8.4.2 of Chapter 8.] Of immediate interest here, however, is the fact that $\bar{D}_{intra}$ dips far below the desired value of ≈0.15 ps/nm-km at the far ends of the WDM band. Fortunately, as predicted theoretically and confirmed by numerical simulation, it should not be necessary to make a correction in every span. Rather, it should suffice if the period for the correction is short relative to the characteristic dispersion length for the solitons, which in this case is ≈1600 km. (A thorough set of numerical studies [95, 96] of the effects of random variation in D on DMS pulse behavior has shown essentially the same stabilization with frequent enough "pinning" of $\bar{D}$ to the correct value.) Thus, making correction just once per round-trip in the 600-km loop should be good enough. In a real system, one could then employ one or another of several possible devices providing a dispersion curve complementary to that shown in Fig. 6.21. Since such a device was lacking at the time of the experiments, the next best thing was done: Depending on the wavelength range being tested, one or two small coils of SMF were added to the loop to provide the best approximate correction. Although that makes $\bar{D}_{intra}$ too large for optimum performance of the middle channels, they are still present, and still functioning, albeit at somewhat worse error rate. Of greater importance, since only the channels in the immediate neighborhood of the channel being tested have a significant effect on it, the test carried out in this way still shows perfectly well how the system performs.

Finally, Fig. 6.24 shows the measured BER vs. wavelength for 109 channels, stretching from ≈1549 to 1594 nm, at the fixed distances of 9000 and 18,000 km. The exact number of channels represented here (109) is a bit arbitrary and is limited more by accidental details of the experiment, such as the increasing difficulty in controlling the gain flatness at the extreme ends of the band, than by anything fundamental. The gradually rising error rates toward the short wavelength end of

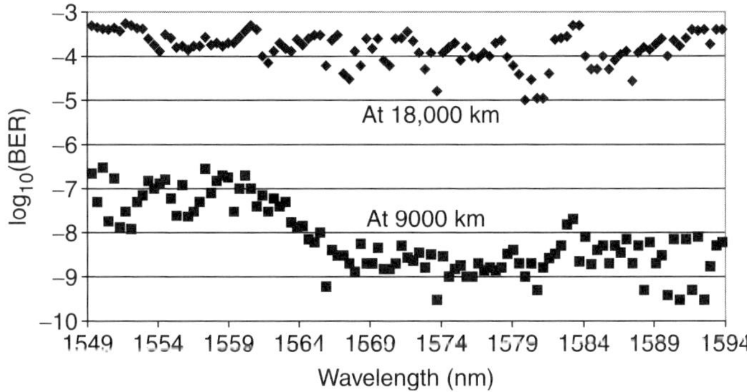

FIGURE 6.24 Measured BER vs. wavelength for 109 channels of 10 Gbit/s each at distances of 9000 and 18,000 km.

the band probably stem from decreased penetration of the Raman pump light, and hence increased ASE noise, with decreasing wavelength. (The longer wavelength pump light extracts power from the shorter pump wavelengths through Raman interaction between the two.)

So what do the data of Fig. 6.24 prove? As far as the effect of collisions on BER performance is concerned, very little beyond what would have been proved by the same sorts of data but with a total of no more than about eight or so channels, as can be seen from a glance at Fig. 6.8. On the other hand, they do confirm the assertion, made previously, that substantial correction to $\bar{D}_{intra}$ needs to be made no more often than once per 600 km or so. And finally, of course, they demonstrate that even under the less favorable conditions of a terrestrial system, with dispersion-managed solitons and all-Raman amplification, one can have a terabit-capacity system whose range is nearly half the circumference of the Earth. Such has never before been demonstrated.

6.2.3. Calculation of the Theoretical BER Curve for Single-channel Transmission

The somewhat involved story behind the theoretical curve in Fig. 6.22 of BER resulting only from ASE-induced amplitude errors is as follows. First, the guiding filters were Fabry–Perot etalons, used in reflection, with front and back mirror reflectivities of $R=0.04$ and $R=0.7$, respectively, and whose free spectral ranges exactly matched the channel spacing of 50 GHz. Their transmission characteristics are shown in Fig. 6.24. As discussed in Chapter 3, Section 3.3, the use of such filters

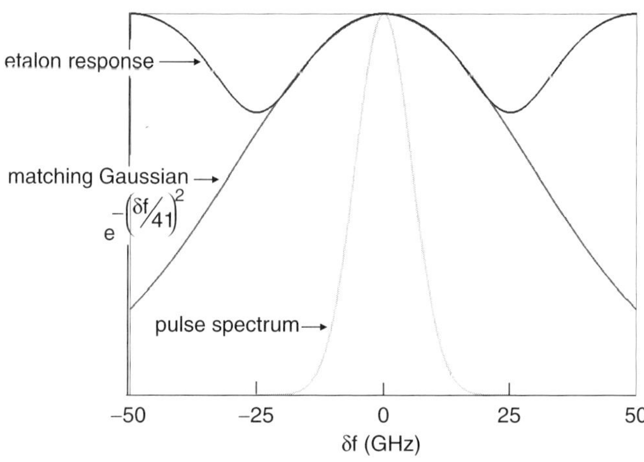

FIGURE 6.25 Transmission of guiding filters compared with that of a matching Gaussian filter and with a pulse spectrum.

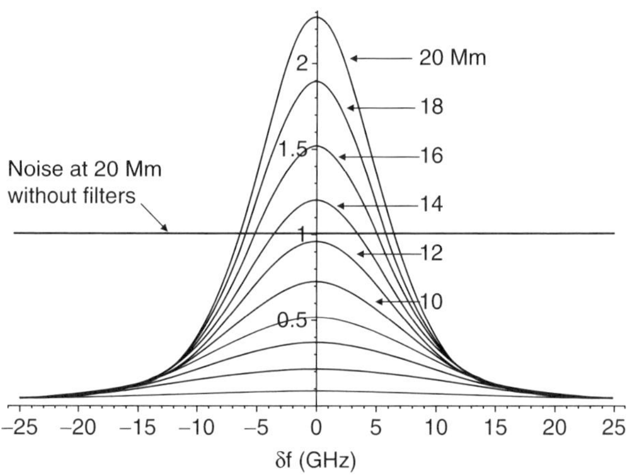

FIGURE 6.26 Computed noise spectrum for the indicated distances, normalized to that which would obtain at 20 Mm without filters.

imposes a loss on the solitons, which must be compensated by a small excess gain. In the meantime, the filters narrow the noise spectrum. The noise spectrum resulting from use of one every 300 km of the filters of Fig. 6.25, computed as a function of transmission distance Z, is shown in Fig. 6.26. Computation of the BER rate must take that noise spectrum into account. Substituting m, S_1+m, $\sqrt{m}$, and $\sqrt{S_1+m}$

for $\mu_0, \mu_1, \sigma_0,$ and σ_1, respectively, into Eq. (3.31), the Q factor becomes

$$Q(S_1, m) = \frac{S_1}{\sqrt{2S_1 + m} + \sqrt{m}}, \tag{6.14}$$

where S is the signal-to-noise ratio, $m = 2 \times BW\,(\text{GHz})/10$ is the number of modes of the radiation field getting to the detector, and BW is the noise bandwidth created by the filters. From Fig. 3.12, we find that the path-average noise (equipartition energy) $= 0.0135Z$ fJ (Z in Mm), so for a path-average signal pulse energy of 25 fJ, without the filters, we have $S_1 = 25/(0.0135Z) = 1850/Z$. To account for the exponential rise in the noise and the ever-decreasing bandwidth from the filters (both as inferred from Fig. 6.26), however, we must write

$$S_1 = \frac{1850}{Z\exp(0.044Z)} \quad \text{and } m \cong \frac{40}{Z}. \tag{6.15}$$

In the Gaussian approximation, the bit error rate is then given by

$$BER(Z) = \frac{1}{2}\text{erfc}\left[\frac{1}{\sqrt{2}}Q\left(\frac{1850}{Z\exp(0.044Z)}, \frac{40}{Z}\right)\right]. \tag{6.16}$$

The dashed line in Fig. 6.22 is the $\log_{10}$ of the $BER(Z)$ computed from Eq. (6.16).

Chapter 7

Polarization and Its Effects

7.1. Apologia

Thus far in this book, we have tended to gloss over polarization and effects of the fiber's birefringence on it. That is, for the most part, we have been content to note the thorough polarization averaging, created by the fiber's random residual birefringence, over those distances for which the nonlinear term has significant effects. We have used that averaging to justify further neglect of polarization. Nevertheless, there are certain important polarization phenomena that require examination. One of these is called *polarization-mode dispersion*, PMD for short, which makes the transit time for a pulse dependent on its polarization history and produces some dispersive wave radiation in the process. The other is the fact that colliding solitons in WDM alter each other's polarization states. Therefore, we now examine the linear birefringence of fibers and its statistical properties, the birefringence induced nonlinearly by the pulse itself, and the effects of both on transmission. A review of the basic properties of PMD in linear transmission lines is given in Gordon and Kogelnik [97].

7.2. Polarization States and the Stokes–Poincaré Picture

If z is the propagation direction, and $\hat{x}$ and $\hat{y}$ are unit vectors in the x and y directions, respectively, a unit normalized polarization vector can be written as $\hat{\mathbf{u}} = (r\hat{x} + s\hat{y})$, where r and s are complex numbers with $|r|^2 + |s|^2 = 1$, so that $\hat{\mathbf{u}}^* \cdot \hat{\mathbf{u}} = 1$.

A corresponding normalized real field component at frequency ω has the form

$$\text{Re}(\mathbf{u}) = \text{Re}(r\hat{x} + s\hat{y})e^{i\psi} \; : \; \psi = kz \quad \omega t + \Phi. \tag{7.1}$$

If the phases of r and s are both changed by the same amount $\delta\Phi$, then Φ in Eq. (7.1) simply changes to $\Phi + \delta\Phi$. With no loss of generality we can therefore express r and s in polar form by $r = \cos(\theta)\exp(-i\phi)$ and $s = \sin(\theta)\exp(i\phi)$, where $0 \leq \theta \leq \pi/2$ so that $\cos(\theta) = |r|$ and $\sin(\theta) = |s|$. Thus the real field can be written as

$$\text{Re}(\mathbf{u}) = \hat{x}\cos(\theta)\cos(\psi - \phi) + \hat{y}\sin(\theta)\cos(\psi + \phi). \tag{7.1a}$$

One way to visualize the polarization state of the field is to plot the motion of the vector Re($\mathbf{u}$) in an x–y coordinate system as ψ varies from 0 to 2π. The resulting figure is generally an ellipse. The state of polarization of the field is determined by the shape of the ellipse and by the direction of motion of the vector around the ellipse. It is independent of the constant Φ in the phase ψ. In general, a change from ϕ to $-\phi$ reverses the direction of motion around the same ellipse, since Eq. (7.1a) is invariant to a sign change of both ϕ and ψ. When $\phi = 0$ or $\phi = \pi/2$, the ellipse degenerates into a straight line making an angle θ or $-\theta$, respectively, with the x axis. This represents linear polarization. When $\phi = \pi/4$, the axes of the ellipse are the x and y axes, and they have lengths of $2\cos(\theta)$ and $2\sin(\theta)$, respectively. When $\theta = \pi/4$, so that $\cos(\theta) = \sin(\theta) = 1/\sqrt{2}$, the axes of the ellipse are rotated by $\pi/4$ from the x and y axes, and they have lengths of $2|\cos(\phi)|$ and $2|\sin(\phi)|$, respectively. When both ϕ and ψ are equal to $\pi/4$, the ellipse degenerates to a circle and we have circular polarization.

A more useful tool for visualizing the state of polarization as it varies during transmission is the real three-dimensional Stokes vector. A Stokes vector $\hat{\mathbf{S}}$ of unit length is derived from the normalized polarization vector in Eq. (7.1) or (7.1a). It has the Cartesian components

$$S_1 = |r|^2 - |s|^2 = \cos(2\theta), \tag{7.2a}$$

$$S_2 = 2\text{Re}(r^* s) = \sin(2\theta)\cos(2\phi), \tag{7.2b}$$

$$S_3 = 2\text{Im}(r^* s) = \sin(2\theta)\sin(2\phi). \tag{7.2c}$$

From Eqs. (7.2a)–(7.2c), however, it can be seen that the angles 2θ and 2ϕ are, respectively, the polar and azimuthal angles of the vector $\hat{\mathbf{S}}$ in a spherical coordinate system with the S_1 axis as the polar axis.[1] A unit sphere centered on such a coordinate system, known as the Poincaré sphere, can be extremely useful for the

[1] In the Poincaré representation, the S_3 axis is often referred to as the polar axis. Nevertheless, Eqs. (7.2a)–(7.2c) require S_1 as the true polar axis and yield the conventional Poincaré representation.

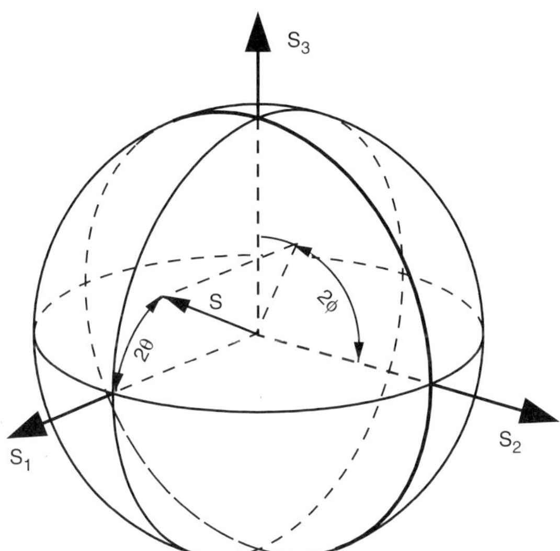

FIGURE 7.1 The Poincaré sphere, with unit Stokes vector $\hat{\mathbf{S}}$ represented in it. Note that the S_1 axis is the polar axis, and *not* S_3. Stokes vectors for linear polarization are confined to the S_1, S_2 plane, while those out of that plane correspond to elliptical polarizations; Stokes vectors corresponding to circular polarization are confined to the S_3 axis.

visualization of $\hat{\mathbf{S}}$ (see Fig. 7.1). It is worth noting that S_1 is proportional to the difference in the powers that would emerge from linear polarizers in the x and y directions, S_2 is likewise proportional to the difference in the powers that would emerge from linear polarizers rotated from the previous two by an angle of $\pi/4$, thus bisecting the x and y directions, while S_3 is the difference between the powers that would emerge from left and right circular polarizers. Hence the Stokes vector can be measured directly, and instruments to do so are now fairly common. One can verify that the (S_1, S_2) plane represents plane-polarized fields ($2\phi = 0$ or π), while the S_3 axis represents circularly polarized fields. Note that the Stokes vectors corresponding to the orthogonal polarization states $\hat{x} \sim (r = 1, s = 0)$ and $\hat{y} \sim (s = 1, r = 0)$ lie antiparallel along the S_1 axis. More generally, two polarization vectors $\hat{\mathbf{u}}_a$ and $\hat{\mathbf{u}}_b$ are orthogonal if $\hat{\mathbf{u}}_a^* \cdot \hat{\mathbf{u}}_b = 0$. Thus $(s^*\hat{x} - r^*\hat{y})$ represents the state orthogonal to $(r\hat{x} + s\hat{y})$ and the Stokes vectors of these two polarization states always lie antiparallel along the same axis.

7.3. Linear Birefringence of Transmission Fibers

7.3.1. Birefringence Element and Its Effects

Consider the effect on $\hat{\mathbf{S}}$ of a short length dz of birefringent fiber. There are many reasons for such birefringence, principal among them being a slight ellipticity

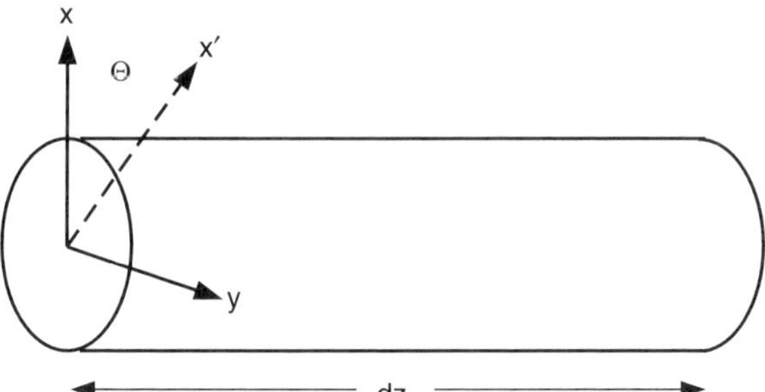

FIGURE 7.2 Element of fiber with birefringence axis x'.

of the fiber, or some strain in it. Suppose first that the principal states of the birefringence are the x and y linear polarizations ($\theta = 0$ in Fig. 7.2). If δk is the corresponding difference in wavenumber, then, in the course of traversing the piece of fiber, a phase shift $\delta\phi = \delta k dz$ will develop in the quantity rs^*. Consequently, the Stokes vector precesses through the angle $\delta\phi$ around the S_1 axis, marking out a cone. The corresponding generalization is that for any birefringence, the Stokes vector precesses through an angle $\delta\phi = \delta k dz$ around the axis in Stokes space that corresponds to the two principal states of the birefringence. If β is a vector whose length is δk and which lies along this axis of birefringence, then $\hat{\mathbf{S}}$ precesses around β according to the equation

$$\frac{d\hat{\mathbf{S}}}{dz} = -\beta \times \hat{\mathbf{S}}. \tag{7.3}$$

Figure 7.3 illustrates this behavior. When $\delta\phi$ reaches 2π, $\hat{\mathbf{S}}$ has swept out a complete cone. An optical element of this sort is called a full-wave plate. In a fiber, the needed length $2\pi/\delta k = \lambda(k/\delta k)$ is called the beat length. It is typically very short, a few meters to tens of meters. If the birefringence axis varies in a random way along the fiber, the Stokes vector soon comes to have a random direction on the Poincaré sphere.

Along with the wavenumber birefringence just discussed comes a birefringence in the inverse group velocity. Let $\mathbf{b} = d\beta/d\omega$. If β does not change direction with frequency (in practice, the change tends to be negligibly small), then we have

$$b = \frac{d\beta}{d\omega} = \frac{d\delta k}{d\omega} = \delta v_g^{-1}, \tag{7.4}$$

where b, for example, is the magnitude (length) of the vector $\mathbf{b}$, and the last equality is so because $dk/d\omega = v_g^{-1}$ holds for each of the two principal states of polarization. Referring again to our piece of fiber with x–y birefringence, the average time delay for the energy of field in the polarization state $(r\hat{x} + s\hat{y})$ is

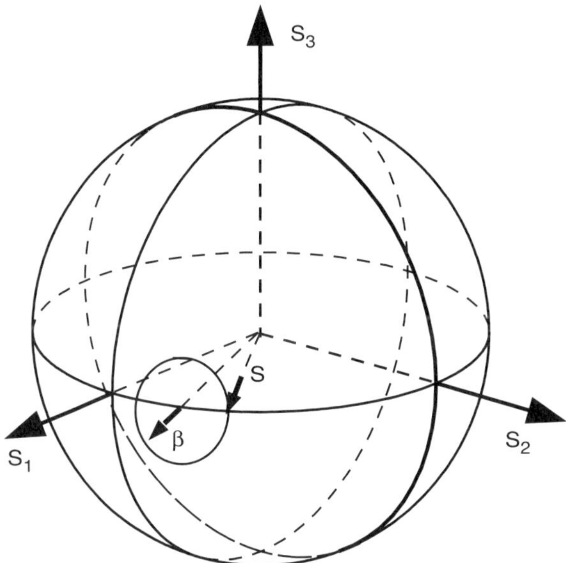

FIGURE 7.3 The birefringence element β causes $\mathbf{S}$ to precess in a cone around it.

proportional to $(|r|^2 - |s|^2)$ and thus to S_1. In general, the time delay is proportional to the projection of $\hat{\mathbf{S}}$ on the birefringence axis, and so we get

$$dt_d = \frac{1}{2}\hat{\mathbf{S}} \cdot \mathbf{b}\, dz. \tag{7.5}$$

7.3.2. Calculus for Long Fibers

In a typical transmission fiber both the strength and axis of the birefringence vary along the length of the fiber, causing the Stokes vector to wander more or less randomly around the Poincaré sphere. The motion of $\hat{\mathbf{S}}$ in going between any two points in the fiber can be expressed formally by

$$\hat{\mathbf{S}}(z_2) = \mathbf{M}(z_2, z_1) \cdot \hat{\mathbf{S}}(z_1), \tag{7.6}$$

where $\mathbf{M}$ is a rotation matrix, a (3×3) form of a Mueller matrix, and $\mathbf{M}(z_1, z_2)$ is the inverse of $\mathbf{M}(z_2, z_1)$. For a long fiber of length Z, the total delay caused by the birefringence can be written as

$$t_d = \frac{1}{2}\int_0^Z \hat{\mathbf{S}}(z) \cdot \mathbf{b}(z) dz = \frac{1}{2}\hat{\mathbf{S}}(0) \cdot \int_0^Z [\mathbf{M}(0,z) \cdot \mathbf{b}(z) z = \frac{1}{2}\hat{\mathbf{S}}(0) \cdot \mathbf{T}, \tag{7.7}$$

where we have used the inverse Mueller matrix to project the vector dot product at location z back to the input of the fiber, and the last equality defines the vector $\mathbf{T}$.

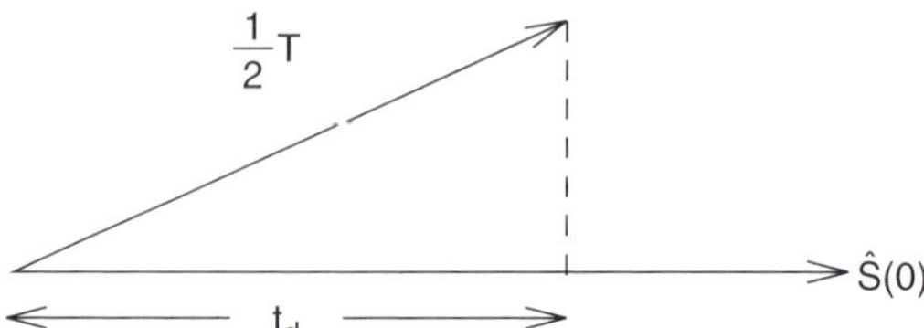

FIGURE 7.4 Relative time delay, in long fiber, of pulse with polarization vector $\hat{\mathbf{S}}$.

We call $\mathbf{T}$ the polarization time-dispersion vector [98].[2] Its length is the difference between the maximum and minimum delay times, and the Stokes vectors pointing in its positive and negative directions represent the principal states of polarization for which the delay times are, respectively, longest and shortest. It behaves very much like the local birefringence, except that its magnitude and direction on the Poincaré sphere are more rapidly frequency dependent (see Fig. 7.4). In the case of linear propagation, any input pulse can be resolved into a linear combination of components in the two principal states, which have different delays. As a result, one sees the output pulse width vary as a function of the input polarization, its mean delay time being given by Eq. (7.7). In the case of soliton propagation, if the PMD is not too big, the nonlinear effects hold the pulse together, so that the pulse width does not change, but its mean delay time also obeys Eq. (7.7). Soliton propagation is further discussed below.

7.3.3. *Growth of* T *with Increasing Fiber Span Length*

We can think of a long fiber as concatenation of two shorter fibers, joined at some point $z = z_1$. Accordingly, the vector $\mathbf{T}$ for the whole fiber can be split into two pieces as

$$\begin{aligned}\mathbf{T} &= \int_0^{z_1} \mathbf{M}(0,z)\cdot\mathbf{b}(z)\,dz + \mathbf{M}(0,z_1)\cdot\int_{z_1}^{Z} \mathbf{M}(z_1,z)\cdot\mathbf{b}(z)\,dz \\ &= \mathbf{T}_1 + \mathbf{M}(0,z_1)\cdot\mathbf{T}_2,\end{aligned} \tag{7.8}$$

where $\mathbf{T}_1$ and $\mathbf{T}_2$ are the $\mathbf{T}$ vectors for the two sections taken individually. We are primarily interested in fiber spans that are long with respect to the characteristic distance (typically only a few meters) for the reorientation of $\mathbf{b}$. In that case, the

[2] Prior to the appearance of the paper by Mollenauer and Gordon [98], however, the polarization dispersion vector and its statistical properties had already been thoroughly explored by Poole and co-authors (see, for example, Refs. [99] and [100]). Note that the vector Ω in those papers is the same as $\mathbf{M}(Z,0)\cdot\mathbf{T}$ here, where Z is the length of the fiber.

direction of $\mathbf{T}_2$ is at random with respect to that of $\mathbf{T}_1$; furthermore, since there is no correlation between $\mathbf{M}(0,z_1)$ and $\mathbf{T}_2$, the second term in the preceding sum is still oriented at random with respect to the first. This division can be iterated, giving

$$\mathbf{T} = \mathbf{T}_1 + \sum_{i=2}^{N-1} \mathbf{M}(0,z_i)\cdot\mathbf{T}_i, \tag{7.9}$$

until the $\mathbf{T}$ vectors of the individual sections cease being uncorrelated. From this exercise it should be obvious that the growth of T (the magnitude of $\mathbf{T}$) is a random walk process, where T is expected to grow as $z^{1/2}$. Another way of seeing this is to look at the quantity

$$\begin{aligned} T^2 = \mathbf{T}\cdot\mathbf{T} &= \int_0^Z \int_0^Z [\mathbf{M}(0,z)\cdot\mathbf{b}(z)]\cdot[\mathbf{M}(0,z')\cdot\mathbf{b}(z')]dz'dz \\ &= \int_0^Z \int_0^Z \mathbf{b}(z)\cdot\mathbf{M}(z,z')\cdot\mathbf{b}(z')dzdz'. \end{aligned} \tag{7.10}$$

Beyond the length over which $\mathbf{b}(z)$ is correlated with $\mathbf{M}(z,z')\cdot\mathbf{b}(z')$ this double integral is expected to grow noisily but roughly linearly with Z. To take a simple example, imagine a fiber, initially with a constant linear birefringence, which is cut into short sections of length l_{sect} and put back together after each section has been rotated through a random angle around its cylindrical axis. Then, referring to Eq. (7.10), since $\mathbf{b}(z)$ is oriented along the local birefringence axis, $\mathbf{M}(z,z')$ reduces to the identity matrix so long as z and z' are both in the same section. There is no correlation between the directions of $\mathbf{b}(z)$ in the different sections, so in getting the expected value of T^2, the first integral over z' reduces simply to $b^2 l_{sect}$, and the second integral produces the factor Z.

An expected value of T is usually inferred from measurements made over a wide band of optical frequencies. As the fiber is cut back, the data [101] indeed tend to fit a curve of form

$$T(z) = D_p z^{1/2}. \tag{7.11}$$

D_p is known as the polarization-mode dispersion parameter. For the highest quality transmission fibers available at present, $D_p \leq 0.05$ ps/km$^{1/2}$.

7.3.4. Statistical Properties of T

The PMD parameter D_p for a length of fiber is found from averaging the values of T obtained over a rather broad range of frequencies. The distribution of T values so obtained tends to be Maxwellian. Changing the temperature of a fiber

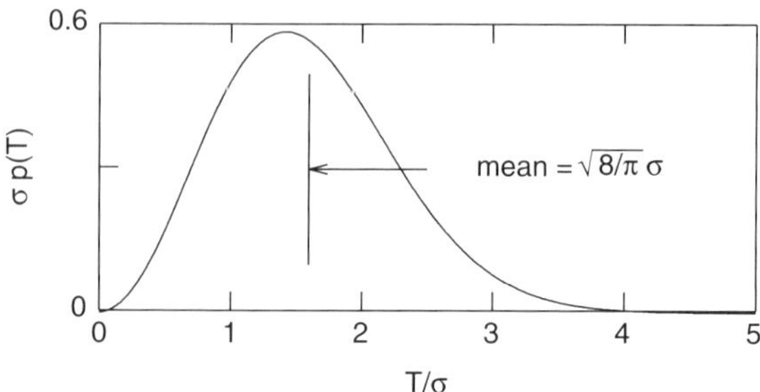

FIGURE 7.5 Maxwellian probability distribution of T.

leads to similar results. This distribution can be understood as follows. The value of T is sensitive to the wavenumber birefringence of the fiber, since this is what causes precession of the Stokes vector, and so determines the values of the Mueller matrices. We have seen that $\mathbf{T}$ can be considered to be the sum of a large number of independent vectors $\mathbf{T}_i$ from fiber sections, each projected to the beginning of the fiber by the appropriate Mueller matrix. As these Mueller matrices are changed by changing frequency, say, one would expect the components of $\mathbf{T}$ to have nearly independent Gaussian distributions.

If each of the three components of $\mathbf{T}$ in Stokes space (T_1, T_2, T_3) has an independent Gaussian distribution with standard deviation σ, then the distribution of $\mathbf{T}$ has spherical symmetry, and the magnitude of $\mathbf{T}$ will have a Maxwellian distribution, as illustrated in Fig. 7.5. The probability that T lies between T and $T+dT$ is

$$p(T)\,dT = \sqrt{\frac{2}{\pi}}\frac{1}{\sigma^3}T^2 e^{-1/2(T/\sigma)^2}\,dT. \tag{7.12}$$

The most probable value of T/σ is $\sqrt{2} \approx 1.414$, its mean value is $\sqrt{8/\pi} \approx 1.596$, and its standard deviation is $\sqrt{3} \approx 1.732$. By definition D_p is the mean value of T divided by the square root of the length of the fiber, so that $\sigma = D_p\sqrt{\pi Z/8}$. As an example, if $D_p \sim 0.1$ ps/km$^{1/2}$ and $Z = 10{,}000$ km, then $\sigma \sim 6.3$ ps. As a final bit of housekeeping here, note from Eq. (7.7) that the delay time for a pulse is half of the component of $\mathbf{T}$ along the direction of $\hat{\mathbf{S}}(0)$. Thus the expected standard deviation of the pulse delay is $\sigma_{delay} = \sigma/2 = D_p\sqrt{\pi Z/32}$.

7.4. Soliton Propagation

We now consider soliton pulse propagation in a randomly birefringent fiber. First, it is known that the nonlinear coefficient n_2 is a function of polarization, varying, relative to its polarization average, from 9/8 for linear polarization to 3/4 for circular polarization (on average, linear polarization is twice as likely as circular polarization). The changes in n_2 as a soliton's polarization state varies during propagation causes some radiative loss of energy. However, this loss is much smaller than is a similar loss due to PMD and may be ignored. To avoid unnecessary complication, we shall therefore continue to treat n_2 as though it were polarization independent.

We now come to consider the residual birefringence. On the often-used assumption that the local form of the birefringence is not very important overall, we shall invoke a well-worn model similar to that used above in which the fiber is composed of short sections of constant linear birefringence, whose axes are assumed randomly directed, and whose magnitude may also have some random distribution. Consider the effect of one such section. We can denote the slow and fast axes of this piece of fiber by the orthogonal unit vectors $\hat{x}$ and $\hat{y}$. The entering soliton will be in some linear combination of these two polarization states and will in general have the form

$$\mathbf{u}(z,t) = (r\hat{x} + s\hat{y})\,\mathrm{sech}(t), \tag{7.13}$$

where the polarization state vector has unit length, as before, satisfying $(|r|^2 + |s|^2 = 1)$. After traversing the section of length l, this soliton will have changed to the form

$$\mathbf{u}(z+l,t) = e^{i\theta}\left[re^{i\phi}\hat{x}\,\mathrm{sech}(t-\epsilon) + se^{-i\phi}\hat{y}\,\mathrm{sech}(t+\epsilon)\right], \tag{7.14}$$

where the angle $\phi = (1/2)\delta k\; l$ arises from the local wavenumber birefringence δk, while $\epsilon = (1/2)bl$ is half the time delay birefringence of the fiber section. For good fibers ϵ is a very small number. Using this, we can expand the sech functions to first order in ϵ $[d\,\mathrm{sech}(t)/dt = -\,\mathrm{sech}(t)\tanh(t)]$ and, with a bit of manipulation, arrive at

$$\begin{aligned}\mathbf{u}(z+l,t) \approx e^{i\theta}\Big[&\left(re^{i\phi}\hat{x} + se^{-i\phi}\hat{y}\right)\mathrm{sech}\left(t - \epsilon\left(|r|^2 - |s|^2\right)\right)\\ &+ \epsilon 2rs\left(s^* e^{i\phi}\hat{x} - r^* e^{-i\phi}\hat{y}\right)\mathrm{sech}(t)\tanh(t)\Big].\end{aligned} \tag{7.15}$$

We have split the terms proportional to ϵ into two parts, one having the same polarization state as the soliton, which we put back together with the soliton, and the other having a polarization state orthogonal to the soliton. We see that the soliton's polarization state and PMD time delay have changed in accordance with our previous general discussion. The field scattered into the orthogonal polarization state, proportional to $\mathrm{sech}(t)\tanh(t)$, can be shown to be a dispersive field, and so represents a loss mechanism for the soliton. The fractional energy loss due to this section of fiber is the ratio of the time integrals of $|u|^2$ in the two orthogonal polarization states, which yields $\delta E/E = (4/3)|rs|^2\epsilon^2$, independent of ϕ. If we assume that the birefringent time delay stays small with respect to the pulse width of the soliton, then the scattered fields will be uncorrelated, and the total loss will be the sum of the energies scattered from each section. To find the expected loss, it is appropriate to do an average over polarizations. From Eq. (7.2) one can see that $4|rs|^2$ is the sum of squares of the S_2 and S_3 components of the unit Stokes vector $\hat{\mathbf{S}}$, and therefore its polarization average is 2/3. Using this, and evaluating ϵ, we get

$$\alpha_{pmd} l = \frac{\delta E}{E} = \frac{1}{18} b^2 l^2, \tag{7.16}$$

where α_{pmd} is the exponential energy loss coefficient. In comparison, the time delay for this section, $\delta t_d = \epsilon(|r|^2 - |s|^2)$, is proportional the the S_1 component of $\hat{\mathbf{S}}$, so its polarization average is zero, but its variance is

$$\sigma_d^2 = \frac{1}{3}\epsilon^2 = \frac{1}{12} b^2 l^2. \tag{7.17}$$

The loss due to PMD is therefore related to the variance of the time delay by $\alpha_{pmd} l = (2/3)\sigma_d^2$. Since the loss and the variance of the time delay both grow linearly with the number of sections, the loss for a length z of fiber will be given by

$$\alpha_{pmd} z = \frac{2}{3}\sigma_d^2(z) = \frac{\pi}{48} D_p^2 z. \tag{7.18}$$

This equation is written in soliton units. It can be made dimension-free by dividing its right side by t_c^2, or $(\tau/1.7627)^2$; removing the common factor of z as well, one gets

$$\alpha_{pmd} = 0.2034 D_p^2/\tau^2. \tag{7.19a}$$

To take a good fiber as an example, with $D_p = 0.1\,\text{ps/km}^{1/2}$, then a pulse with $\tau = 20$ ps yields $\alpha_{pmd} = 5\times 10^{-6}$/km, or 0.005/Mm. Although the loss rate in soliton energy due to PMD appears small in this case, note that the total energy lost in 10 Mm is 5% of the soliton's energy. Since this lost energy is scattered into dispersive waves, it can add significantly to the total noise, especially in a broadband transmission line. On the other hand, sliding-frequency filters (see Chapter 3, Section 3.4) control the growth of this noise just as effectively as they control the growth of spontaneous emission noise.

It should also be noted from Eq. (7.18a) that the loss rate from PMD increases rapidly with decreasing pulse width. For example, for the maximum pulse width ($\tau \sim 2$ ps) permitting a single channel rate of 100 Gbit/s, and again for $D_p = 0.1\,\text{ps/km}^{1/2}$, α_{pmd} rises to 0.5/Mm, a value large enough to cause very serious problems. This is yet another reason why, for the attainment of very large net bit rates, dense WDM is by far the better choice.

If the value of D_p gets too big, solitons can become unstable. For distances of the order of z_c, essentially linear propagation occurs, so a soliton has a chance of being split into its two principal state components. Some years ago, a criterion for stability was established by numerical simulation, using a kind of worst-case scenario [102]. In soliton units, the result was simply

$$D_p \leq 0.27 t_c / z_c^{1/2}. \tag{7.19}$$

Using Eq. (7.19) with the equality sign to establish the largest allowable D_p, and setting $D_p z^{1/2} \approx t_c$, note that a linear pulse would split into two pieces spaced by t_c in a distance $(0.27^{-2} \approx 14) \times z_c$. Under those same conditions, however, nonlinear effects hold the soliton together over an indefinitely long distance of propagation. From Eq. (2.15a), we recall that $t_c/z_c^{1/2}$ scales with the dispersion constant D, so in standard units Eq. (7.19) becomes

$$D_p \leq 0.3 D^{1/2}. \tag{7.19a}$$

Note that for $D_p = 0.1$ ps/km$^{-1/2}$, Eq. (7.19a) is satisfied for D as small as about 0.11 ps/nm-km. [For an example of what can happen when the criterion of Eq. (7.19a) is violated, see Wai *et al.* [103].]

It is interesting that even at the stability border, the loss calculated from Eq. (7.18) seems very small. If we use Eq. (7.19) with an equals sign, then from Eq. (7.18) we get $\alpha_{pmd} = 0.048/z_c$, which does not seem big enough to signal impending doom for the soliton. The answer to this conundrum is probably in the statistics. The loss was calculated on the basis of polarization averaging, while the split-up of a soliton can occur in any section of the transmission fiber a few z_c long if the **T** vector for that section is big enough. Simulations also support this conclusion. So long as the soliton transmission is stable, the loss (per z_c) is quite small.

7.5. Polarization Scattering by Soliton–soliton Collisions

As already noted in the introduction to this chapter, colliding solitons in WDM alter each other's polarization states. Although this polarization scattering was described earlier by Manakov [104, 105], its consequences were not generally appreciated until they became manifest in a later experiment [106]. In the experiment, each of several 10-Gbit/s WDM channels was subdivided into two polarization (and time) division multiplexed, 5-Gbit/s subchannels. As had been known for some time [107], such polarization division multiplexing works well,

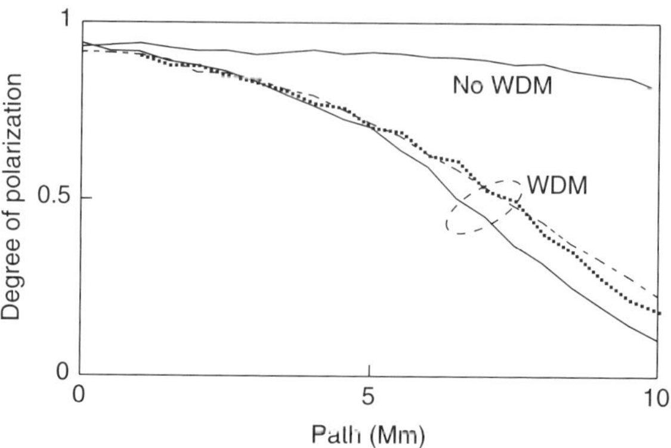

FIGURE 7.6 Experimentally measured degree of polarization for a given wavelength channel as a function of distance. No WDM: Only one channel present on the transmission line. WDM: Two channels present on the line. Channel separations: Solid line, 0.6 nm; dashed line, 1.2 nm; dotted line, 1.8 nm. In all cases, the channels were initially co-polarized.

at least in the absence of WDM. Indeed, in the experiment, with only one such polarization-multiplexed, 10-Gbit/s channel present, the orthogonality of the 5-Gbit/s subchannels was well maintained over transoceanic distances, and the transmission was error-free. As soon as a second WDM channel was added, however, polarization scattering from the collisions destroyed the orthogonality of the subchannels, and the error rate became high for all but very short distances. To confirm that polarization scattering was to blame, the degree of polarization of each 10-Gbit/s channel (this time with all pulses initially co-polarized) was measured as a function of distance. The results are summarized in Fig. 7.6. With only one channel present (no WDM), as expected, the degree of polarization (DOP) was only slightly reduced in 10 Mm, from the mild effects of spontaneous emission [106]. With just one other channel present, however, the degree of polarization of either channel was reduced nearly to zero in the same distance.

The origin of the polarization scattering in WDM transmission is not hard to understand. It comes about because the magnitude of the cross-phase modulation (XPM) between waves of different frequencies is dependent on their relative polarizations, being twice as large for co-polarized waves as for orthogonally polarized waves. One can think of this as a nonlinearly induced effective birefringence, keeping in mind that the birefringence seen by each of the waves is different from that seen by the other. We discuss this effect in more detail later. Consider now a collision between solitons in two different frequency channels of a WDM system. Label the solitons a and b. The XPM phase shift given to the component of soliton a co-polarized with soliton b is twice as large as that given to the component of soliton a orthogonal to soliton b. The result is a change in the polarization state of soliton a. In terms of the three-dimensional Stokes vector representation of a soliton's polarization state, the first-order result is a precession of soliton a's Stokes vector around that of soliton b. Soliton b is similarly influenced by soliton a, so their Stokes vectors precess around one another. The XPM occurs only while the solitons overlap, and after a completed collision, the differential phase shift (which, as we shall show, equals the precession angle of the Stokes vectors) is approximately equal to $L_{coll}/(1.76z_c)$. Note that there is no change in the polarizations of either soliton if they are in the same or orthogonal polarization states.

To analyze the polarization scattering in detail, we describe the optical field envelope in a fiber, in the manner of Eq. (7.13), using

$$\mathbf{u} = u_x\hat{x} + u_y\hat{y}. \tag{7.20}$$

Here u_x and u_y are the x and y components of the field, normalized so that $\mathbf{u}^* \cdot \mathbf{u} = |u_x|^2 + |u_y|^2$ is the optical power in the fiber. Instead of the Stokes vector of unit length used in Section 7.2, we shall here use an instantaneous Stokes vector with the components $(S_1, S_2, S_3) = (|u_x|^2 - |u_y|^2, 2\mathrm{Re}(u_x^* u_y), 2\mathrm{Im}(u_x^* u_y))$, so that the length of the Stokes vector is the optical power. Using a unitary transformation (the Jones matrix), we can mathematically eliminate the rapid motion of the polarization of the optical field caused by the fiber's wavenumber birefringence. Then, taking into account that the collision length for the solitons of interest here is short enough ($L_{coll} \le 200$ km) that the effects of polarization dispersion on the relative polarizations of the two fields can be neglected, and that the nonlinear term is averaged over all polarizations, we are left with a much simplified version of the propagation equation, known as the Manakov equation [104, 105]:

$$-i\frac{\partial \mathbf{u}}{\partial z} = \frac{1}{2}\frac{\partial^2 \mathbf{u}}{\partial t^2} + (\mathbf{u}^* \cdot \mathbf{u})\mathbf{u}. \tag{7.21}$$

Now, if $\mathbf{u}$ is composed of two fields with distinctly separable frequency ranges (e.g., pulses in a WDM system), we can isolate the terms of Eq. (7.21) in each frequency range. If the two frequencies are identified by subscripts a and b, then for the frequency of $\mathbf{u}_a$, we find the equation

$$-i\frac{\partial \mathbf{u}_a}{\partial z} = \frac{1}{2}\frac{\partial^2 \mathbf{u}_a}{\partial t^2} + (\mathbf{u}_a^* \cdot \mathbf{u}_a)\mathbf{u}_a + (\mathbf{u}_b^* \cdot \mathbf{u}_b)\mathbf{u}_a + (\mathbf{u}_b^* \cdot \mathbf{u}_a)\mathbf{u}_b. \tag{7.22}$$

The equation for the frequency of $\mathbf{u}_b$ is obtained by interchanging indices. There are three nonlinear terms on the right-hand side of Eq. (7.22). The first is the self-phase modulation term, while the second and third constitute the polarization-dependent XPM. For co-polarized fields, the second and third terms become identical [compare with Eq. (5.3)], while for orthogonally polarized fields, the third term goes to zero.

By expanding Eq. (7.22) into its field components, it is straightforward to show that the XPM terms of Eq. (7.22) modify the Stokes vector of the a field according to the equation

$$\frac{\partial \mathbf{S}_a}{\partial z} = \mathbf{S}_a \times \mathbf{S}_b. \tag{7.23}$$

This equation and that for $\mathbf{S}_b$ (exchange subscripts) show that the nonlinear term causes the Stokes vectors of the two fields to precess around one another, so that $\partial(\mathbf{S}_a + \mathbf{S}_b)/\partial z = 0$. It is important to note from Eq. (7.23) that when the pulses are either exactly co-polarized or exactly orthogonally polarized, there is no scattering. Thus, for example, in the experiment of Fig. 7.5, there was no scattering between the initially co-polarized channels, until the fiber's PMD itself gradually opened up the angle between their Stokes vectors.

To apply Eq. (7.23) to solitons, we have only to integrate it over the course of a collision. In first order, if we neglect the simultaneous precession of soliton b, then the integration of Eq. (7.23) gives an effective precession angle of $\mathbf{S}_a$ around $\mathbf{S}_b$. Single solitons of Eq. (7.21) have the general form

$$\mathbf{u}(z,t) = \hat{\mathbf{u}} A \operatorname{sech}[A(t+\Omega z)] \exp\left[iz\left(A^2 - \Omega^2\right)/2 - i\Omega t\right], \tag{7.24}$$

where $\hat{\mathbf{u}}$ is a unit normalized polarization vector ($\hat{\mathbf{u}}^* \cdot \hat{\mathbf{u}} = 1$). To evaluate the precession angle most easily, let $\mathbf{u}_a$ be a stationary soliton ($\Omega_a = 0$), and let $\mathbf{u}_b$ be a soliton substantially displaced in frequency ($|\Omega_b| \gg 1$). Then the integration over the collision involves the integration over z of $A_b^2 \operatorname{sech}^2[A_b(t+\Omega_b z)]$, which gives a precession angle of $2A_b/\Omega_b$. This formula uses soliton units. Noting that the full spectral width of the soliton in Eq. (7.24) is $\Delta\omega = (2/\pi) \times 1.763A$, we can re-express the polarization angle in the dimensionally independent form

$$\Delta\theta_a \sim 1.78 \frac{\Delta\nu_b}{\Delta\nu_{ab}}, \tag{7.25}$$

where $\Delta\theta_a$ is the precession angle for the stationary soliton, $\Delta\nu_b$ is the full (spectral) width at half maximum of the passing soliton, and $\Delta\nu_{ab}$ is the frequency separation of the two solitons. Note the inverse dependence of the change in polarization angle on channel separation ($\Delta\nu_{ab}$). Because of the fact that the number of collisions in a given distance is in direct proportion to $\Delta\nu_{ab}$, however, the net spread in polarization vectors tends to be independent of channel separation, as observed experimentally (Fig. 7.5). Note also that Eq. (7.25) is valid whether or not the colliding solitons are equal in amplitude. Since the bandwidth of a soliton is proportional to its amplitude, the collision of two unequal solitons will yield unequal precessions for the two. In the WDM experiments described in Chapter 5, the ratio of the channel spacing to the soliton's spectral FWHM was about five, which would make the precession angle per collision about 0.35 radians. This is a small enough angle to make the preceding theory applicable, and yet is large enough that just a few collisions are enough to prohibit the use of polarization division multiplexing.

The argument just presented gives results applicable to a WDM communications system, where the channel spacing is much greater than the soliton bandwidth, so that the precession angle in a collision is small. Manakov [104, 105] was able to show that Eq. (7.21) supports polarized solitons in the strict sense. A collision of two solitons therefore does not give rise to any scattered radiation. The solitons emerge from the collision with their energies and velocities unchanged, but their polarizations change as well as their positions and phases. A translation of Manakov's result gives the following Stokes–Poincaré picture. If colliding solitons a and b have amplitudes A_a and A_b, and normalized Stokes vectors $\hat{\mathbf{S}}_a$ and $\hat{\mathbf{S}}_b$, then the vector $\mathbf{A} \equiv A_a\hat{\mathbf{S}}_a + A_b\hat{\mathbf{S}}_b$ remains the same before and after the collision. As a result of the collision, the two Stokes vectors $\hat{\mathbf{S}}_a$ and $\hat{\mathbf{S}}_b$ precess around the axis defined by $\mathbf{A}$ through an angle ϕ whose tangent is given by $\tan\phi = 2A\Omega/(\Omega^2 - A^2)$, where Ω is the difference between the solitons' frequencies and A is the length of the vector $\mathbf{A}$. With a little practice in geometry, one can show that when $\Omega^2 \gg A^2$, the exact result reduces to the approximate one just given.

Another consequence of the polarization scattering from collisions, more fundamental than the simple prohibition of polarization division multiplexing, is a jitter in pulse arrival times, mediated by the fiber birefringence. Mollenauer and Gordon [98] described a qualitatively similar birefringence-mediated jitter, as initiated by the (relatively small) noise-induced scatter in polarization states. As the spread in polarization states from collisions tends to be much larger (note that it eventually tends to spread the Stokes vectors over a large fraction of the Poincaré sphere), the jitter is correspondingly greater, and in a typical case can easily add at least a few tens of picoseconds to the total spread in arrival times over trans-oceanic distances. This represents a significant reduction in safety margin for individual channel rates of 10 Gbit/s or more. Nevertheless, note that in the WDM results reported in Section 5.9, reduction, while undoubtedly present, was not fatal.

Chapter 8

Hardware and Measurement Techniques

8.1. Soliton Sources

For best performance, soliton transmission requires a reliable source of unchirped pulses, whose intensity envelopes are, to good approximation, either of Gaussian or sech^2 shape, and whose repetition rate can be easily adjusted to the desired bit rate. Many of the early experiments, however, were carried out using mode-locked, electrically driven semiconductor lasers, which tend to produce chirped and asymmetrical pulses with widths that are hard to control. They also require very careful matching of the cavity's $c/2L$ resonance frequency (where L is the cavity length) to match the microwave drive frequency, since even very small mismatches tend to cause large changes in the pulse characteristics. Harmonically mode-locked erbium fiber ring lasers can provide unchirped, essentially Gaussian pulses, but they require even more complex adjustment, viz., the round-trip times of an internal stabilizing Fabry–Perot etalon and of the fiber ring must be simultaneously adjusted to match, respectively, $1\times$ and an exact integral multiple of the bit period. In sum, both forms of mode-locked laser were much too complex, expensive, and fussy to be viable candidates for use in practical systems, especially where dense WDM may require 100 or more such devices in one and the same transmitter. By contrast, the scheme alluded to in earlier chapters to use a sinusoidally driven $LiNbO_3$ Mach–Zehnder modulator to carve pulses from the output of a cw laser tends to meet all requirements simply and at relatively low cost.

8.1.1. The Pulse Carver

The $LiNbO_3$ Mach–Zehnder modulators used for the pulse carver come in two forms: "X-cut" and "Z-cut." Note from Fig. 8.1 that the X-cut version automatically provides for the creation of equal but opposite phase shifts in the two arms of the interferometer from the single drive electrode, so that the carved pulses are always unchirped. In the Z-cut version, however, unchirped pulses are obtained only when the two drive electrodes are supplied with voltages that are equal in amplitude but π out of phase with each other; that arrangement is assumed in the following description when a Z-cut device is used. Perhaps it should also be noted here that the optical inputs to the modulators must be linearly polarized, with the optical E field aligned along the Z axis.

The basic scheme of operation of the pulse carver is simple: As shown in Fig. 8.2, the drive voltage for the $LiNbO_3$ modulator is a pure sinusoid whose frequency is equal to half the bit rate and whose peak-to-peak amplitude exactly matches the voltage difference between successive nulls of the modulator [84]. With the modulator bias adjusted for peak transmission without the sinusoidal drive, two identical pulses are produced per cycle of the drive voltage, one on its upswing and the second on its downswing. That is, with the modulator biased as just stated, the transfer function is

$$I/I_0 = \cos^2(\pi V/V_0), \tag{8.1}$$

where I and I_0 are the transmitted and input intensities, respectively, V is the applied voltage (exclusive of the bias), and V_0 represents the voltage difference between successive nulls. Applying the voltage $V(t) = (V_0/2)\cos(\pi t/T)$, where

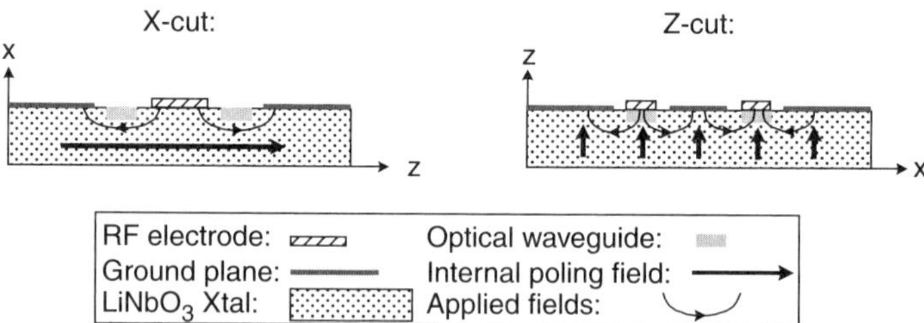

FIGURE 8.1 Cross-sections of X-cut and Z-cut $LiNbO_3$ Mach–Zehnder modulators. Note that in the X-cut version, the single drive electrode provides fields that aid and oppose the large internal field in the two interferometer arms, respectively, while the two electrodes of the Z-cut version must be oppositely driven to achieve the same effect. RF, Radio frequency; Xtal, crystal.

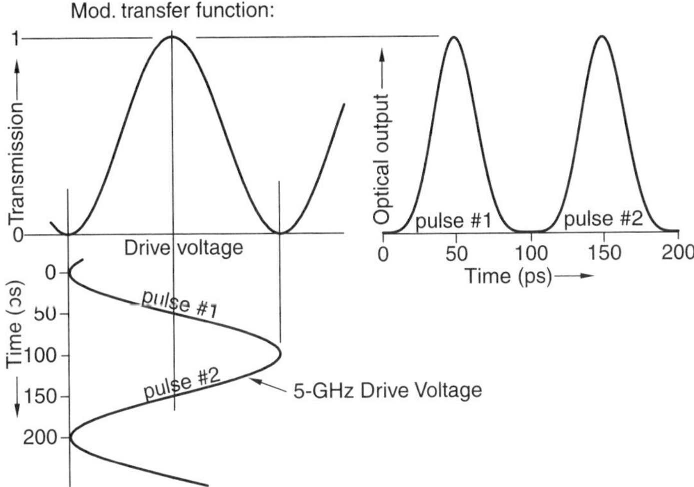

FIGURE 8.2 Basic principle of the pulse carver. A sinusoidal drive voltage whose frequency is half the bit rate swings the modulator transmission between two adjacent nulls. The two nearly Gaussian pulses created per cycle of the drive voltage are shown on the right. (The particular drive frequency and time scales indicated here are, of course, for a 10-Gbit/s source.)

T is the bit period ($T = 100$ ps in the figure), the transmitted intensity becomes

$$I(t) = I_0 \cos^2(\tfrac{\pi}{2} \cos(\pi t/T)). \tag{8.2}$$

Any simple graphics program will confirm that Eq. (8.2) yields pulses whose shape makes a nearly perfect fit (save in its far wings) to a Gaussian whose intensity FWHM is $\cong T/3$, just as shown on the right side of Fig. 8.2.

Figure 8.3 is a schematic of the 10-Gbit/s pulse carver used for the dense WDM experiments described in Chapter 6, Section 6.2. It uses an X-cut modulator for the simplicity afforded by the single-port drive and the automatic production of chirp-free pulses. (When a Z-cut modulator is used, the drive must be split into two equal parts, which are then made π apart in phase through the use of connecting cables having a precisely controlled differential in their lengths. Although not exceptionally difficult, this arrangement adds unnecessary complexity and cost.) The microwave amplifier in the lower right-hand corner of the figure operates in saturation to provide a fixed output voltage for a range of input voltages. Note that its output is split into three parts, with one part being used to feed the power amplifier that drives the modulator. The power amplifier is also over-driven, so that its output can be controlled very reliably and simply by adjustment of its DC power supply voltage. [At 5 GHz, the required voltage swing is typically no

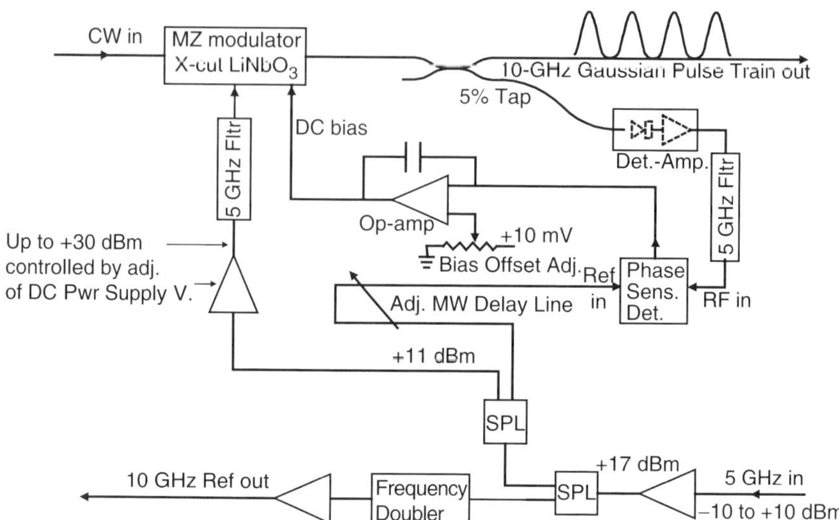

FIGURE 8.3 Schematic of a 10-Gbit/s pulse carver. MZ, Mach–Zehnder; MW, microwave; SPL, splitter.

more than about 12 V peak-to-peak, or about 4.24 V rms, which corresponds to 360 mW of microwave power into the ≈50-ohm load presented by the modulator. Thus, the 30-dBm (1 W) output power capability of the amplifier shown in Fig. 8.3 represents a bit of overkill.] To guarantee a pure sinusoidal drive to the modulator, the amplifier's output is filtered to eliminate all higher harmonics of its 5-GHz output. It will be noted that one great advantage of this particular pulse carver scheme is that it is not dependent on details of the modulator's frequency response. Rather, the only requirement is that sufficient drive voltage be available for a single frequency at just half the bit rate. Thus, this scheme can be extended to higher bit rates, such as 40 Gbit/s, without much difficulty.

Most of the other components to the right of the modulator are part of a scheme to provide automatic bias control. The control scheme is based on the fact that unless the bias is correct, the pulses are unevenly spaced in time, so that the pulse train contains a significant Fourier component at 5 GHz. Thus, as the bias is moved through the correct setting, the 5-GHz component at first decreases, then passes through zero at the correct setting, and then grows again, but with the opposite phase with respect to that of the 5-GHz modulator drive. To measure the 5-GHz component, as shown in the figure, a sample of the pulse train is split off, detected, amplified, passed through a 5-GHz filter, and finally sent to the radio frequency (RF) input port of a phase-sensitive detector (variously known as a "double-balanced mixer"), where it is compared with (beat against) a properly phased sample of the 5-GHz drive voltage. Since the two inputs to the phase-sensitive detector are at exactly the same frequency, its output is a DC voltage whose polarity depends on the relative phase of the two inputs, and which nominally becomes zero for zero input into the right-hand (RF input) port. Thus, the output of the phase-sensitive detector affords the desired error signal, which, when greatly multiplied by an inexpensive "op-amp," locks the system to exactly the DC bias required for proper operation of the modulator. Since the balance of the phase-sensitive detector is often not quite perfect, it typically tends to produce a DC output of a few millivolts even when the RF input is exactly zero; this imbalance is offset by the bias adjustment to the op-amp shown in the figure. It should also be noted that for the wrong adjustment of the microwave (MW) delay line (for the 5-GHz reference), the system will lock in to a mode producing the inverse of the pulse train shown in Fig. 8.2; clearly, the best adjustment is in the middle of the range providing the correct pulse train. Finally, it should be noted that this scheme for bias adjustment, which has been used with great success for many years, avoids the cumbersome "dithering" of alternative schemes and tends to provide a faster and more precise lock to the correct bias.

It will be noted from Fig. 8.3 that a 10-GHz clock for synchronizing the pulse pattern generator is derived by frequency doubling of the 5-GHz reference input. In a real transmission system, however, that arrangement must be reversed, since there, the fundamental clock input will always be at the fundamental (10 GHz) bit rate. In that case, a simple "toggle flip-flop" chip can be used to divide the 10-GHz clock input frequency by two, to yield the 5 GHz required for the modulator drive. That arrangement should be just as cheap, simple, and reliable as the one shown in Fig. 8.3.

Certain details of the optical interconnection of the pulse carver to the cw laser source and to the following data modulator (as in Fig. 6.20, for example) are also important. First, since both modulators require inputs with the correctly aligned linear polarization, it is highly desirable that correctly aligned polarization-maintaining fibers be used for all interconnections. Although ordinary fibers provided with polarization controllers can be used in principle, thermal and mechanical drift makes it nearly impossible to maintain correct polarization settings, and the number of adjustments required for a dense WDM system is tediously large. Second, the arrangement for an entire transmitter can be simplified if both pulse carver and data modulators are contained in the same package and are directly optically coupled together. At the time of this writing (2005), a check of the World Wide Web shows that at least one manufacturer now supplies just such convenient combinations, with features custom tailored for dispersion-managed soliton transmission, no less! One of those special features is an integral detector at the modulator's output. To see how that might be possible, see Fig. 8.4.

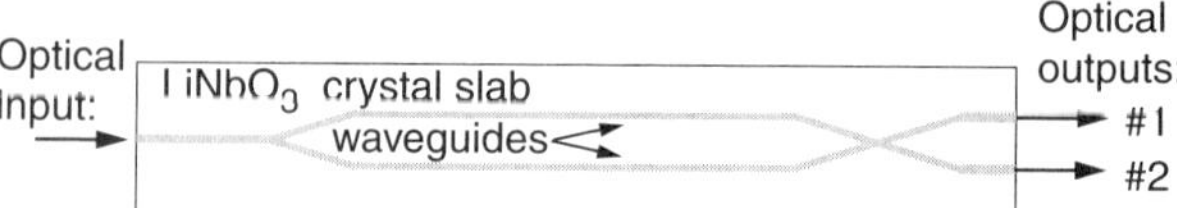

FIGURE 8.4 Possible input and output configurations of a $LiNbO_3$ modulator. Traditionally, only two of the four ports of the Mach–Zehnder interferometer were brought out, so that each end of the modulator tended to look like the left-hand side of the figure. But there is no reason that two ports cannot be brought out at either end, as shown on the right-hand side of the figure. In that case, one of such a pair can be the working output, while the other can be directly coupled to a detector for monitoring purposes, or as part of the bias control scheme of Fig. 8.3.

8.2. The Temporal Lens

In Chapter 3, it has already been pointed out that in the presence of significant jitter in pulse arrival times, the ideal detection system bases its decisions on the integral of all the energy arriving at the detector in each bit period. Unfortunately, however, such "integrate-and-dump" detectors do not yet seem to exist. Rather, it is common for the detection system to base its decision on a sample integrated over only a small fraction of the bit period, supposedly about 20 ps wide in the typical 10-Gbit/s system, with the consequent imposition of severe penalties from timing jitter. In the absence of effective integrate-and-dump detectors, however, there is another way to largely avoid such penalties, and that is through the use of a so-called temporal lens [108–111]. The basic scheme of the temporal lens is shown in Fig. 8.5. As can be seen, the device consists of nothing more than a phase modulator, appropriately driven in synchronism with the locally recovered clock, and followed by an essentially linear dispersive element (usually a coil of fiber). The phase modulator gives each incoming pulse a frequency shift proportional to its jitter displacement, and the dispersive element then serves to translate, or

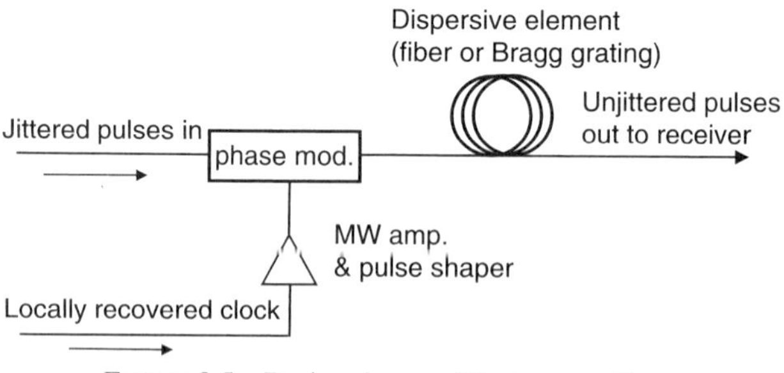

FIGURE 8.5 Basic scheme of the temporal lens.

"focus," each pulse onto the mean arrival time (modulo the bit period). The net effect is then to considerably widen the effective acceptance window for arrival times.

In the ideal mode of operation, the phase shift, $\phi(t)$, produced by the modulator is a series of truncated parabolas, centered about the middle of each bit period, such that the corresponding frequency shift [the time derivative of $\phi(t)$] is directly proportional to time as measured from the center of each period (see Fig. 8.6). At microwave frequencies, however, the voltage waveform for this ideal mode is very hard to produce, so that in practice, it is usually necessary to settle for a simple sinusoid, peaked at the centers of the successive bit periods. Nevertheless, this compromise also works very well, as can be seen from Fig. 8.7, where the fraction of the pulse energy used for decision making is plotted as a function of arrival time, for three cases: (1) a perfect integrate-and-dump detector, (2) the sinusoidally driven temporal lens, and (3) the ordinary detector with an ≈20-ps integration time. Note that while the acceptance window for the ordinary detector is scarcely wider than the 33-ps pulse itself, the temporal lens opens up that window to at least 60 ps, a width not much smaller than that of the ideal integrate-and-dump detector. The rather dramatic effect of the temporal lens is shown in yet another way in Fig. 8.8, which shows eye diagrams from a dense WDM transmission experiment, as seen with and without the use of the lens.

Any device used at the receiver needs to have an essentially polarization-independent response. Unfortunately, however, as already noted in Section 8.1, $LiNbO_3$ modulators are very strongly polarization dependent. The most straightforward way around this problem is to use a polarization diversity scheme. The one

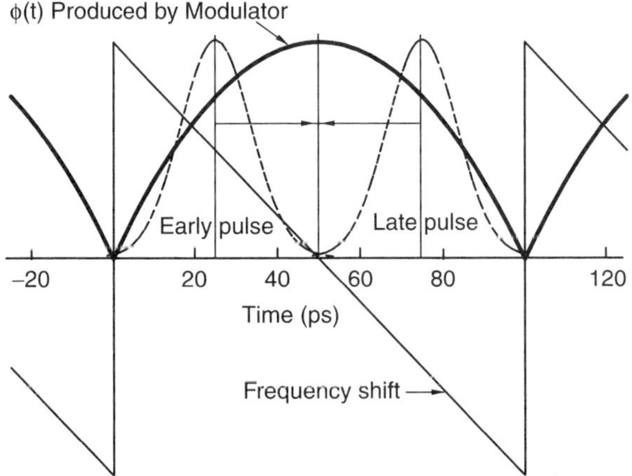

FIGURE 8.6 Temporal details of the ideal mode of operation.

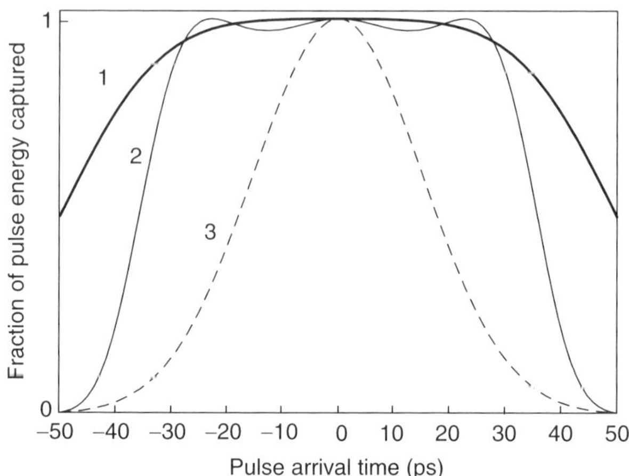

FIGURE 8.7 Fraction of the energy of a 33-ps-wide Gaussian pulse available for decision making, plotted as a function of arrival time, for the following cases: (1) a perfect integrate-and-dump detector, (2) a sinusoidally driven temporal lens, and (3) the ordinary detector with an ≈20-ps integration time.

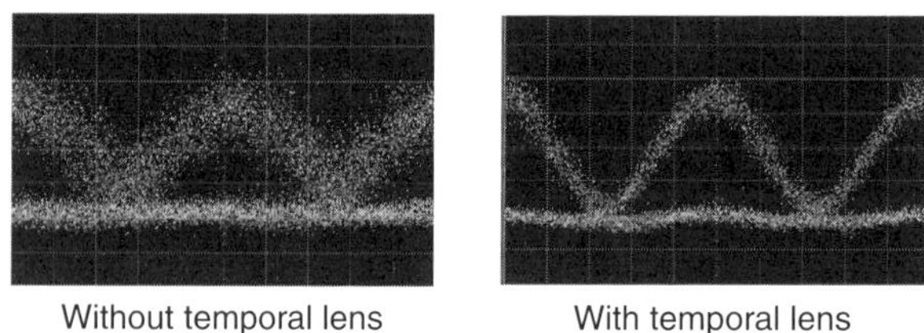

FIGURE 8.8 Eye diagrams at 7200 km of one channel of a dense WDM experiment at 10 Gbit/s and 50-kHz channel separation, using dispersion-managed solitons, with and without use of the temporal lens. Although the measured BER for the situation on the left was $\approx 10^{-6}$, that for the right was better than 10^{-9}.

we have used [108,109] (see Fig. 8.9) performed extremely well, as the polarization dependence of the device was nearly immeasurable.

8.2.1. *Analytical Treatment*

The discussion here of the temporal lens has thus far been largely qualitative, and the potential of the temporal lens to change the shape of the pulses has been ignored. The following quantitative treatment is intended to fill that void. It is based on the ordinary differential equations outlined in Chapter 2, Section 2.1.4,

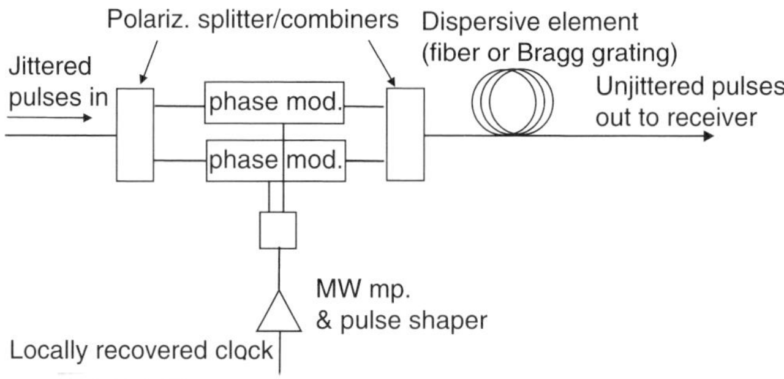

FIGURE 8.9 Polarization diversity scheme for the temporal lens. The first polarization splitter separates the input of arbitrary polarization into its two, orthogonal, linearly polarized components. Each of those components is then sent through its own phase modulator. The outputs of the two phase modulators are then combined by a second polarization combiner before being sent on to the dispersive fiber. Note that the input and output (polarization-maintaining) fiber leads, as well as the coaxial cables to the modulators, must constitute three carefully matched pairs in length.

and, for simplicity, it assumes the ideal model of Fig. 8.6. Also, since the problem is essentially a linear one, we shall treat all elements as lossless. Nevertheless, the treatment here should be adequate to convey all of the really important ideas.

At the input to the phase modulator, we assume the unchirped pulse

$$u_0(t) = u_0 \exp\left[-\frac{1}{2}\eta_0(t-\delta t)^2\right], \tag{8.3}$$

where δt is the displacement of the center of the pulse from $t=0$ (the middle of the bit period). It will be recalled from Section 2.1.4 that $1/\sqrt{\eta}$ is a measure of the pulse width; that is, the pulse intensity FWHM $\tau = \sqrt{4\ln 2}/\sqrt{\eta}$.

The quadratic phase shift imposed by the modulator can be written in terms of a chirp parameter β_{mod}, so that after the modulator, we have

$$u_1(t) = u_0 \exp\left[-\frac{1}{2}\eta_0(t-\delta t)^2 - i\left(\frac{1}{2}\beta_{mod}t^2\right)\right], \tag{8.4}$$

which can be rewritten as

$$u_1(t) = u_0' \exp\left[-\frac{1}{2}(\eta_0 + i\beta_{mod})(t-\delta t)^2 - i(\beta_{mod}\delta t)t\right]. \tag{8.4a}$$

[Here u_0' differs from u_0 only because it has absorbed the constant phase term $\exp(i\beta_{mod}(\delta t)^2/2)$.] Note that although the pulse still has the same width (τ_0), it now has a linear chirp β_{mod} and a shift in mean frequency of $\delta\omega = -\beta_{mod}\delta t$. When the pulse is then put through a linear element of dispersion

$$\Delta = 1/\beta_{mod}, \tag{8.5}$$

it will then be moved in time by $\delta\omega\,\Delta = -\delta t$, exactly as needed for focusing onto the center of the bit period. Note that although Λ and β_{mod} have the same algebraic sign, the pair can be either positive or negative. (The curves of Fig. 8.6 correspond to $\beta_{mod} > 0$.)

The potential of the temporal lens to modify the widths of the pulses emerging from it constitutes a second kind of temporal focusing that should be analyzed. From Eq. (2.10), we know that when the pulses are subject to a linear element of dispersion Δ, the quantity $q = \eta_0/(\eta + i\beta_{mod})$ must obey the equation

$$q_2 = q_1 + i\Delta\eta_0, \tag{8.6}$$

where q_2 and q_1 represent the values after and just before the dispersive element, respectively. From Eq. (8.4a), we have $q_1 = \eta_0/(\eta_0 + i\beta_{mod})$. Thus, combining Eqs. (8.5) and (8.6), we have

$$\frac{1}{\eta_2 + i\beta_2} = \frac{1}{\eta_0 + i\beta_{mod}} + \frac{i}{\beta_{mod}}. \tag{8.7}$$

With a bit of manipulation, Eq. (8.7) can be rewritten as

$$\eta_2 + i\beta_2 = \beta_{mod}^2/\eta_0 + -i\beta_{mod}. \tag{8.7a}$$

Finally, substituting $\beta_{mod} = g\eta_0$ into Eq. (8.7a), we obtain the final pulse width

$$\tau_2 = \tau_0/g. \tag{8.8}$$

It is important to know the peak-to-peak swing in phase, $\delta\phi_{max}$, required from the modulator. Let us assume an inital pulse width $\tau_0 = T/3$ (T is the bit period), so $1/(\eta_0) = T/(3\sqrt{4\ln 2})$. Thus we have

$$\delta\phi_{max} = \frac{1}{2}\beta_{mod}(T/2)^2 = (g/8)(\eta_0 T^2) = \frac{3\sqrt{4\ln 2}}{8} = (3.1192\ldots)g, \tag{8.9}$$

or nearly π radians at $g = 1$.

It is also now easy to calculate the required dispersion, Δ, for "focus." For example, for $g = 1$, where $\beta_{mod} = \eta_0$, we have $\Delta = 1/\beta_{mod} = 1/\eta_0 = \tau_0^2/(4\ln 2)$, so for $\tau_0 = 33$ ps, $\Delta = 393\,\text{ps}^2 \rightarrow -315$ ps/nm, a dispersion easily supplied by a modest coil of DCF.

8.2.2. *Some Practical Details of Operation*

With the more practical sinusoidal drive, the optimum window shape (see Fig. 8.7) is obtained when the peak frequency shifts (occurring at $t = \pm T/4$) are the same as those of the parabolic drive at the same times. In that case, it can easily be shown that the required peak-to-peak drive for the sinusoid is $(2/\pi)\times$ that for the parabolic drive. Thus, for $g=1$, where previously we found that the required peak-to-peak parabolic drive is $\approx\pi$, the corresponding requirement for the sinusoidal drive is just 2 radians peak-to-peak. The voltage swing required to produce that phase shift at 10 GHz in most $LiNbO_3$ modulators is typically on the order of 3–6 V, corresponding to RF powers in the very modest range of ~20–80 mW into the 50-ohm load posed by the modulator. Thus, it is still quite practical to double that drive to obtain $g=2$ and to halve the required dispersion. An experimental test seems to show that the temporal lens works as well or even better under that higher drive condition. In practice, the modulator drive voltage is adjusted until the best focus is obtained, as seen from an eye diagram on a sampling scope.

It is also important to know when the phase modulator drive is correctly synchronized with the mean pulse arrival time. The required information is easily obtained from the optical spectrum of the pulse train. That is, if the adjustment is correct, the spectrum after the modulator is perfectly symmetrical about the original spectral position, but when the drive is late (early), the spectrum is strongly skewed to the low-frequency (high-frequency) side of the original spectral center. (The rule given here is for $\beta_{mod}>0$; for $\beta_{mod}<0$, the association between late–early/low–high is reversed.) It should also be noted, however, that there are really two phase settings that result in perfect symmetry of the spectrum, the second one an incorrect adjustment, corresponding to a defocusing of the pulses away from the center of the bit period. The appearance of the eye diagram makes it obvious whether the choice is correct or not, however.

Finally, it should be noted that when a coil of fiber is used as the dispersive element, the clock phase at the output of the fiber, relative to that at its input, will tend to drift significantly with changing temperature. Thus, the temporal lens requires its own clock recovery at the modulator, independent from the later clock recovery used for the decision circuit of the detector.

8.2.3. The Need for an Integrate-and-dump Detector

While it has proved to be a wonderful tool in the laboratory, the temporal lens tends to be too complex and expensive for practical acceptance in a dense WDM system. (Each channel would need its own proper lens.) By contrast, the integrate-and-dump detector, which, as we have seen, could potentially provide just as good or even better performance (Fig. 8.7), as could an electronic device on a single chip, would be much cheaper. With the extremely high-speed electronics now being developed for 40 Gbit/s, there is no reason why a high-performance integrate-and-dump chip could not be developed for 10 Gbit/s. Integration over entire bit periods could be obtained by splitting the output of the detector between two identical circuits processing successive bit periods. In each circuit, the detector voltage would be converted into a current source feeding an integrating capacitor. Following integration over the first full bit period, interrogation and capacitor discharge would follow in the second. (In the meantime, integration over that second bit period would be carried out by the other circuit.) Outputs of the two decision circuits would then be combined to yield the net integrate-and-dump detector output.

8.3. Clock Recovery

Clock recovery is much easier when the signals are in RZ (return-to-zero) format, as they are for soliton transmission, than when they are in NRZ (non-return-to-zero) format. The reason is that an RZ signal train always has a large Fourier component at the frequency of the bit rate, while NRZ has none of the same for random data. Thus, for RZ signals, there are two straightforward and relatively easy ways to effect clock recovery. The first, shown schematically in Fig. 8.10, consists of a current- or voltage-controlled oscillator, phase-locked to the incoming RZ data through a phase-sensitive detector. The oscillator, for example, might be based on a sub-millimeter-diameter sphere of the ferromagnetic material YIG (yttrium–iron–garnet), whose microwave resonance frequency is directly proportional to a applied magnetic field. With a permanent magnet providing the major part of the required field, a small coil carrying just milliamperes of current can complete

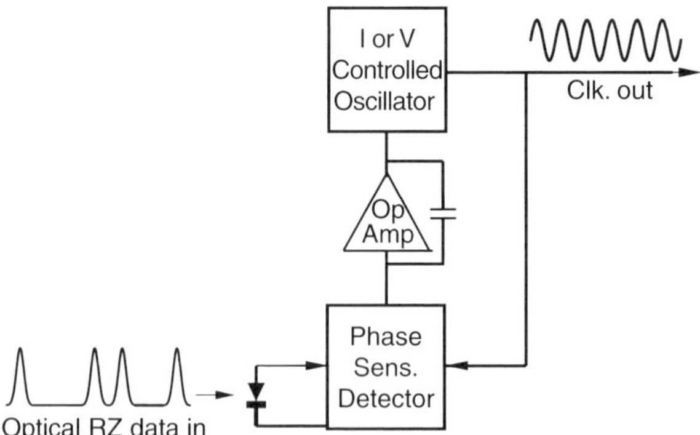

FIGURE 8.10 Clock recovery for RZ data based on a current- or voltage-controlled oscillator in a phase-locked loop. As shown here, a detected sample of the incoming data is compared in a phase-sensitive detector with the output of the oscillator, thereby creating an error signal that serves to lock the oscillator's phase to that of the fundamental Fourier component of the incoming data stream.

the tuning. A detected sample of the incoming data is beat against the oscillator's output in a phase-sensitive detector to generate an error signal that controls the frequency, and ultimately the exact phase of the oscillator, through an amplified version of the phase-sensitive detector's output. This overall arrangement, known as a phase-locked loop, has the fundamental advantage of being able to respond to a range of clock rates, and with proper design, it can be very stable and dependable. Hence, this is the version that is almost universally chosen for commercial telecommunication systems. Compact, relatively inexpensive, pre-engineered versions of the hardware shown in Fig. 8.10 are readily available.

The second form of clock recovery is strictly passive and consists of nothing more than a very high-Q-factor resonator, tuned to the clock frequency, which serves as a filter to extract the fundamental Fourier component from the incoming RZ data (see Fig. 8.11). The resonator shown there is based on the TE_{011} mode, used for a long time in wave meters because of its potentially very high Q factor. One reason for the high Q factor is that in the TE_{011} mode, surface currents in the cylinder walls make circular paths in the plane normal to the cylinder axis. Thus, no current must flow across the (potentially lossy) joint between the cylinder walls and the end caps. Rather, as shown in the figure, a small gap is purposely introduced between the walls and the end caps to help suppress unwanted modes. The loaded Q factors of several 10-GHz resonators like that of Fig. 8.11, whether machined of copper or aluminum, had measured values consistently in excess of 10,000.

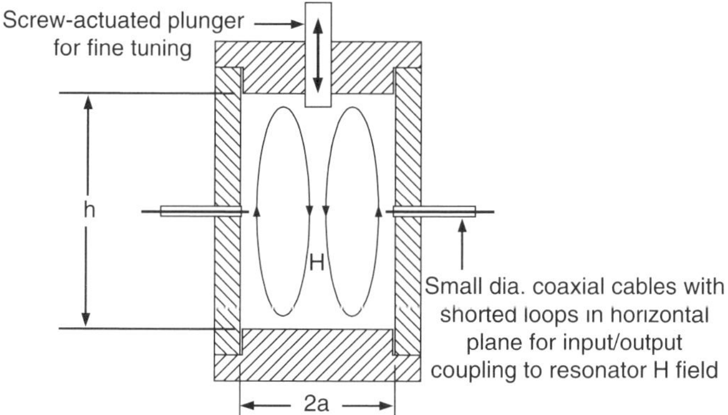

FIGURE 8.11 Microwave resonator for clock recovery of RZ data, based on the high-Q TE_{011} wave-meter mode of a right cylindrical cavity. For that mode, the shape of the H field paths are as shown; the E fields (not shown) make circular paths in the plane normal to the cylinder axis. Formulas for calculating the resonant frequency from the diameter $2a$ and height h of the cylinder are given in the text.

The resonance frequency f_0 is given rather accurately by

$$f_0 = \frac{c}{1.64a}\sqrt{1+0.6745(a/h)^2}, \tag{8.10}$$

where a is the cylinder radius, h is its height, and c is the speed of light. Note that Eq. (8.10) yields $f=10$ GHz from the combination of $a=19$ mm and $h=55$ mm.

Since the Q factor is defined as the ratio of the energy stored in the resonator to the loss per radian, and since voltage scales as the square root of power, if the input to the resonator is removed at $t=0$, at resonance its output voltage rings down as

$$V(t) = V_0 \sin(2\pi t/T)\exp(-\pi t/(TQ)), \tag{8.11}$$

where $T=1/f_0$ is the bit period. Since the characteristic ring-down time of TQ/π represents more than 3000 bit periods at $Q=10{,}000$, the resonator's output maintains the clock over much longer strings of zeros than are ever encountered in practice. Furthermore, the phase of the resonator's voltage output represents the input phase as averaged over a similar number of bit periods. Thus, the standard deviation in time-phase jitter at the resonator's output should be reduced from that at input by a factor of $\approx\sqrt{\pi/Q}$, or about 1/56th for $Q=10{,}000$. That factor is sufficient to reduce even the worst of the typically encountered jitter to a truly negligible level (less than a few tenths of a picosecond).

The principal advantages of the high-Q resonator as a clock recovery device lie mainly in its simplicity, its low cost, and its ability to follow rapid changes in input phase. Its greatest disadvantages lie in the fact that it must be tuned manually to maintain resonance with changing clock frequency, and that the phase of its output is strongly dependent on detuning in the neighborhood of resonance. Those disadvantages tend to make it unsuitable for commercial transmission, and thus restrict its use to the laboratory only. Even there, the potential for detuning from changes in both ambient temperature and humidity require constant vigilance of its adjustment.

8.4. Dispersion Measurement

Dispersion is clearly an all-important parameter in soliton transmission. Therefore, it is equally important to be able to measure the dispersion of real fibers, both the path-average values that determine soliton pulse energies and the local values that play such an important role in dispersion management. This section describes several important techniques for the measurement of both kinds of dispersion.

8.4.1. Measurement of Path-average Dispersion in Ordinary Maps

In an ordinary system (one not using PGD devices), measurement of the path-average dispersion parameter $\bar{D}(\lambda)$ of a recirculating loop is fairly straightforward if a complete WDM transmitter and its associated set of source lasers are available. For each mean wavelength of the measurement, pairs of source wavelengths are selected, symmetrically surrounding the mean wavelength, and typically spaced apart from each other by a few nanometers, such that the expected $\bar{D}$ of a few tenths of a ps/nm-km will produce relative time displacements of a few hundreds of picoseconds in just one pass through the 400- to 600-km-long loop. Such measurements are easily and accurately made on a sampling scope, especially if the pattern generator is set to create a few ones (pulses) surrounded on both sides by several empty bit slots. Note that it is necessary to take the *difference* in relative time displacements between measurements made in two places, first at the immediate entry to the loop, and the second at its immediate output, so that the effect of extraneous dispersion is not included. The path-average dispersion parameter is then just that net time difference divided by the path length and the wavelength separation of the two lasers. The measurements reported in Fig. 2.2 were made in just that way.

8.4.2. Measurement When PGD Devices Are Involved

In Chapter 6, Section 6.1.4, we found that the use of PGD dispersion compensators creates two distinctly different kinds and values of path-average dispersion, viz., $\bar{D}_{inter}$ and $\bar{D}_{intra}$. Clearly, the former can be measured as just described, with only the source laser spacings reduced to the minimum (the nearest channel spacing), to compensate for that fact that $\bar{D}_{inter}$ tends to be many times greater than the normal $\bar{D}$. For measurement of $\bar{D}_{intra}$, however, we have a more difficult problem, since now both wavelengths of the measurement must lie within a given channel. Additionally, both must lie well within the linear region of the group-delay curves of the PGD device (see Fig. 6.13). Thus, the two wavelengths must be separated by less than about 0.2 nm, a difference too small to be maintained accurately. Furthermore, since the time displacements on just a single pass through the loop tend to be very small, much more accurate results can be obtained by observing the displacements following substantial numbers of round-trips.

To circumvent the problem of the very small wavelength separation required of the sources, the measurements reported in Fig. 6.23 used the following trick. The source for each measurement was just a single laser, tuned exactly to the center of a particular channel. In the meantime, however, a phase modulator,

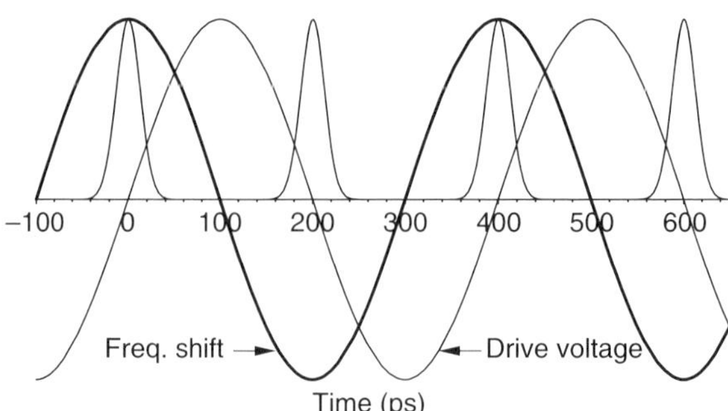

FIGURE 8.12 Scheme for using a phase modulator to create a source of two closely spaced wavelengths for measurement of $\bar{D}_{intra}$ in a system using PGD devices. As shown here, the modulator produces equal but opposite frequency or wavelength shifts on alternate pulses of a repeated 1010... pattern.

driven sinusoidally at one-fourth the bit rate (2.5 GHz), was imposed between the transmitter and the loop, and it was phased so as to produce the maximum possible frequency shifts of alternating sign on pulses of a repeated 1010... pattern (see Fig. 8.12). The wavelength difference between pulses induced by the modulator was not measured directly. Rather, the pulses were first sent through a fiber of accurately known dispersion (a length of standard SMF) and the induced time displacements were noted. The ratio of the relative time displacements in the loop to those of the calibration fiber, multiplied by its dispersion, yielded the results shown in Fig. 6.23.

8.4.3. The Dispersion Optical Time Domain Reflectometer

Almost from the very beginnings of the era of dispersion management, it became clear that something was terribly wrong with at least some dispersion-shifted fibers. In dense WDM, four-wave mixing products were often much greater than they should have been. Or all too often, the measured dispersion of a fiber segment would bear little relation to the dispersion of the longer spool from which it was cut. Such observations created a strong incentive to find a nondestructive way to measure the chromatic dispersion of a fiber as a function of distance along its length. While methods based on modulational instability [112] and on a search for the phase-matching condition in four-wave mixing [113] were proposed and tested, those methods really represented attempts to measure λ_0, the wavelength of zero dispersion, rather than D itself. Aside from the fact that λ_0 is often far removed from the wavelength range of interest, those methods further required extensive data-taking over a considerable wavelength range, so that the measurements for just one fiber took a long time. By contrast, the method we describe here, invented in early 1996 [114], and improved by 1998 [115], measures the map $D(\lambda_i, z)$ directly and with great accuracy, where λ_i is the wavelength of one of the two fixed-wavelength sources used, and the data for a given map can be taken (including many repetitions for signal averaging), processed, and displayed in just a few seconds. The method was immediately seized upon by colleagues at what was then Lucent Denmark as a desperately needed way to monitor their efforts to improve the uniformity of various dispersion-shifted and dispersion-compensating fibers, and it was soon used in the field to survey already installed fibers in long-haul transmission lines. At least one commercial version also became available.

The method is based on the fact that four-wave mixing sidebands oscillate in amplitude with a spatial frequency that is directly proportional to $D(z)$. As an intense pulse containing the two wavelengths λ_1 and λ_2 propagates down the fiber, the oscillating sidebands it generates are observed at the fiber's input through Rayleigh backscattering. Thus the sideband intensity as a function of z is converted into an equivalent return time, in the well-known principle of optical time domain reflectometry (OTDR). Thus, our new instrument soon came to be known as the dispersion OTDR, or more simply, as the DOTDR.

Basic Principles of Operation

As already noted, the method involves launching, simultaneously, strong, sub-microsecond pulses at λ_1 and λ_2 into the fiber, so that they may generate FWM product fields at the Stokes and anti-Stokes wavelengths, λ_S and λ_A, sequentially in each part of the fiber. Consider the field at λ_S, for example. Because of the phase mismatch δk [see Eq. (4.19)], the corresponding power $P_S(z)$ oscillates with the spatial frequency

$$F_S = 1/\Lambda_S = (\delta k)/(2\pi) = cD(\lambda_1)[\delta\lambda/\lambda]^2. \tag{8.12}$$

Measuring the frequency $F_S(z)$ then measures $D(\lambda_1, z)$, with spatial resolution Λ_S. Λ_S is typically in the sub-kilometer range.

The intensity oscillations, however, can be observed only in Rayleigh backscattering at the input end of the fiber. There the signal will fluctuate in intensity at a temporal frequency

$$f_{sig}(t) = (c/2n)F_S(z), \tag{8.13}$$

where n is the effective index of refraction of the fiber and where t is the round-trip time from the fiber input to point z and return, i.e.,

$$t = (2nz)/c. \tag{8.14}$$

Note that for $n = 1.46$, δt is $9.73\,\mu$s for each kilometer of path. Combining Eqs. (8.12) and (8.13), we finally obtain

$$D(\lambda_1, z) = \left(2n/c^2\right)(\lambda_1/\delta\lambda)^2 f_{sig}[t = (2n/c)z]. \tag{8.15}$$

The frequencies dictated by Eq. (8.15) are typically in the range of some tens to a few hundreds of kilohertz.

From Eq. (4.18), and from the known loss and scattering properties of the fiber, one can estimate the strength of the Rayleigh backscattered signal. For the case where the input pulses at λ_1 and λ_2 are co-polarized, and where there is no significant initial signal at λ_S and λ_A, the signal power at λ_S can be computed as

$$P_S(z) = 4\left(\frac{\lambda}{Dc\delta\lambda^2}\right)^2\left(\frac{n_2 P_1^0}{A_{eff}}\right)^2 P_2^0 \sin^2(\delta kz/2) \times \beta\delta z \exp(-4\alpha z). \tag{8.16}$$

[A similar expression yields the signal power, $P_A(z)$, at λ_A.] In Eq. (8.16), P_1^0 and P_2^0 are the pulse powers at the fiber input, β is the Rayleigh backscattering coefficient [see Eq. (3.18) and Table 3.1], δz is the fiber length occupied by the pulses at any given time, and, consistent with the notation used earlier in this book, α is the fiber's loss coefficient, A_{eff} is the effective area of the fiber core, and n_2 is the nonlinear index coefficient. [The factor of four in the exponential loss term stems from the combined facts that the quantity $P_1^2 P_2$ declines as $\exp(-3\alpha z)$, while the Rayleigh backscattering at z suffers an additional loss factor of $\exp(-\alpha z)$ in returning to the fiber input.] Note that the factor $\sin^2(\delta kz/2)$ in Eq. (8.16) provides the desired intensity oscillations of $P_S(z)$. For pulse input powers (P_1^0 and P_2^0) on the order of 1 W, Eq. (8.16) predicts, and actual measurements confirm, adequate signal strength for measurement of spans up to many tens of kilometers long.

Finally, at high powers, the nonlinear contribution to δk might become significant. From Eq. (4.19a), for the process $2\omega_1 \rightarrow \omega_2 + \omega_S$, that contribution is

$$\delta k_{nl} = k_{NL}(2P_1 - P_2), \tag{8.17}$$

with a similar expression (just reverse the subscripts 1 and 2) for the process $2\omega_2 \rightarrow \omega_1 + \omega_A$, and where $k_{NL} = (2\pi n_2)/(\lambda A_{eff})$ is that already defined by Eq. (1.11). Note that for the process $2\omega_1 \rightarrow \omega_2 + \omega_S$, δk_{nl} is zero if $P_2 = 2P_1$. (For the process $2\omega_2 \rightarrow \omega_1 + \omega_A$, δk_{nl} is zero if $P_1 = 2P_2$.) Even if the appropriate one of those conditions is not precisely met, however, note that δk_{nl} will tend to be only a small fraction of the linear δk for all but D very close to zero, since k_{NL} is numerically just $\sim 2.5\ \text{W}^{-1}\,\text{km}^{-1}$ for $A_{eff} = 50\,\mu\text{m}^2$ and $\lambda \sim 1550$ nm.

Technical Details of the Instrument

Figure 8.13 shows a schematic of the instrument. A multiwavelength semiconductor laser, originally designed as a source for dense WDM, provides any desired pair of wavelengths from a grid with spacing 0.6 nm and covering most of the erbium amplifier C band. The laser contains an extra-cavity semiconductor optical amplifier switch, which is used to create 0.8-μs pulses at a 2- to 3-kHz repetition rate. The following erbium-doped fiber amplifier (EDFA) increases the pulse power to ~1 W, before the pulses travel through the circulator and a wavelength division multiplexer into the fiber under test. Rayleigh backscattered light exits the third port of the circulator. A high-finesse, tunable etalon with bandpass of 0.1 nm selects the signal either at λ_S or at λ_A. The signal is then detected and its AC components are electronically amplified and filtered to remove useless high-frequency components. It is then processed by a personal computer (PC) equipped

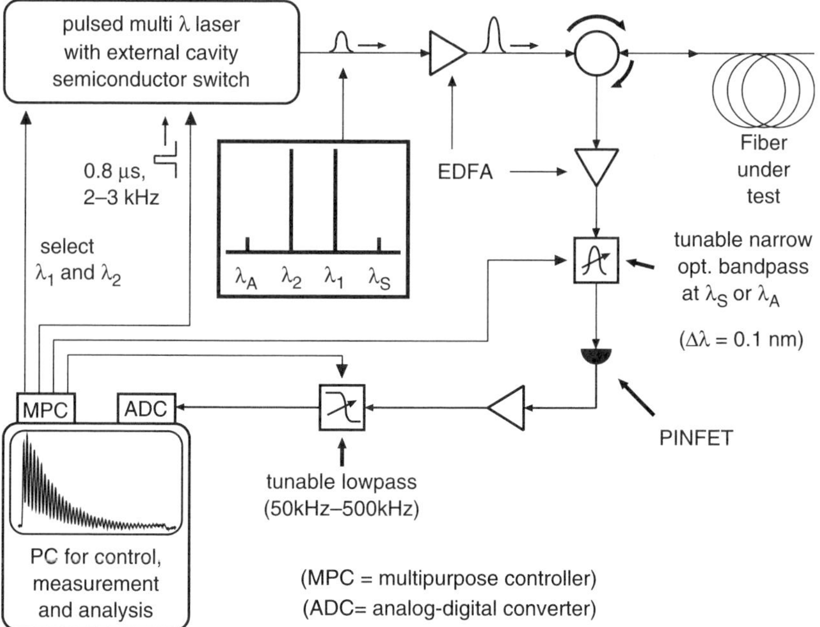

FIGURE 8.13 Schematic of the dispersion OTDR. Dual-wavelength pulses created by the multiwavelength laser are amplified by an EDFA to as much as 1-W power levels before being sent on to the fiber under test. The Rayleigh backscattered signals are directed by the circulator to a second EDFA before wavelength selection by an extremely narrow-band optical filter. For details of electronic processing of the signals following detection, see text. PINFET, Photodiode plus field-effect transistor.

with a data acquisition board. The data board contains 12-bit analog-to-digital converters having a maximum interlaced sampling rate of 20 MHz.

The computer is programmed to take a user-specified number of measurements and display the averaged signal. Typically, thousands of scans can be taken and averaged in just a couple of seconds, for a large increase in signal-to-noise ratio. A special, fast, and efficient algorithm was devised to calculate the dispersion and the dispersion parameter as a function of distance. The scheme of the algorithm is sketched in Fig. 8.14. The algorithm converts the linear oscillation given by the signal into a circular oscillation by taking its fast Fourier transform (FFT), eliminating the negative part of the symmetric spectrum, and then taking the inverse FFT. As shown in the figure, the resulting signal is a curve that spirals around in the complex plane. The absolute value of $D(z)$ is directly proportional to the curvature of this complex signal. Finally, the accumulated dispersion is computed as the integral of D. Even for the longest fibers the algorithm needs only a fraction of a second to compute the dispersion maps.

Strong four-wave mixing in the semiconductor amplifier switch of the source laser produces significant seed fields at the Stokes and anti-Stokes wavelengths. The seeds provide two important benefits. First, without the seed, from time to

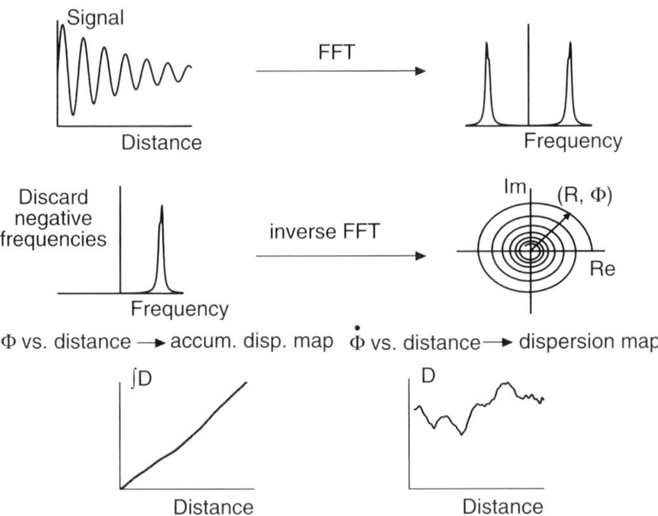

FIGURE 8.14 Computer algorithm for processing the signal. The signal is first transformed into the frequency domain, and its negative frequencies eliminated. The inverse transformation then yields a spiral in the complex plane, with time (or the equivalent distance) a parameter along the spiral. $\Phi(z)$ and its z derivative become the accumulated dispersion and $D(z)$, respectively.

time there arise situations where the Stokes vector rotates about a center at or near the origin of the complex plane, so that the required oscillations in the signal power tend to disappear. With the seed, however, the center of rotation is always well away from the origin, so that this catastrophe can never happen. Second, with the seed field, the instrument's sensitivity is increased, since in that case, the detected signal results from the beat between the relatively large seed field and the smaller FWM field generated in the fiber. With the seed field, the measurement range in dispersion-shifted fiber is extended to beyond 50 km, even without the help of Raman gain.

Results of a Few Sample Measurements

To begin with the simplest case, Fig. 8.15 shows the FWM signal and corresponding measured $D(z)$ for a 22-km length of standard SMF. Since D of standard SMF depends almost entirely on material properties, the measured $D(z)$ is essentially constant, as expected. Note the corresponding constant frequency of the FWM signal.

Figure 8.16 shows the FWM signal and corresponding measured $D(z)$ for a 24-km length of dispersion-shifted fiber. Note the considerable variation of D with z in this case, and the corresponding change in FWM frequency. To demonstrate the high degree of consistency of the measurements, this fiber was measured first from one end and then from the other. Note how very closely the two curves of $D(z)$ fit one another.

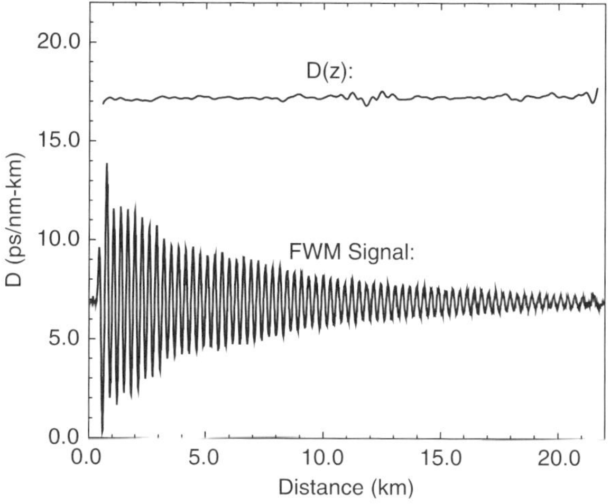

FIGURE 8.15 FWM signal and $D(z)$, as measured by the dispersion OTDR for a 22-km length of standard SMF.

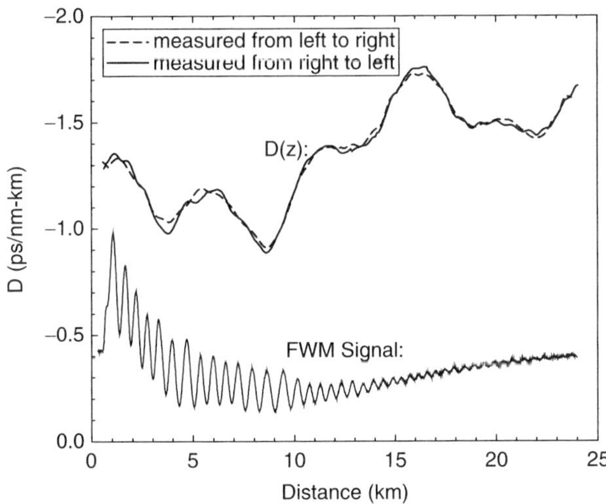

FIGURE 8.16 FWM signal and $D(z)$, as measured by the dispersion OTDR for a 24-km length of dispersion-shifted fiber.

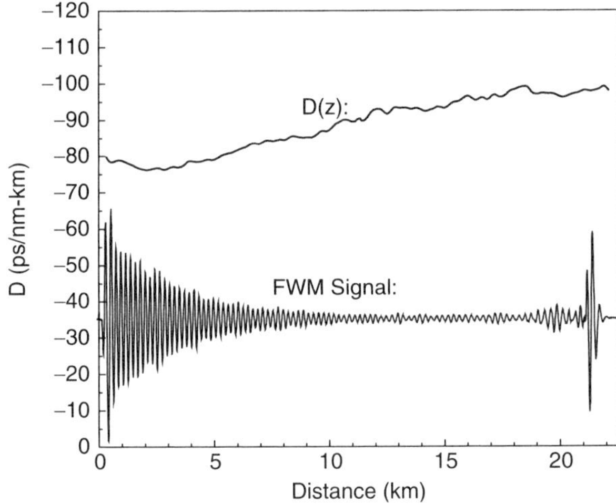

FIGURE 8.17 FWM signal and $D(z)$, as measured by the dispersion OTDR for a 22-km length of DCF. Backward Raman pumping produced the signal rise at the far end of the span.

The dispersion OTDR is equally effective with fibers of very high dispersion. Figure 8.17 shows the FWM signal and corresponding measured $D(z)$ for a 22-km length of DCF. There are several interesting points here. First, to make this measurement, the wavelength spacing $\delta\lambda$ had to be reduced to the smallest value available (0.6 nm), in order to keep the FWM oscillation frequency low enough.

Second, since the loss rate in this particular coil of DCF was exceptionally high at 0.51 dB/km, Raman gain from backward pumping was used to maintain the signal levels for a good S/N ratio over the entire 22-km length. Third, the high oscillation frequency shown here corresponds to a spatial resolution length of about 200 m.

Figure 8.18 shows a measurement of 75 km of a dispersion-shifted fiber with a loss coefficient of 0.21 dB/km. Since the average value of D was small and the distance long, a wavelength spacing $\delta\lambda = 2.4$ nm was chosen to have a good compromise between spatial resolution and signal strength. First the measurements were made on the entire span, with Raman pumping from both ends, and then on sub-spans of 25-km length each, without pumping. Note that without pumping, a measurement of the entire 75 km would not be possible, since in that case the signal power would decrease by an untenable 63 dB over the length of the span. From the nearly perfect fit of the two resultant sets of curves, we infer that the measurement uncertainty is small. For example, the accumulated dispersion deviates by less than ± 0.1 ps/nm over the first 25 km, by less than ± 0.2 over the next 25 km, and by less than ± 0.3 km over the last 25 km. This means that the accumulated dispersion has an exceedingly small relative uncertainty of $\pm 0.3\%$ over these distances. Finally, note that D varies from about -2 to -0.3 ps/nm-km, a ratio of ≈ 6 to 1! This huge variation shows how important it is for the construction of

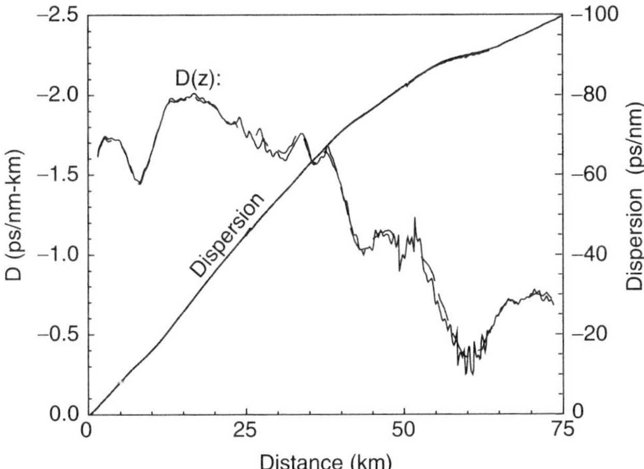

FIGURE 8.18 $D(z)$ and the net dispersion $[\int_0^z D(z)dz]$ as measured by the dispersion OTDR for a 75-km length of dispersion-shifted fiber, with both backward and forward Raman pumping used to keep the S/N ratio high enough over the entire span. Measurements of each of the three 25-km sub-spans, made separately and without Raman gain, are appropriately superimposed (dashed lines) on the 75-km measurements; note the excellent fit.

dispersion-managed systems to be able to measure the actual dispersion maps of the fibers.

8.5. Accurate Measurement of Pulse Widths Using a Detector with Finite Response Time

The attempt is often made to measure the temporal width and shape of a pulse with a fast detector and a sampling scope. The response of the detector and the following electronics often adds significant distortion and error, however, especially when the detector response time is a significant fraction of the pulse width. But if a microwave spectrum analyzer is available, the measurement can be made essentially independent of the detector response. The idea is first to calibrate the detector by measuring the microwave spectrum of the detector's response to a train of known, reference pulses, whose widths are preferably short on the scale of the detector response. The ratio of that measured response to the expected Fourier transform of the reference pulses is then stored, to be used as a set of normalizing factors. The spectral intensities of a given measurement are divided by those normalizing factors, and the adjusted spectrum is then fit to a spectrum of the suspected pulse shape and width. Admittedly, the technique has its limitations: Since the intensity detector loses all phase information, the measurements do not yield a true inverse Fourier transform. It cannot distinguish between symmetrical and asymmetrical pulse shapes. It yields no information about whether or not the pulse is chirped. Finally, it yields the mean behavior of an entire pulse train, and not of individual pulses. But when one can be reasonably sure on other grounds that the pulse is symmetrical, it can yield accurate measurement of the pulse intensity envelope shape and width. For that reason, and on account of its sensitivity, it was used successfully in a number of the first long-haul soliton transmission experiments [116–119].

It should be noted that the Fourier transform of a pulse's intensity envelope is significantly different from the pulse spectral densities that have been referred to earlier in this book. The latter represents the square of the absolute value of the Fourier transform of the amplitude function $u(t)$, a spectral distribution that is considerably narrower than the former. Table 8.1 and Fig. 8.19 attempt to make that very clear for the common cases of Gaussian and sech pulses.

Since the measurement involves the behavior of long pulse trains, the resultant spectrum reflects the influence of timing jitter, and it was used for the first successful measurement of Gordon–Haus jitter. The analysis for any jitter with a Gaussian distribution is as follows: The detector current $I(t)$ (proportional to the

TABLE 8.1 Pulse Shapes, Their Spectral Densities, and Fourier Transforms of Their Intensity Envelopes Compared[a]

Pulse shape	Intensity ($\lvert u(t)\rvert^2$)	Spectral density ($\lvert\tilde{u}(f)\rvert^2$)	Fourier transform of $\lvert u(t)\rvert^2$
Gaussian	$e^{-(1.665t/\tau)^2}$	$e^{-(2\pi f\tau/1.665)^2}$	$e^{-(\pi f\tau/1.665)^2}$
sech	$\mathrm{sech}^2(1.763t/\tau)$	$\mathrm{sech}^2(\pi^2 f\tau/1.763)$	$(\pi^2 f\tau/1.763)\mathrm{csch}(\pi^2 f\tau/1.763)$

[a]As usual, τ is the pulse intensity FWHM.

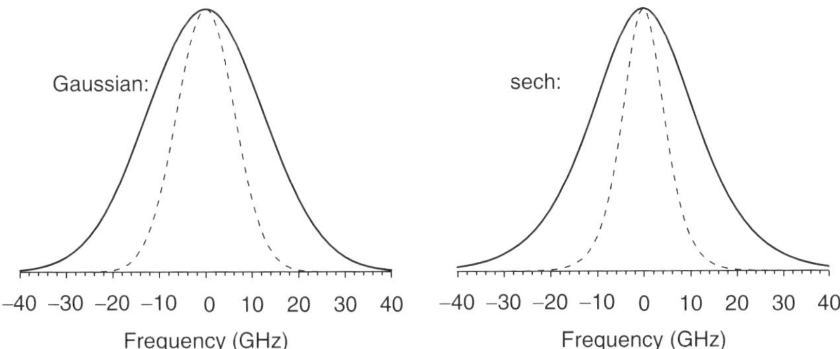

FIGURE 8.19 Fourier transform of intensity envelope (solid curve) and spectral density (dashed curve) for unchirped, $\tau = 30$ ps (intensity FWHM) Gaussian (left) and sech (right) pulses.

optical intensity) can be written as $\sum_{n=1}^{N} I_1(t - t_n)$, where I_1 is the contribution from a single pulse and N is the number of pulses in the train. Its Fourier transform is then

$$\tilde{I}(f) = \int I(t)\exp(i2\pi ft) = \tilde{I}_1(f)\sum_{n=1}^{N}\exp(i2\pi ft_n), \tag{8.18}$$

where f is the microwave frequency. Now, let $t_n = nT + \tau_n$, where T is the expected period between pulses and τ_n is the random Gaussian timing jitter. Furthermore, note that for a long pulse train, $\tilde{I}(f)$ is significantly large only for the harmonic series $f_m \equiv m/T$. Then Eq. (8.18) becomes

$$\tilde{I}(f_m) = \tilde{I}_1(f_m)\sum_{n=1}^{N}\exp(i2\pi f_m\tau_n). \tag{8.18a}$$

Now (as has been noted earlier several times in this book), for a random Gaussian variable x, with variance σ^2, one has $\langle\exp(i\lambda x)\rangle = \exp\left(-\frac{1}{2}\lambda^2\sigma^2\right)$. Thus, the

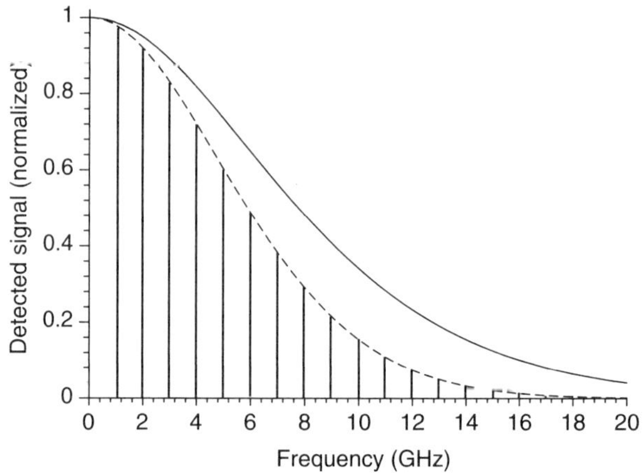

FIGURE 8.20 Microwave spectra of 1-GHz pulse repetition rate pulse trains. Solid curve: Theoretical spectral envelope for unjittered, $\tau = 50$ ps, sech^2 intensity envelope pulse train. Dashed curve: Expected spectral envelope for the same pulse train, but jittered, with $\sigma_{GH} =$ 20 ps. The vertical lines show the discrete spectrum of the pulse train.

microwave spectral intensity of the detector's output is

$$\langle \tilde{I}(f_m) \rangle = N\tilde{I}_1(f_m)\exp\left[-\frac{1}{2}(2\pi f_m \sigma)^2\right]. \tag{8.19}$$

Figure 8.20 shows the effect of $\sigma = 20$ ps Gordon–Haus jitter on the microwave spectrum of the intensity envelope of a train of $\tau = 50$ ps pulses. The figure approximately reproduces one typical result from a real experiment [119]. (In the real experiment, the pulse repetition rate was only 200 MHz, so the density of harmonics was 5× as great, but that has no effect whatsoever on the shape of the spectral envelope.)

Finally, it should be noted that this technique has become something of a fixture in recirculating loop experiments as a monitor of the health of the recirculating pulse train. In a transmission with a bit rate of 10 Gbit/s, for random data the harmonic series is at 10, 20, 30, ... GHz, so there tend to be only a few harmonics within the range of the microwave spectrum analyzer (typically 0–22 GHz, although 0–50-GHz instruments have recently become available). Nevertheless, simple monitoring of the strength of the 10-GHz- or especially the 20-GHz-harmonic as a function of time (hence as a function of distance) as each transmission progresses can yield immediate feedback on whether or not the solitons are being maintained at proper energy levels, and on whether or not they are significantly affected by timing jitter. (Harmonic strength that does not droop is best.)

8.6. Flat Raman Gain for Dense WDM

The purpose of this section is simply to show how relatively easy it is to obtain flat Raman gain for dense WDM. We begin with the pumping scheme used for an all-Raman amplified recirculating loop used for dense WDM experiments that immediately preceded those described in Chapter 6, Section 6.2. It was a 480-km loop containing six 80-km spans of dispersion-shifted fiber (DSF), each compensated by DCF. The pumping scheme is shown in Fig. 8.21. Note that there are only six lasers for each span, three for the DCF and three for each 80-km length of DSF. For the main span, a pair of longer wavelength pump lasers are polarization-multiplexed together before being combined by a WDM coupler with a short-wavelength pump laser. For the DCF, the arrangement is reversed, with two shorter wavelength lasers polarization-multiplexed together before being wavelength-multiplexed with the longer wavelength laser. It should be noted that the shorter wavelength lasers play a dual role here, since they act as pumps for the longer wavelength Raman pumps, thereby helping those longer pump wavelengths to penetrate deeper into the fiber, as well as to supply gain directly to the shorter signal wavelengths. The lasers were all simple Fabry–Perot cavity lasers with no external tuning element, so the lasing well above threshold tended to occur in a band of many modes centered at the peak of the laser's gain band. Since the $c/(2L)$ mode spacing of such lasers is typically ~30 GHz, or about 0.2 nm at 1450 nm, the 6- to 8-nm-wide lasing bands typically contained on the order of 30 or more frequencies (see Fig. 8.22). It is very important that the laser's power (~200 mW) be spread out in that way, since the threshold for stimulated

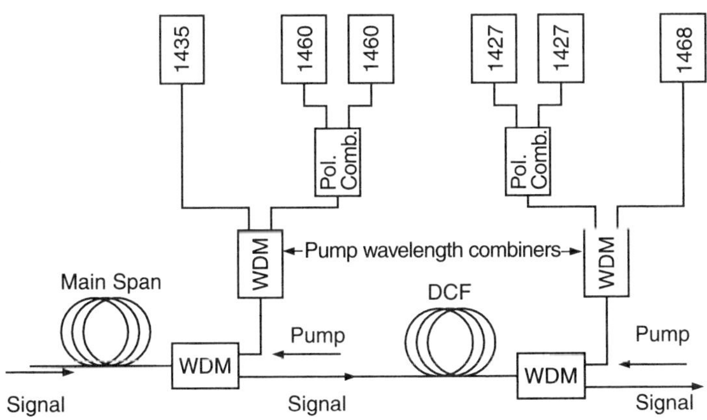

FIGURE 8.21 All-Raman pumping scheme for 80-km spans of dispersion-shifted fiber compensated by DCF.

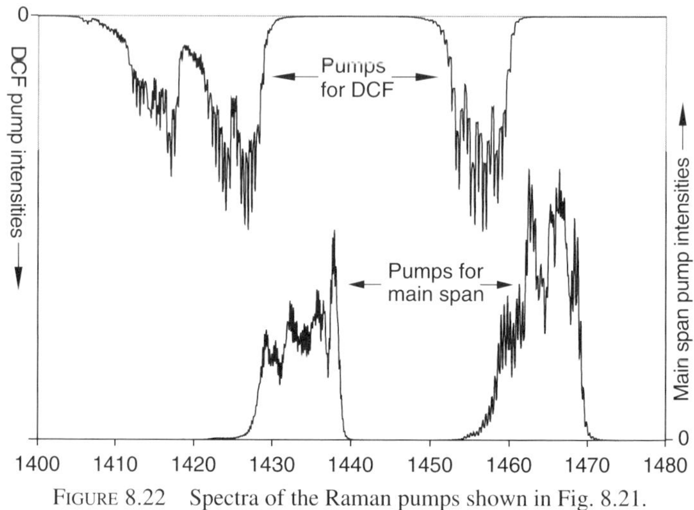

FIGURE 8.22 Spectra of the Raman pumps shown in Fig. 8.21.

Brillouin backscattering in the fibers of interest here is typically no more than about 10–15 mW for light contained in the Brillouin scattering bandwidth of just a few megahertz. (Stimulated backscattering, if it were allowed to occur, would act like a mirror to prevent the Raman pump light from significantly penetrating the fiber.) The spread of the laser's output over many frequencies also makes it easy to thoroughly depolarize the pump light simply by passing it through an appropriate length of PM fiber whose axis is skewed with respect to the original linear polarization. That is, acting as a multiple wave plate, the PM fiber rapidly spreads the polarization states of the various modes over the Poincaré sphere.

It is also handy that the center of the lasing band can be temperature tuned over a considerable range ($\sim\pm 10$ nm), because that much tuning is very helpful for the achievement of flat gain. (Thus, the wavelengths listed for the lasers in Fig. 8.21 tend to be nominal only.) With a bit of such tweaking of the mean wavelengths of the individual lasers, the resultant gain band can be remarkably flat (see Fig. 8.23). Note that over a gain band a bit wider than the erbium C band, the net gain variation is no greater than about ± 0.5 dB out of a total gain (for the six spans of the loop) of 132 dB.

More lasers are needed as the WDM band becomes wider. Figure 8.24 shows the arrangement used for pumping of the loop for the WDM experiments described in Chapter 6, Section 6.2. Note that there are now eleven lasers altogether for each 100-km span, comprising four lasers for each of the two 50-km sub-spans, and three more for the DCF coil. Otherwise, the details are much as described above

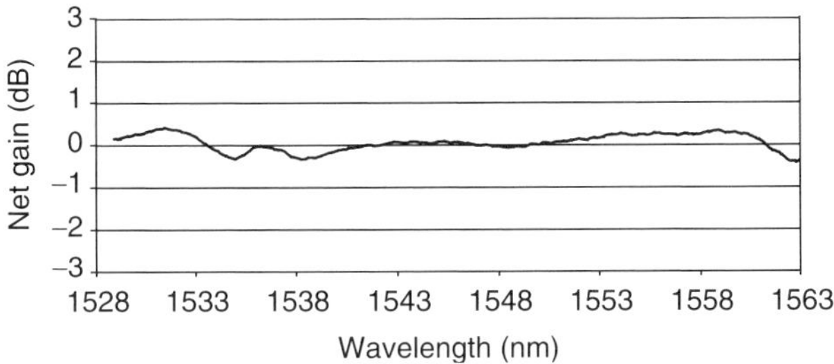

FIGURE 8.23 Net gain (gain − loss) vs. wavelength for six spans using the Raman pumping scheme shown in Fig. 8.21. The total gain here is 132 dB.

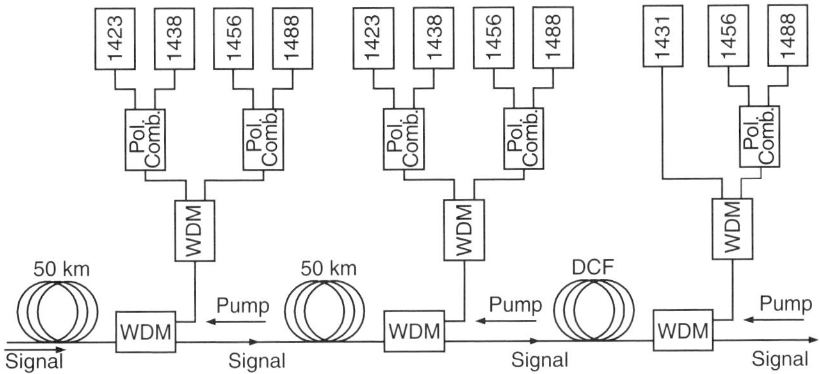

FIGURE 8.24 All-Raman pumping scheme for 100-km spans of dispersion-shifted fiber, mid-span back pumped, and compensated by DCF.

for the scheme of Fig. 8.21—that is, lasers of similar wavelength first polarization-multiplexed together, followed by wavelength multiplexing of the shorter with the longer wavelength groups. Also, as before, the listed wavelengths are nominal only.

Figure 8.25 is a curve of the net gain (this time out of 170 dB) of the six-span loop versus wavelength for a temperature tuning of the lasers designed to achieve the maximum possible WDM bandwidth. Although the gain variation is now bigger ($\approx\pm1.5$ dB), the WDM band is twice as wide ($\approx$60 nm). The situation shown here is *not* that obtained for the experiments in Section 6.2, however; there, the lasers were tuned somewhat closer together, and all to longer wavelengths, to best support the somewhat narrower WDM band ($\approx$1549–1594 nm) of Fig. 6.24. As a result of the closer pump wavelength spacing, the gain dip in the middle of the WDM band was not as deep as that seen in Fig. 8.25.

For any given WDM band, the set of wavelengths and powers for the various pump lasers was initially determined from a calculus designed to achieve minimum

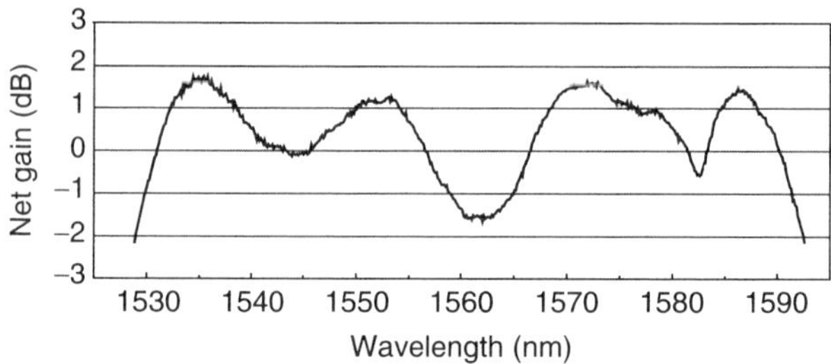

FIGURE 8.25 Net gain (gain − loss) vs. wavelength for six spans using the Raman pumping scheme shown in Fig. 8.24. The total gain here is 170 dB.

possible gain variation. Although beyond the scope of this book, that calculus has been fully described elsewhere [120]. It took into account the Raman interactions among all of the pump and signal wavelengths and the change in fiber-effective cross-section with wavelength, such that the signal gain was always calculated from correct profiles of the pump powers as functions of distance. Although the computed parameters almost always produced a gain profile close to that predicted, the final configuration almost always involved a certain amount of empirical tweaking.

8.6.1. Gain Flatness Achievable from a Continuum of Pump Wavelengths

We have just seen how a handful of pump wavelengths can produce a Raman gain profile having remarkably small ripple. A very natural question then is, if the number of pump wavelengths could be greatly increased, perhaps even to form an effective continuum, could the gain flatness be improved, and if so, by what factor? There is a straightforward answer to that question in the context of the so-called "Raman smart pump"[121]: Imagine a pump laser whose optical frequency is swept back and forth periodically so rapidly that all parts of a counter-propagating signal would tend to experience the same spatially averaged effect from the pump, and where the pump power spectral density, $P_p(\nu_p)$, is determined by the relative amount of time spent at each frequency. Note that this scheme completely eliminates the potential for complicating Raman interaction among the pump frequencies, since each is spatially separated from all others as it travels down the fiber. The total gain $g(\nu_s)$ (ν_s is the signal frequency) can then be very simply written as

$$\ln g(\nu_s) = \int G(\nu_p - \nu_s) P_p(\nu_p) d\nu_p, \tag{8.20}$$

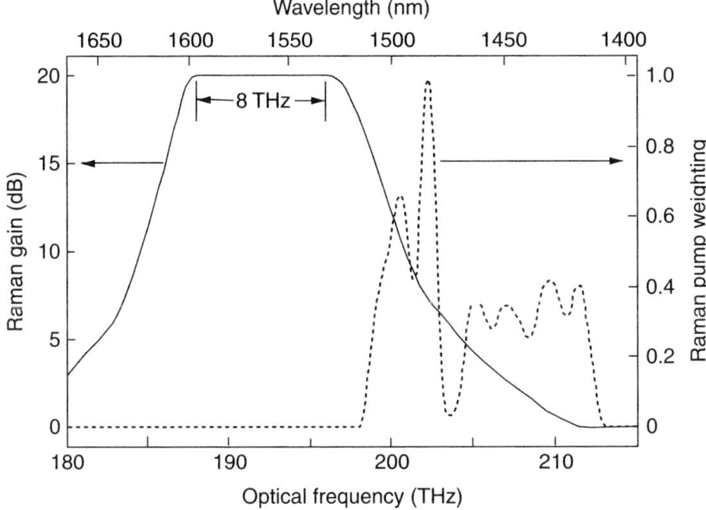

FIGURE 8.26 Illustrating the potential performance of a Raman smart pump. Dashed curve: Computed pump spectral intensity. Solid curve: The resultant Raman gain. Gain ripple in the flat region is less than 0.01 dB peak-to-peak.

where $G(\nu_p - \nu_s)$ is the Raman gain coefficient like that shown in Fig. 3.3, and where $P_p(\nu_p)$ has already been multiplied by the appropriate exponential decay factor and integrated over z. In principle, Eq. (8.20) can be solved for the distribution $P_p(\nu_p)$ required for a given $g(\nu_s)$ (such as flat gain over a particular frequency range), by making use of the fact that the Fourier transform of the convolution of the product of two terms is just the simple product of the Fourier transforms of the individual terms. For flat gain, however, the need to specify reasonable skirts to the flat region provides a significant complication. Thus, in practice, it has turned out to be better to solve the problem "brute force" by repeatedly adjusting the relative powers of a large number of discrete pump frequencies on a uniform grid until the desired gain distribution is attained. (Although it sounds tedious, the necessary number of iterations can be carried out on a fast PC in no more than a minute or two.) Figure 8.26 shows the results of one of many such computations [120, 121]. Note the apparently dead flat gain region of approximately 8-THz width, or enough space for 160 WDM channels at the standard spacing of 50 GHz. An expanded view of that region reveals a peak-to-peak gain ripple of less than 0.01 dB, representing only 0.05% variation out of the nominal 20-dB gain. Similar computations reveal that the 15-THz-wide span of pump frequencies required here can be reduced to less than 12 THz, with only an approximately twofold increase in gain ripple, and that a 5-THz-wide flat gain region requires only a 10-THz span of pump frequencies. Still further computations reveal that just about any reasonable gain profile could be produced by the smart pump. Thus, for example, one could also have a gain linearly rising with frequency at just the right rate to compensate for the Raman interaction between signal frequencies, or an even more complex gain profile to compensate for gain ripple produced elsewhere.

The catch here is the fact that, to date at least, no one seems to know how to produce the high-powered (many hundreds of milliwatts) pump lasers capable of being rapidly (at rates on the order of 1 MHz) tuned over the frequency ranges required for the smart pump. Nevertheless, even with present technology, it might be possible to produce a lower powered version to provide convenient correction to system gain ripple. In any event, the idea of the smart pump is too intriguing to be ignored.

8.6.2. A "Radical" Proposal

In the very conservative telecommunications industry, tradition dies hard. The issue here is the "amplifier huts" used to house the amplifiers and other hardware needed periodically along a long-haul transmission line. A hardcover training manual, dated 1938, for employees of the AT&T "long-lines" department, shows several photographs of such "huts," which are in fact most substantial one- and two-story brick buildings! In those days, the huts housed vacuum tube amplifiers for the open-air lines and coaxial cables carrying signals primarily at sub-megahertz frequencies. Another photo shows the interior of a hut, complete with thousands of pounds of lead-acid storage batteries and rack after rack of amplifiers and other gear. With the advent of optical transmission, except for replacement of the vacuum tube amplifiers with semiconductor-based optical repeaters, apparently things did not change very much. Clearly, all of that real estate represents a costly investment that is equally costly to maintain. It is no wonder, then, that a major holy grail of the industry was to make the spacing between such "repeater" locations as great as possible. That tradition and the mind-set that goes with it tend to remain to this day. Hence the designation "radical" for the modest proposal we make here.

The proposal is that in the context of all-optical transmission, the huts represent a costly and unnecessary anachronism. They should be replaced with underground manholes, no more than about 2-m deep, that would house little more than a few Raman pump modules (see Fig. 8.27). Since the pump modules would consume just a few tens of watts each (including power for thermoelectric cooling), they could be powered from an electrical cable of modest size, buried along with the fibers, and carrying DC voltage just low enough (say 1 kV) to allow for easy transformation to the few volts required for the modules.

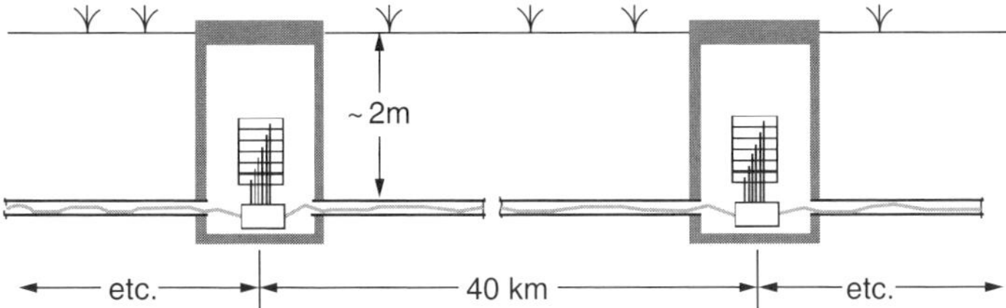

FIGURE 8.27 Manholes to replace amplifier huts. The arrangement shown here would span the typical 400- to 600-km distance between node points.

There would be many advantages, both technical and economic, from such an arrangement. First, the low cost of the manholes would enable spacings of no more than about 40 km. By thus allowing for backward Raman pumping with such close spacing, as we have already seen, the noise penalty would become essentially zero, and nonlinear penalties would tend to be reduced. Additionally, since the required Raman pump power scales in direct proportion to the distance between points of pump injection, the cost of the modules would go down and their reliability would increase. The temporary loss of any one module would tend not to be fatal, since the modest gain lost (about 8 dB) could be easily made up by increased output from a few surrounding modules. Second, the thermal environment a few meters underground tends to be stable and benign (cool). This would make cooling of the pump modules easier, and fiber dispersion would no longer be subject to significant thermal variation. (Major temperature changes are particularly problematic when DCF modules are involved.) Also, of course, no air conditioning would be required! Finally, the arrangement would tend to allow for more flexible and efficient dispersion map design: Map periods would no longer be rigidly fixed to the present 80- to 100-km hut spacing, thereby facilitating proper dispersion-managed soliton transmission at bit rates >10 Gbit/s. Lossy and expensive DCF modules might be eliminated altogether, so maps would use the more efficient design employing transmission spans with D values of alternating signs. In short, terrestrial system design would tend to gain much, if not all, of the flexibility enjoyed in undersea systems, but with vastly greater ease of monitoring, maintenance, and access for repair.

Appendix A

A Sample Maple Program for the ODE Method

A.1. Introduction

The following pages are a printout of a program for efficient numerical computation of pulse behavior in a dispersion map by the ODE method described in Chapter 2, Section 2.1.4. The program was written by Bell Labs colleague Jürgen Gripp in the late 1990s and is based on the Maple mathematics software package. It has since proved to be an extremely handy tool for exploration of the behavior of dispersion-managed solitons, and for map design. (Many of the graphs presented in Chapter 2 were generated with this program.)

Following is an outline of the program sections:

- Section 1 (Initialization) simply sets up a few general computing parameters and defines a few physical constants.
- Section 2 (Fundamental Input Parameters) allows the user to specify details of the dispersion map, viz., the number of different fiber segments in the map, and for each segment, its length, its D value, effective core area, loss coefficient, and Raman gain coefficient. Usually, only the length of the + (or −) D fiber is specified, and the program then calculates the length of the other type to provide a user-specified $\bar{D}$. In the convention used here, the fiber segments are numbered 1, 2, 3, ..., from left to right, i.e., in order of increasing distance z. Provision is also made for specifying splice losses, and the free spectral range of an etalon filter if one is used. Here also the user specifies an unchirped pulse width.

- Section 3 (Calculate Derivative Parameters) calculates quantities such as z_c, the characteristic dispersion length [Eq. (2.8)], the total map length, and the map strength parameter S [Eq. (2.17)].
- Section 4 (Calculate Intensity Profile with and without Raman Gain) establishes analytic formulas for calculating the Raman pump and signal powers as a function of distance. In the sample program shown here, the Raman pumping is all in the "backward" direction, i.e., the pump light enters the right end of each segment and travels left, counter to the signals. (The program can and has been modified to allow for forward Raman pumping, and for lumped amplifiers as well. For the case of lumped amplifiers, the intensity profiles are, of course, just decaying exponentials.)
- Section 5 (Define and Plot Intensity Profile for a Given Raman Pump Power) computes actual pump and signal powers as functions of z from a user specified set of input pump powers. (The user must keep changing the set of pump powers until he gets the desired signal power profile.) This section sets the stage for computation of the quantity $K(z)$ [Eq. (2.12)] in the next section.
- Section 6 (Calculate Pulse Evolution) iteratively applies the Maple routine "dsolve" to ODE Eq. (2.6), and adjusts the signal power and input pulse chirp parameter until the pulse at the output of the map essentially matches that at its input.
- Section 7 (Calculate Graphs) does just that, as functions of z, for whatever subset one wants from the pulse's energy, width, bandwidth, chirp parameter, as well as q and q_{NL}. Other requested Maple output is shown in italics.

As was shown in Chapter 2, Section 2.2.4, the location of either of the zero chirp points is a function of the intensity profile, and hence is unknown before the ODE has been solved. Thus, for whatever map starting point is chosen, the program must try different values of the chirp parameter for that starting point as well as different pulse energies, until a satisfactory match is found between the pulses at map output and map input. As long as the starting point is not too close to the beginning or end points of a $+D$ or $-D$ fiber span, the program always converges very quickly to a solution. On the other hand, if it is at or too close to an end point, the program tends to become unstable and dsolve cannot find a solution. Thus, it is usually helpful to start the program at or near the mid-point of a span, as was done in the sample shown here, where the length of DCF (which in reality is all one piece) was divided in two. Note that the Raman pump powers at the inputs to segments 1 and 4 (the two halves of the DCF coil) are in a ratio determined by the loss in the first-pumped segment of the DCF.

The guiding etalon filter is in transmissive mode, and R is the intensity reflection coefficient common to both mirrors. The program as written always places the filter at the very far end of the map. Although the curve of transmission vs. frequency of a typical etalon guiding filter tends to be shallow and almost sinusoidal (see Fig. 3.28), consistent with the assumptions of the ODE method, the program replaces the actual curve with a Gaussian whose curvature matches that of the etalon at the transmission peak (see Fig. 6.25). In the program, the filter also does not impose a frequency-dependent delay, or phase shift, as real filters do. Thus, in the program, the only effect of the filter is to produce a step change in the bandwidth of the pulse's (Gaussian) spectrum (see, for example, Fig. 2.20). Nevertheless, because of the fact that the pulse bandwidth is usually less than about 30% of the etalon's free spectral range (which is always made equal to the WDM channel spacing), and because the dispersive effects of the filter tend to be small and of odd (mostly third) order, the program computes the principal effects of the filter with reasonable accuracy.

Finally, the first two lines of section 2 (beginning "DeleteOld") allow the user to choose between clearing out the old data when beginning a new program, or saving data in order to create multiple-curve plots. For example, it is often useful to run the program for a given map, but with a sequence of different values of the unchirped pulse width; in that case, choose the option "DeleteOld := No" by commenting out the other. In that way, each successive run will produce yet another curve in each of the plots.

A.2. The Maple Program

Initialization

```
restart:
Digits := 12:
colorcode := [red, green, blue, magenta, cyan, green, black, blue]:
with(plots) : with(linalg):
setoptions(titlefont=[TIMES, ROMAN, 14], axesfont= [TIMES, ROMAN, 14], labelfont=[TIMES, ROMAN, 14]):
pi := evalf(Pi):                              # establishing a numerical value for Pi
lambda := 1550.0:                             # wavelength (nm)
c :=     2.9979e5:                            # speed of light (km/s)
n2 :=  2.6e-20:                               # nonlinear index (m^2/W)
Cs :=   2.0 * log(2.) * lambda^2 / pi / c:    # proportionality factor for dispersion length (m*s*1e27)
```

Fundamental Input Parameters

```
#DeleteOld := No;
DeleteOld := Yes;                             # plot several calculations with colorcode
Tau:=  30:                                    #unchirped pulsewidth (ps)
FSR :=  50:                                   # free Spectral Range of etalon filter (GHz)
D_TW   := 7:
D_DCF  := -100:
D_bar  := 0.15:                               # path average dispersion parameter
##### the following lines depend on the design of the dispersion map ###########
Num := 4:                                     # number of different fiber segments in the map
Length  := 100+L_DCF:                         # total map length
L_DCF1  := -50*(D_TW - D_bar) / D_DCF;
L_DCF2  := -50*(D_TW - D_bar) / D_DCF;
L_DCF  := L_DCF1 + L_DCF2;
L_TW1 := 50;
L_TW2 := 50;
LineTkns := [2, 3, 2, 3]:                     # line thickness in plots
Len := [L_DCF1, L_TW1, L_TW2, L_DCF2];        # lengths of individual pieces (km)
Disp := [D_DCF, D_TW, D_TW, D_DCF]:           # dispersion parameter (ps/nm-km)
Aeff := [19, 50, 50, 19]:                     # fiber effective core area (um^2)
Alpha := [0.7, 0.21, 0.21, 0.7]:              # loss coefficients (dB/km)
Gain := [160, 52, 52, 160]:                   # Raman gain coefficients (um^2/W-km)
Apply_G1 := [1, 0, 0, 0]:                     # where Raman pump 1 enters
Apply_G2 := [0, 1, 0, 0]:                     # where Raman pump 2 enters
Apply_G3 := [0, 0, 1, 0]:                     # where Raman pump 3 enters
Apply_G4 := [0, 0, 0, 1]:                     # where Raman pump 4 enters
Spl_loss := [0.0, 0.0, 0.0, 0.0, 0.0]:        # splice loss (dB)
```

```
            DeleteOld := Yes
            L_DCF1  :=3.42500000000
            L_DCF2  :=3.42500000000
            L_DCF   :=6.85000000000
            L_TW1   :=50
            L_TW2   :=50
      Len := [3.42500000000, 50, 50, 3.42500000000]
```

Calculate Derivative Parameters

```
alpha_s := Alpha * ln(10.0) / 10.0:                # loss (1/km)
alpha_p := alpha_s * (1550.0 / 1465.0)^4:          # loss (1/km)
zc := map(x -> Tau^2 / Cs / x, Disp):              # characteristic dispersion length (km)
spl_loss := Spl_loss * ln(10.0) / 10.0;
Spl_pos[1] := 0:
spl_pos[1] := 0:
for n from 1 to Num do:
  len[n] := Len[n] / zc[n];                        # normalized piece length
  Spl_pos[n + 1] := Spl_pos[n] + Len[n]:
  spl_pos[n + 1] := spl_pos[n] + len[n]:
od:
Length:= Spl_pos[Num + 1];                         # total span length (km)
D_bar:= dotprod(Disp, Len) / (Length-L_DCF);       # average dispersion (ps/nm-km)
S  := add(abs(len[n]), n=1..Num)/2;                # map strength
```

Length := *106.850000000*
D_bar := *0.150000000000*
S := *2.72100353831*

Calculate Intensity Profile with and without Raman Gain

```
Ip1[Num + 1] := (n, z) -> 1:
for n from Num by -1 to 1 do:       # calculate 1. pump intensity profile
  Ip1[n]:=(n,z)->(exp(-spl_loss[n+1]*Apply_G1[n]))*Ip1[n+1](n+1,Spl_pos[n+1])
         * exp(Apply_G1[n]*alpha_p[n]*(z-Spl_pos[n+1])):
od:
Ip2[Num + 1] := (n, z) -> 1:
for n from Num by -1 to 1 do:       # calculate 2. pump intensity profile
  Ip2[n]:=(n,z)->(exp(-spl_loss[n+1]*Apply_G2[n]))*Ip2[n+1](n+1,Spl_pos[n+1])
         * exp(Apply_G2[n]*alpha_p[n]*(z-Spl_pos[n+1])):
od:
Ip3[Num + 1] := (n, z) -> 1:
for n from Num by -1 to 1 do:       # calculate 3. pump intensity profile
  Ip3[n]:=(n,z)->(exp(-spl_loss[n+1]*Apply_G3[n]))*Ip3[n+1](n+1,Spl_pos[n+1])
         * exp(Apply_G3[n]*alpha_p[n]*(z-Spl_pos[n + 1])):
od:
Ip4[Num + 1] := (n, z) -> 1:
for n from Num by -1 to 1 do:       # calculate 4. pump intensity profile
  Ip4[n]:=(n,z)->(exp(-spl_loss[n+1]*Apply_G4[n]))*Ip4[n+1](n+1,Spl_pos[n+1])
        * exp(Apply_G4[n]*alpha_p[n]*(z-Spl_pos[n+1])):
od:
Is[0] := (n, z) -> 1:
for n from 1 to Num do:       # calculate signal intensity profile (without Raman gain)
  Is[n]:=(n,z)->(exp(-spl_loss[n]))*Is[n-1](n-1,Spl_pos[n])
        *exp(-alpha_s[n]*(z-Spl_pos[n])):
od:
Is_RG[0] := (n, z) -> 1:
for n from 1 to Num do:       # calculate signal intensity profile (with Raman gain)
  Is_RG[n] := (n, z) -> (exp(-spl_loss[n])) * Is_RG[n - 1](n - 1, Spl_pos[n])
         * Is[n](n, z) / Is[n](n, Spl_pos[n])
    * exp(Gain[n] / 8e4 * (Apply_G1[n] * Wpump1 * Ip1[n](n, Spl_pos[n])
                 + Apply_G2[n] * Wpump2 * Ip2[n](n, Spl_pos[n])
                 + Apply_G3[n] * Wpump3 * Ip3[n](n, Spl_pos[n])
                 + Apply_G4[n] * Wpump4 * Ip4[n](n, Spl_pos[n]))
            / alpha_p[n] * (exp(alpha_p[n] * (z - Spl_pos[n])) - 1)):
od:
```

Define and Plot Intensity Profile for Given Raman Pump Power

```
Wpump1:= 140.0;                                  # pump power in mW
Wpump2:= 237.0 ;                                 # pump power in mW
Wpump3:= 237.0 ;                                 # pump power in mW
Wpump4:= 81.0;                                   # pump power in mW
zmax := 100:                                     # step size for calculation of average signal power
P_Ip1:=[]: P_Ip2:=[]: P_Ip3:=[]: P_Ip4:=[]:      # initialize Raman pump plots as empty sets
P_Is:=[]:                                        # initialize signal plot without Raman pump as empty set
P_Is_RG:=[]:                                     # initialize Raman pump plots as empty sets
n:= 'n':                                         # unassign variable n (required for sum)
W_ave:=sum(add(Is_RG[n](n,z/zmax*(Spl_pos[n+1]-Spl_pos[n])+Spl_pos[n]),z=1..zmax)
     /zmax * (Spl_pos[n+1] - Spl_pos[n]), n=1..Num) / Length;
for n from 1 to Num do:                          # add pieces to sets of plots
  P_Ip1:= [op(P_Ip1), plot(10*log10(Ip1[n](n,z)), z=Spl_pos[n]..Spl_pos[n+1], color= blue)]:
  P_Ip2:= [op(P_Ip2), plot(10*log10(Ip2[n](n,z)), z=Spl_pos[n]..Spl_pos[n+1], color= green)]:
  P_Ip3:= [op(P_Ip3), plot(10*log10(Ip3[n](n,z)), z=Spl_pos[n]..Spl_pos[n+1], color = cyan)]:
  P_Ip4:= [op(P_Ip4), plot(10*log10(Ip4[n](n,z)), z=Spl_pos[n]..Spl_pos[n+1], color= red)]:
  P_Is:=  [op(P_Is), plot(10*log10(Is[n](n,z)), z=Spl_pos[n]..Spl_pos[n+1], color = red,thickness=2)]:
  P_Is_RG:=[op(P_Is_RG),plot(10*log10(Is_RG[n](n,z)), z=Spl_pos[n]..Spl_pos[n+1], y=-8.0..2.0,
    color=blue,thickness=4)]:
od:
P_ave:= plot([z, 10*log10(W_ave), z=0..Length], color=blue):
display(P_Is_RG, P_ave);                         # plot signal with and without Raman pump
#display(P_Ip1, P_Ip2,P_Ip3);                    # plot Raman pumps
Is_end_SF:=Is_RG[Num](Num, Spl_pos[Num]):        # relative signal power at end of span
```

Wpump1 := 152.0
Wpump2 := 236.0
Wpump3 := 236.0
Wpump4 := 76.0
W_ave := 0.755827251687

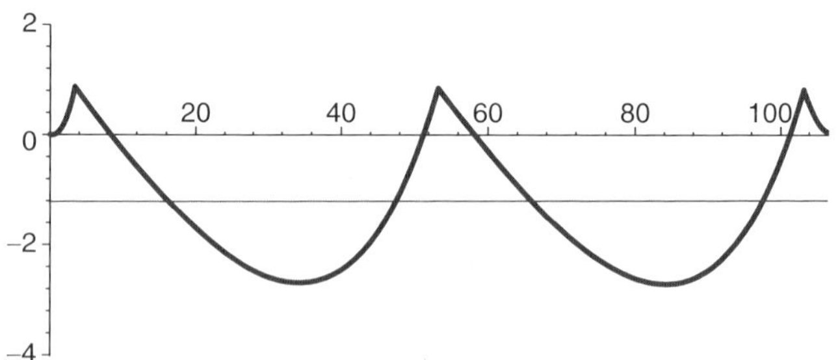

Calculate Pulse Evolution (this adjusts pulse energy to get periodic pulse width) with Target Mirror Reflectivity

```
################################################################################
final_R := 0.0:
initial_chirp := 0:     # adjust prechirp to get wanted mirror refl. (next paragraph)
delta_chirp := 0.03:
R0  := 0.0:             # initial mirror rcflcctivity
K0 := -0.012:           # initial pulse energy (negative if first dispersion negative!)
R := 1e6:
eps_R := 0.001:
loop_no := 1:
while (abs(final_R - R) > eps_R) do:
print(loop_no, `initial chirp`, initial_chirp);
################################################################################
dK := 0.01:             # step size
eps := 0.0001:          # precision
Iq_end[3] := 2 * eps:
while (abs(Iq_end[3]) > eps) do:
 for m in [-1, 1, 0] do:
  K := K0 + m * dK:
  ################# calculate prechirp ##########################
  if (abs(initial_chirp) > 0.001) then
   Isignal:=z->Is_RG[1](1, zc[1]*z)/Is_RG[1](1,0);
   ode:={diff(Rq(z),z)=K*Isignal(z)*2*Rq(z)*Iq(z)*(Rq(z)/(Rq(z)^2+Iq(z)^2))^1.5,
    diff(Iq(z),z)=1-K*Isignal(z)*(Rq(z)^2-Iq(z)^2)*(Rq(z)/(Rq(z)^2+Iq(z)^2))^1.5,
        Rq(0)=1,Iq(0)=0};
   q_ini:=dsolve(ode,{Rq(z),Iq(z)}, type=numeric,output=listprocedure):
   R_ini:=subs(q_ini,Rq(z)):       # Re(q) is proportional to 1/BW^2
   I_ini:=subs(q_ini, Iq(z)):      #Im(q) is proportional to the dispersion
   f_chirp := -I_ini / (R_ini^2+I_ini^2):
   z_zero:=-initial_chirp:
   eps_z_zero:=abs(initial_chirp / 100.0):
   while (abs(f_chirp(z_zero)-initial_chirp) > eps_z_zero) do:
    z_zero:=z_zero*initial_chirp / f_chirp(z_zero);
   od:
  else
   z_zero:= 0;
   R_ini(z_zero):= 1:
   I_ini(z_zero):= 0;
  fi:
  Rq_ini:= R_ini(z_zero);
  Iq_ini:= I_ini(z_zero);
  ############ end of prechirp calculation #####################
```

```
    for n from 1 to Num do:
      Isignal:=z->Is_RG[n](n,zc[n]*(z-spl_pos[n])+Spl_pos[n]) / Is_RG[n](n, Spl_pos[n]);
      ode:={diff(Rq(z),z)=K*Isignal(z)*2*Rq(z)*Iq(z)*(Rq(z) / (Rq(z)^2 + Iq(z)^2))^1.5,
        diff(Iq(z),z)=1-K*Isignal(z)*(Rq(z)^2-Iq(z)^2)*(Rq(z) / (Rq(z)^2 + Iq(z)^2))^1.5,
            Rq(spl_pos[n]) = Rq_ini, Iq(spl_pos[n]) = Iq_ini};
      q[n]:= dsolve(ode, {Rq(z),Iq(z)}, type=numeric, output=listprocedure):
      Req[n]:= subs(q[n], Rq(z)):       # Re(q) is proportional to 1/BW^2
      Imq[n]:= subs(q[n], Iq(z)):       # Im(q) is proportional to the dispersion
      if (n < Num)
       then K:= K*exp(-spl_loss[n+1])*Isignal(spl_pos[n+1]) / Isignal(spl_pos[n])* Aeff[n] / Aeff[n+1]*
Disp[n] / Disp[n+1];
      fi:
      Rq_ini:=Req[n](spl_pos[n+1]):
      Iq_ini:= Imq[n](spl_pos[n+1]):
    od:
    Iq_end[m+2]:= Iq_ini - I_ini(z_zero):
  od:
  print (`K0=`, K0);
  if evalf(abs(Iq_end[3])) > eps then
    A := (Iq_end[3] - Iq_end[1]) / 2.0:
    B := Iq_end[1] + Iq_end[3] - 2 * Iq_end[2]:
    ff := 2 * Iq_end[2] * B / A^2:
    if abs(ff) < 1e-5
      then K0 := K0 - dK * Iq_end[2] / A:
      else K0 := K0 - dK * A / B * (1 - sqrt(1 - ff)):
    fi:
    dK := -dK * Iq_end[2] / A:
  fi:
 od:
 Rq_end:= Rq_ini:
 Iq_L := Iq_ini:
W0 := sqrt(2.*ln(2.)) / 2 / pi^(3/2) / n2 / c*lambda^3 * Aeff[1]*K0*Disp[1] / Tau* 1e-24:
 print (`Initial pulse energy W0 (pJ) =`, W0);
 print (`Average pulse energy (pJ) =`, W0 * W_ave);
 ############
 dR  := 0.005:      # step size
 eps := 0.0001:     # precision
 b   := (Tau / 1665.1)^2:
 deltaRq[3] := 2 * eps:
 while (abs(deltaRq[3]) > eps) do:
  for m in [-1, 1, 0] do:
    R  := R0 + m * dR:
    Q  := 4. * R / (1-R)^2:                   # Q factor of etalon (1)
    xmax:= 5. / sqrt(b):
    d00 := (1177.4/Tau)^2*(1 / (evalf(int(x^2*exp(-b*x^2)/(1+Q*sin(x/FSR/2)^2), x=0..xmax)/
int(exp(-b*x^2)/(1+Q*sin(x / FSR / 2)^2),x=0..xmax))) - 2*b);
    d0  := d00 / Rq_end;
    deltaRq[m + 2] := R_ini(z_zero) - Rq_end - d0,
  od:
```

```
  if (abs(deltaRq[3]) > eps) then
   A := (deltaRq[3] - deltaRq[1]) / 2.0:
   B := deltaRq[1] + deltaRq[3] - 2 * deltaRq[2]:
   ff := 2 * deltaRq[2] * B / A^2:
   if abs(ff) < 1e-5
    then R0 := R0 - dR * deltaRq[2] / A:
     else R0 := R0 - dR * A / B * (1 - sqrt(1 - ff)):
   fi:
   dR := -dR * deltaRq[2] / A:
  fi;
  print (`R=`, R);
od:
############
if (round(loop_no) = 1) then
  ini_ch[1] := initial_chirp:
  fin_R[1] := R:
  if (R < final_R) then
   initial_chirp := initial_chirp + delta_chirp;
  else
   initial_chirp := initial_chirp - delta_chirp;
  fi:
else
  ini_ch[loop_no] := initial_chirp:
  fin_R[loop_no] := R:
  initial_chirp := ini_ch[loop_no] + (ini_ch[loop_no] - ini_ch[loop_no - 1])*
(final_R - fin_R[loop_no])/ (fin_R[loop_no] - fin_R[loop_no - 1]);
fi:
loop_no := loop_no + 1;
od:
initial_chirp := ini_ch[loop_no - 1];
print ("========================  done  ===========================");
############
```

1, initial chirp, 0

K0=, −0.012

K0=, −0.0124149891830

K0=, −0.0124150633251

Initial pulse energy W0 (pJ) =, 0.0397153883315

Average pulse energy (pJ) =, 0.0285782889419

R=, 0.

R=, 0.000921377853846

R=, 0.000921401031166

initial_chirp := 0

"======================== *done* ================ ==========="

Calculate Graphs

```
if (DeleteOld = Yes)        # set DeleteOld to 'No' to keep old plots
then
  P_E:= []: P_q:= []: P_qNL:= []: P_PW:= []: P_BW:= []: P_CH:= []:
  T_E := "Pulse Energy  -  Average Values (pJ): ":
  T_q := "The complex quantity q":
  T_qNL := qNL  -  required R, D (ps/nm-km): ":
  T_PW := "Pulsewidth  -  Initial Values (ps): ":
  T_BW := "Bandwidth":
  T_CH := "Chirp  -  Initial Values: ":
fi:
PlotNo := nops(P_q) / Num + 1:
print (PlotNo);
P_E := [op(P_E), plot([z, 10 * log10(W0 * W_ave), z = 0..Length], color=colorcode[PlotNo])]:
for n from 1 to Num do:
  if (spl_pos[n] < spl_pos[n + 1])
    then P_range := spl_pos[n]..spl_pos[n + 1];
    else P_range := spl_pos[n + 1]..spl_pos[n];
  fi:
  z_real := (z - spl_pos[n]) * zc[n] + Spl_pos[n]:
  P_E:= [op(P_E),plot(10*log10(W0*Is_RG[n](n,z)), z=Spl_pos[n]..Spl_pos[n + 1],
          color=colorcode[PlotNo], thickness=LineTkns[n])]:
  P_q:= [op(P_q),odeplot(q[n],[Rq(z),Iq(z)], P_range,
          color=colorcode[PlotNo], thickness=LineTkns[n])]:
  P_qNL:=[op(P_qNL),odeplot(q[n],[Rq(z)-R_ini(z_zero),Iq(z)-z-I_ini(z_zero)],
        P_range,color=colorcode[PlotNo], thickness=LineTkns[n])]:
  P_PW:=[op(P_PW),odeplot(q[n],[z_real,Tau*sqrt((Rq(z)^2+Iq(z)^2) / Rq(z))],
        P_range,color=colorcode[PlotNo], thickness=LineTkns[n])]:
  P_BW:=[op(P_BW),odeplot(q[n],[z_real,440 / (Tau * sqrt(Rq(z)))],
        P_range,color=colorcode[PlotNo], thickness=LineTkns[n])]:
  P_CH:=[op(P_CH),odeplot(q[n],[z_real,-Iq(z) / (Rq(z)^2+Iq(z)^2)],
        P_range,color=colorcode[PlotNo], thickness=LineTkns[n])]:
od:
T_E   := sprintf("%s  %s=%.5f", T_E, colorcode[PlotNo], W0 * W_ave):
T_qNL := sprintf("%s  %s=%.3f, %.3f, %.3f",T_qNL,colorcode[PlotNo],R,D_bar,
        0.28288138 * Tau^2 * (spl_pos[Num + 1] - Iq_L) / Length):
T_PW  := sprintf("%s  %s=%.1f", T_PW, colorcode[PlotNo], Tau):
T_CH  := sprintf("%s  %s=%.3f", T_CH, colorcode[PlotNo], initial_chirp):
```

Plot Graphs

```
display(P_E,   title = T_E);        # plot all graphs at once
#display(P_q,   title = T_q);
#display(P_qNL, title = T_qNL);
display(P_PW,  title = T_PW);
display(P_BW,  title = T_BW);
display(P_CH,  title = T_CH)
```

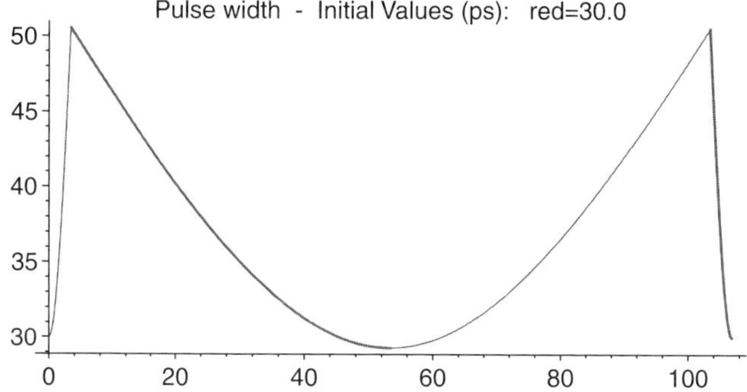
Pulse width - Initial Values (ps): red=30.0
50
45
40
35
30
0
20
40
60
80
100

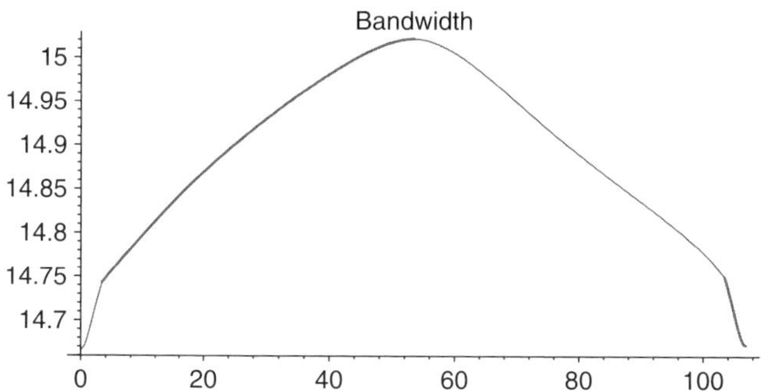
Bandwidth
15
14.95
14.9
14.85
14.8
14.75
14.7
0
20
40
60
80
100

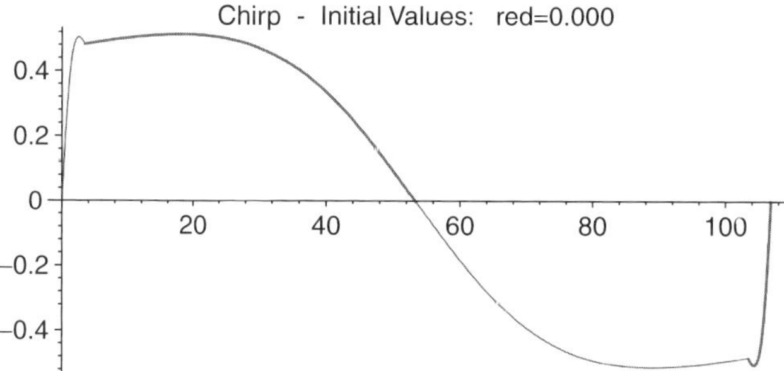
Chirp - Initial Values: red=0.000
0.4
0.2
0
–0.2
–0.4
20
40
60
80
100

Appendix B

A Brief History of Solitons

B.1. Apologia

The general subject of solitons has become a vast one, with ramifications in many fields of science and technology. This very brief history does not even begin to attempt to cover such an extensive field. Rather, it is necessarily focused on the background to the specific technical subject matter of this book—solitons in optical fibers, and specifically, their use in optical communications. It also provides the proper place to discuss some important ideas that did not easily fit into the principal scheme of this book.

B.2. The Beginning: John Scott Russell and His Discovery

The Union Canal winds along the Scottish lowlands, from Falkirk to the Lochrin Basin in Edinburgh. With width and depth of about 4 and 1.5 meters, respectively, other than the fact that there is not one lock or sluice in its 50-km length, there would seem to be little to distinguish it from many another early 19th century canal. But in 1834, it was the site of a momentous discovery by the then 26-year-old recently appointed lecturer at Herriot Watt University in Edinburgh, and later naval architect, John Scott Russell. Scott Russell's experience, which took place at Hermiston, close to the Riccarton campus of Herriot Watt, is best appreciated from his own words [122]:

> I was observing the motion of a boat which was rapidly drawn along a narrow channel by a pair of horses, when the boat suddenly stopped—not so the mass of water in the channel which it had put in motion; it accumulated round the prow of the vessel in a state of violent agitation, then suddenly leaving it behind, rolled forward with great velocity, assuming the form of a large solitary elevation, a rounded, smooth and well-defined heap of water, which continued its course along the channel apparently without change of form or diminution of speed. I followed it on horseback, and overtook it still rolling on at a rate of some eight or nine miles an hour, preserving its original figure some thirty feet long and a foot to a foot and a half in height. Its height gradually diminished, and after a chase of one or two miles I lost it in the windings of the channel. Such, in the month of August 1834, was my first chance interview with that singular and beautiful phenomenon which I have called the Wave of Translation.

Scott Russell's "solitary... Wave of Translation," as an essentially nondispersive wave, was, of course, the water-wave analog of the fundamental soliton discussed in this book. What is truly solitary about the experience, however, is not the wave itself—surely, similar events had occurred time and again, and been ignored; rather, it was Scott Russell's recognition that he had just witnessed something truly important, and worth following up on. He did just that, by building a 30-foot-long water tank in his back yard so that he might generate and make a controlled study of the solitary waves. Through a careful series of measurements, he determined the following empirical formula for the velocity of his solitary waves:

$$v = \sqrt{g(h+k)}, \tag{B.1}$$

"k being the height of the crest of the wave above the plane of repose of the fluid, h the depth throughout the fluid in repose, and g the measure of gravity." Later, as a naval architect, Scott Russell had the idea that his solitary wave could somehow be used to improve the efficiency of sailing vessels. One can only speculate on how pleased he might have been if he could have known that some day, albeit in a very different context, his stationary waves would play an important role in the world of engineering.

Scott Russell tried very hard, but in vain, to find the equation of motion for his solitary waves. His lack of success is understandable, however, since the mathematical basis of classical physics was then just in its infancy, partial differential equations were a novel concept, and nonlinear differential equations were virtually unheard of. In particular, he examined theoretical models by Lagrange, Poisson, Cauchy, Kelland, and Airy, among others, and found them all wanting, with none able to predict his experimentally determined formula [Eq. (B.1)] for the group velocity.

B.2.1. The Equation of Korteweg and de Vries

Many years had to pass before two Dutchmen, Diederik Korteweg and Gustav de Vries, in 1895, published [123] the famous equation that bears their names and describes Scott Russell's solitary waves. In its original form, the Korteweg–de Vries (KdV) equation is

$$\frac{\partial \zeta}{\partial t}+c\frac{\partial \zeta}{\partial z}+\frac{3c}{2h}\zeta\frac{\partial \zeta}{\partial z}+\frac{ch^2}{6}\delta\frac{\partial^3 \zeta}{\partial z^3}=0, \tag{B.2}$$

where $\zeta(z,t)$ is the height of the fluid above its plane of repose, $c=\sqrt{gh}$ is the velocity of a wave of infinitesimal height on the surface of fluid h deep, and δ is a parameter derived from the coefficient of surface tension, the gravitational constant g, and h. For $k \ll h$, Eq. (B.2) has the solution

$$\zeta = k\,\mathrm{sech}^2[\sqrt{k/2}(z-Vt)], \tag{B.3}$$

where $V=c(1+k/2h)$. Note that this velocity is just the small-amplitude expansion of Scott Russell's original formula [Eq. (B.1)].

Equation (B.2) can be reduced to "dimensionless" form with the equivalent of the soliton units of Chapter 1. By first changing to a frame moving at velocity c to eliminate the second term in Eq. (B.2), and then renormalizing by setting $t'=t/T$, $z'=z/Z$, and $u=\zeta/\eta$, one gets

$$\frac{\partial u}{\partial t'}+\frac{3c}{2h}\frac{\eta T}{X}u\frac{\partial u}{\partial z'}+\frac{ch^2}{6}\delta\frac{T}{Z^3}\frac{\partial^3 u}{\partial z'^3}=0. \tag{B.4}$$

The soliton units η, T, Z must satisfy the equations

$$\frac{\eta T}{Z}=\frac{4h}{c} \quad \text{and} \quad \frac{T}{Z^3}=\frac{6}{ch^2\delta}.$$

As in the treatment of the NLS equation, there are three parameters and only two equations, but it is convenient to set $\eta=1$. Dropping the primes, the transformed Eq. (B.4) becomes

$$\frac{\partial u}{\partial t}+6u\frac{\partial u}{\partial z}+\frac{\partial^3 u}{\partial z^3}=0, \tag{B.5}$$

which is the modern, canonical form of the KdV equation. Its solitary solution is

$$u(z,t)=a\,\mathrm{sech}^2(\sqrt{a/2}(z-2at)). \tag{B.6}$$

Except for the facts that the velocity $(2a)$ is just the incremental velocity, i.e., the net velocity with c subtracted off, and that all terms are now in soliton units, the solution Eq. (B.6) is clearly the same as that of Eq. (B.3). [Note that the incremental velocity of Eq. (B.3), multiplied by $T/Z = 4h/c$ for transformation to soliton units, becomes $2k$, i.e., twice the amplitude, just as in Eq. (B.6).]

It is important to note the similarities and the differences between this solitary solution and the soliton of the NLS equation that we have studied in this book. Although the KdV wave has the same shape (sech^2) as the intensity envelope of the NLS soliton, note that its group velocity ($2a$) is proportional to its amplitude, while its width is $\sqrt{2/a}$. Thus a higher amplitude (and thinner) solitary wave, initially behind a lower, wider one, will eventually overtake and pass through the lower wave.

B.2.2. *How Solitary Waves Became Solitons*

We must now fast-forward in time, to 1965, when Norman Zabusky and Martin Kruskal, both then working at Bell Laboratories in Murray Hill, New Jersey, were investigating numerical solutions of the KdV equation. In particular, they noticed that colliding solitary waves of the KdV equation emerged from the collision with all of their original properties intact. They further noticed that sinusoidal waves would evolve into solitary waves plus dispersive radiation. Because of this particle-like behavior of the solitary waves, they decided to name them "solitons" [124]. The name was apt and caught on quickly, especially in the world of mathematical physics. After all, the soliton was perhaps the first "particle" to be discovered with pure mathematics!

Incidentally, in his water tank experiments, Scott Russell had observed that his solitary waves could pass through each other and emerge without change. In his day, however, since the concept of fundamental particles scarcely existed, he could not have been expected to see his waves as such.

B.3. Solitons in Optical Fibers

B.3.1. *Origins of the Nonlinear Schrödinger Equation*

The NLS equation is much newer than the KdV equation.[1] A stationary version of the NLS equation (one without the time-derivative term) first appeared in 1950 within the context of superconductivity [125], and in the 1960s, the stationary version was used to describe the self-focusing of light in nonlinear media [126–128]. A time-dependent version first appeared in 1961, where it was used to describe Bose condensates in solid state physics [129–131]. Nevertheless, Zakharov, in 1967,

[1] We wish to thank Ildar Gabitov for considerable help with this sketch of the origins of the NLS equation.

was the first to use the time-dependent NLS equation to describe the evolution of optical wave packets in nonlinear dielectric media [132]. The one-dimensional, time-dependent version (basically, the one we have studied in this book) also first appeared in 1967 [133]. As already pointed out in Chapter 1 of this book, Zakharov and Shabat [1,2] were the first to show that the general solution of the NLS equation consisted of solitons accompanied by dispersive wave radiation, and in Chapter 4 we have been introduced to some of the powerful and useful tools that can be derived from their inverse scattering theory.

B.3.2. *First Application to Optical Fibers*

In the early 1970s, when the technology of low-loss optical fibers was still in its infancy, Akira Hasegawa was a theoretician at Bell Labs in Murray Hill, NJ. As a specialist in plasma physics, however, Hasegawa knew about the NLS equation, and he knew about Zakharov's application of the NLS equation to the general problem of dielectric media. Hasegawa was thus primed to realize that the NLS equation was appropriate for the calculation of pulse propagation in optical fibers, and that they should therefore support solitons. In a seminal work published in 1973 [134], he and co-author Frederick Tappert showed how the NLS equation applied to single-mode fibers, derived the essential properties of the corresponding solitons, and in supporting numerical simulation, showed that the solitons were stable and robust. It is noteworthy that at the time, fibers having low loss in the region of anomalous dispersion ($\lambda > 1300$ nm; see Chapter 1, Section 1.2.3) did not exist. Traces of heavy metal and OH ions, not thoroughly purged until several years later, created significant absorption in that longer wavelength region, forcing the loss minimum to the much shorter wavelength of about 800 nm. Thus, the paper by Hasegawa and Tappert [134] was also a work of genuine foresight and imagination. [Also because of that loss problem, Hasagewa and Tappert followed up almost immediately with another paper [135] describing "dark" solitons, i.e., sech-shaped holes in a cw background, which could exist in the presence of normal dispersion. For a number of practical reasons, however, the dark solitons have never been used for transmission.]

B.3.3. *Higher Order Solitons*

In another seminal work that appeared soon after the Hasegawa and Tappert papers, Satsuma and Yajima [136], using the inverse scattering method of Zakharov and Shabat, studied the resolution of various initial wave functions into solitons and dispersive wave radiation. Among the input functions they studied was the special case $u(t, z=0) = A\,\text{sech}(t)$. They found that this input pulse contains a number of fundamental solitons, with amplitudes A_j given by the positive values of $2A-1$, $2A-3$, $2A-5$, and so on, accompanied generally by a small amount of radiation. Thus, for $0.5 < A \le 1.5$ there is one soliton, for $1.5 < A \le 2.5$ there are two solitons, and so forth. Recall that the energy of the input pulse is equal to $2A^2$, while the energy of each fundamental soliton is equal to $2A_j$. From this, it is easy to show that when A is an integer N, the input pulse contains just N fundamental solitons with amplitudes 1, 3, 5, ... $2N-1$, and there is no radiation. These input pulses, namely $u(t, z=0) = N\,\mathrm{sech}(t)$, have been heuristically called Nth order solitons.

Satsuma and Yajima also discovered that when $A=N$ the pulse behavior undergoes a pattern of narrowing and splitting with a period, in soliton units, of $z_0 = \pi/2$. Translated into real-world units, this "soliton period" is

$$z_0 = \frac{\pi}{2} z_c = \frac{1}{(1.763)^2} \frac{\pi^2 c}{\lambda^2} \frac{\tau^2}{D}, \tag{B.7}$$

where z_c is the characteristic dispersion length defined by Eq. (1.18). Satsuma and Yajima provided the following analytic solution for the case $N=2$,

$$u(z,t) = \frac{4\exp(-iz/2)[\cosh(3t) + 3\cosh(t)\exp(-i4z)]}{\cosh(4t) + 4\cosh(2t) + 3\cos(4z)}. \tag{B.8}$$

We note that the formalism of Satsuma and Yajima yields the soliton functions as the complex conjugates of ours.

Using the equations given in the beginning of Chapter 4, Section 4.2.2, analytic forms for the $N=2$ and $N=3$ solitons can be derived without great difficulty (hint, take $A_1 = 1$, $A_2 = 3$, $A_3 = 5$, and $\phi_{10} = 0$, $\phi_{20} = \pi$, and $\phi_{30} = 0$). The result for the $N=3$ soliton is

$$u(z,t) = 6e^{iz/2} \times \frac{\cosh(8t) + 8e^{i4z}\cosh(6t) + \left(18e^{i4z} + 10e^{i12z}\right)\cosh(4t) + \left(16 + 40e^{i12z}\right)\cosh(2t) + 16e^{i16z} + 22.5e^{i8z} + 2.5e^{-i8z}}{\cosh(9t) + 9\cosh(7t) + 36\cos(4z)\cosh(5t) + (64 + 20\cos(12z))\cosh(3t) + (36 + 90\cos(8z))\cosh(t)} \tag{B.9}$$

Note that the $N=2$ and $N=3$ soliton functions at $z=0$ are respectively expansions of the quantity $N\cosh[(N^2-1)t]/\cosh[N^2 t] = N\,\mathrm{sech}(t)$. Note also that the arguments of the phase terms are all multiples of $4z$, so that these functions repeat with the aforementioned soliton period of $z_0 = \pi/2$. All of the higher order solitons have these properties.

Figure B.1 graphs the temporal behavior of the power $|u(z,t)|^2$ of the $N=2$ and $N=3$ solitons for several values of z.

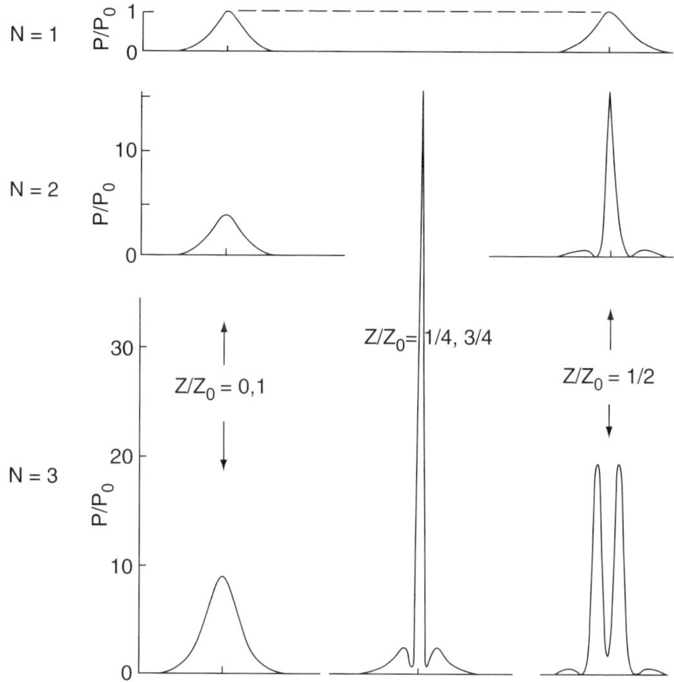

FIGURE B.1 Behavior of the $N=2$ and $N=3$ solitons, compared with that of the fundamental ($N=1$) soliton. Note that the vertical scale for the fundamental soliton has been magnified with respect to that of the other two.

It should be emphasized that for each $N > 1$, the higher order soliton is really a nonlinear superposition of N fundamental solitons (with amplitudes A_j of 1, 3, $5, \ldots, 2N-1$). The substantial pulse narrowing and splitting seen in Fig. B.1 results from periodic interference among the solitons. It is also so that the higher order solitons, unlike the fundamental soliton, are not stable in the face of perturbations. Perturbations which differently affect the frequencies of the underlying solitons, such as third-order dispersion or the ASE noise from amplifiers, cause the underlying solitons to come apart as the pulse propagates. Nevertheless, as we shall soon see, the higher order solitons were of great importance to the first experimental observation of solitons in optical fibers and to a femtosecond pulse source known as the "soliton laser."

B.3.4. First Experimental Observation

[To avoid a certain awkwardness, the account in this section is in the first person of L. F. Mollenauer.]

In the early 1970s, the most common sources of picosecond light pulses were synchronously pumped, mode-locked lasers using a thin stream of solvent containing one or another organic dye as the gain medium. Since there appeared to be no dyes for similar laser action in the near infrared, I spent several years developing substitutes based on certain color centers in alkali halide crystals [137]. By the late 1970s, two color centers, the "F_2^+" center in the host NaCl, and the "$Tl^0(1)$" center in the host KCl, both pumped at 1060 nm by a cw, Nd:YAG laser, could provide continuous tuning over the ranges of ≈1420–1780 and ≈1400–1600 nm, respectively, with cw output powers at band center on the order of 1 W. As these laser-active centers were unstable at room temperature, they and their equally delicate alkali halide hosts had to be maintained in high vacuum on a copper finger cooled to 77 K (liquid N_2 temperature), not a small nuisance! Nevertheless, when synchronously pumped, the corresponding lasers could produce unchirped pulses of about 7-ps width at a repetition rate of 100 MHz. Thus, despite all of the surrounding technical difficulty, as "the only game in town" for such service, they were very welcome at the time.

In very late 1979 or early 1980, colleague Rogers Stolen acquired a 700-m length of one of the very first fibers having low loss in the 1500-nm region, a fiber that was also like standard SMF in its other characteristics. Since we had the picosecond laser source as well, we were eager to try to observe the soliton, pulse compression, and possibly even some of the more complex behavior predicted by Satsuma and Yajima.

The first experimental observation [138] was straightforward. Using microscope objectives, we coupled the mode-locked color center laser's output into the fiber, and the fiber's output into an autocorrelator (see Fig. B.2) for observation of the shapes of the emerging pulses. It was then simply a question of observing the pulse behavior as we gradually raised the power coupled into the fiber. We had already noted that for the 7-ps pulse width and for the assumed fiber dispersion parameter D of 16 ps/nm-km (later corrected to 15 ps/nm-km [139]), the soliton period z_0 should have been about 1260 m (later 1350 m), or very nearly twice the 700-m length of the fiber; thus, we anticipated behavior, at the appropriate power levels, like that shown in the $z/z_0 = 1/2$ column of Fig. B.1. Figure B.3 shows the resultant autocorrelation traces at certain critical peak pulse powers. From left to right, as the input power increases, we see, first, significant dispersive broadening, followed by return to the original pulse width at the power (≈1.2 W) for the fundamental soliton, then, at 4.2× that power (5 W), the narrowing expected at the half soliton period of the $N = 2$ soliton, followed at 9.5× and 18×, the autocorrelation patterns of the expected two- and three-fold splitting at the half soliton period of the $N = 3$ and $N = 4$ higher order solitons, respectively. (Note that the measured power ratios, to within the limits of experimental error, essentially follow the expected

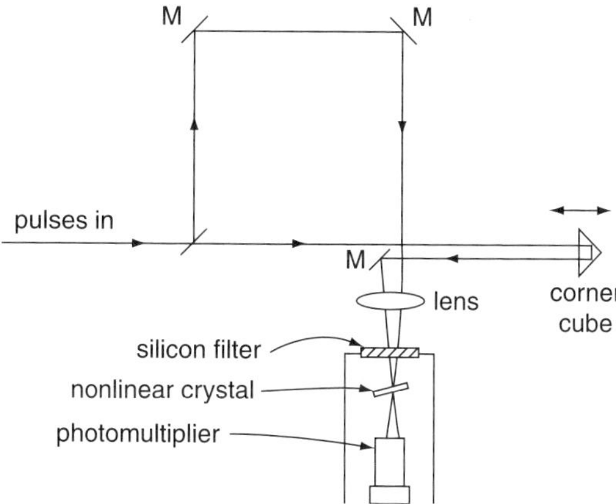

FIGURE B.2 Schematic diagram of the autocorrelator. The input beam is divided into two roughly equal beams, which, after traveling separate paths, are brought together in the nonlinear crystal. Second harmonic light is generated only if pulses from both beams are simultaneously present in the crystal. Thus the strength of the second harmonic (registered by the photomultiplier) reflects the temporal overlap of the pulses in the two converging beams. A measurement of second harmonic intensity as a function of relative delay (created by motion of the corner cube) then yields the pulse shape in autocorrelation. (The silicon filter passes 1500-nm light but keeps out visible room light.)

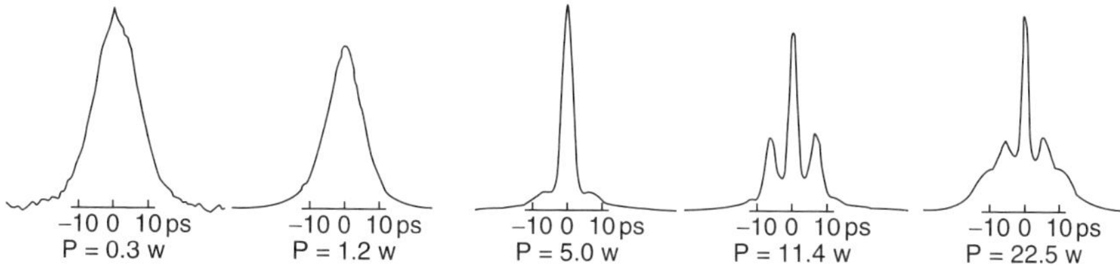

FIGURE B.3 Autocorrelation shapes of pulses emerging from a $z \cong z_0/2$ length of fiber for various input powers. Seen from left to right are first, dispersive broadening, then the anticipated behaviors of the $N=1$, $N=2$, $N=3$, and $N=4$ solitons, respectively. Peak pulse powers at the fiber's input are indicated below each trace. The vertical scales of the various traces have been arbitrarily adjusted to give all approximately equal peak height.

sequence of $1, 2^2, 3^2$, and 4^2.) Since we would have considered it success merely to see the fundamental soliton and some degree of pulse narrowing above the soliton power, we were elated at these results, which were obtained all within the course of one afternoon's measurements.

In hindsight, the fact that the available fiber was almost exactly $z_0/2$ in length represents a wonderful bit of serendipity. That is, had it been significantly less than $z_0/2$, we would have missed the sharply defined pulse behavior that occurs at the half-period point, especially for the $N \geq 2$ solitons. Had it been too long, on the other hand, we probably would have been reluctant to cut it; at that time, that 700-m length of fiber was considered precious.

Finally, it should be noted that, thanks to (co-author) Gordon, our paper [138] also contained a complete translation between the dimensionless world of the then available theoretical literature and the real world, i.e., it established the basis for the "soliton units" described in Section 1.3.2 of Chapter 1. The understanding generated, especially of how the soliton behavior scaled with τ and D, was seminal to all of our further work.

B.3.5. The Soliton Laser

The soliton compression we had just witnessed seemed to offer a promising way to significantly shorten the pulses produced by the mode-locked, color center laser. Indeed, in a follow-up study [140], we were able to narrow the laser's pulses to a small fraction (as small as 0.26 ps) of the initial 7-ps width simply by sending the laser's output, at high power, through a judiciously chosen length of fiber. But there was at least one serious drawback: with increasing degree of compression, the energy left over in the uncompressed "wings" of the pulse represented an ever greater fraction (>70% for compression to 0.26 ps) of the total pulse energy. So then we began to wonder, what would happen if we were to somehow try to feed back the narrowed pulses into the laser? Wouldn't that stimulate the laser itself to produce ever shorter pulses, until some sort of equilibrium were reached? With those thoughts in mind, we built and tested the device shown in Fig. B.4, which

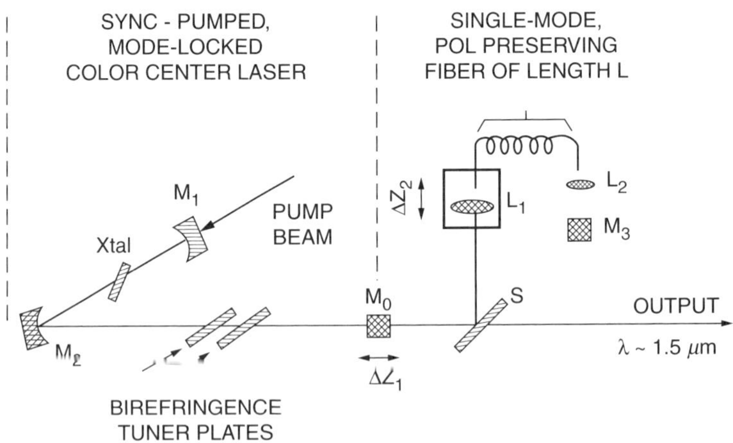

FIGURE B.4 Schematic of the soliton laser.

we called the "soliton laser" [141]. With the external fiber loop length carefully adjusted (Δz_2 in Fig. B.4) to be an integral multiple of the main cavity length, the device immediately began to provide sub-picosecond pulses. Data taken with an ever shorter succession of control fiber lengths seemed to indicate that the device operated on $N = 2$ solitons, with the control fiber length L just a bit shorter than one-half soliton period (see Fig. B.5). Figure B.6 shows autocorrelation traces of the shortest pulses produced in that first round of experiments.

The data of Fig. B.5 were a bit rough, however, since at the time, the soliton laser action started and stopped sporadically as vibration and thermal drift caused the relative lengths of the two coupled cavities to vary in and out of the proper interference condition. Later, however, servo stabilization of the relative cavity lengths by postdoc Fedor Mitschke and myself [142] enabled the laser to settle down to a very low-noise mode of operation where the output pulses exhibited nearly textbook-perfect, sech^2 shapes. That clean and very stable mode of

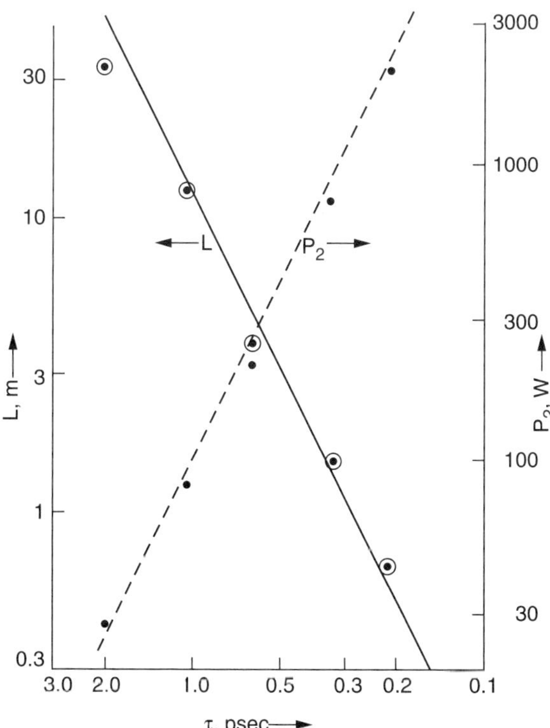

FIGURE B.5 Control fiber length L and peak fiber input powers for the soliton laser as functions of the produced pulse width τ.

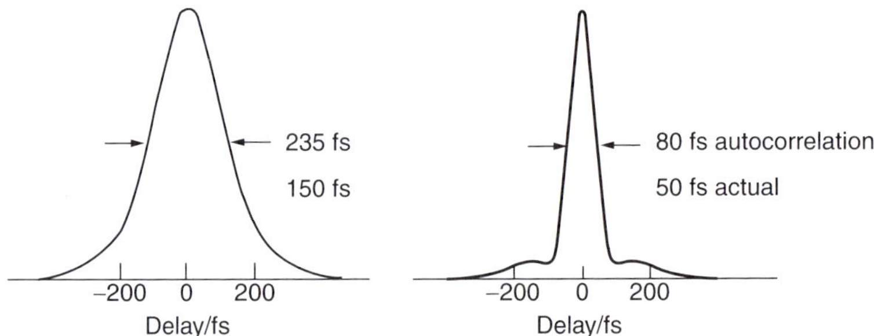

FIGURE B.6 Autocorrelation traces of (left) 150-fs FWHM pulses produced by the soliton laser and (right) the same, after compression to 50 fs by a short length of external fiber.

operation continued down to the shortest pulses produced by the stabilized soliton laser ($\tau = 60$ fs directly, and 19 fs after compression in an external fiber). Careful study of the stabilized laser operating in that best mode showed that the control fiber was actually closer to just $z_0/4$ long, so that the pulses returned to the laser were considerably narrower than the output pulses. We also discovered that after the soliton laser action had begun, pumping of the color center laser could be switched from the synchronous pulsed mode to simple cw pumping, as the soliton laser action was sufficient to maintain mode locking all by itself.

Later, others discovered that soliton pulse narrowing was not absolutely necessary for the production of short pulses in coupled cavity lasers. In the concept of "additive pulse mode locking" [143, 144], self-phase modulation from a purely nonlinear element in the external cavity causes the returned pulses to interfere constructively with the main laser pulses near their peaks, while interfering destructively with them in their wings (thereby producing a narrower pulse). (The nonlinear phase shift would be similar to that shown in Fig. 1.4 of Chapter 1.) It was shown, both theoretically and experimentally, that such additive pulse mode locking could be equally effective in the production of ultra-short pulses.

Nevertheless, our soliton laser really did involve the pulse narrowing of $N = 2$ solitons. It generated a lot of excitement in its day, and it spurred an entire cottage industry for the production of femtosecond pulses from lasers using nonlinear elements to enhance the mode locking. In our own laboratory, it provided the pulse source for at least one important experimental discovery, as related in the following section.

B.3.6. Discovery of the Soliton Self-frequency Shift

Mitschke and I decided to use pulses from the newly stabilized soliton laser to study soliton propagation in the sub-picosecond regime. Launching $\tau \approx 500$ fs pulses into a 392-m length of fiber, we observed nearly 100× broadening at the lowest power levels, since z_c is only ≈4 m for that initial pulse width. With increasing power, we saw pulse narrowing until the power reached the soliton power, P_c, at which point the initial pulse width was restored; thus far, there was nothing new or unexpected. But as the power was increased beyond P_c, as seen in autocorrelation, the pulse developed a satellite, which began to split off very rapidly in time as the power was further increased. At first, this behavior was very puzzling, since there seemed to be no reasonable cause for the great temporal splitting. The optical spectrum, however, revealed that the satellite was really a weak, spectrally narrow, nonsoliton component of the input pulse, which remained at the original optical frequency, while the bulk of the energy was in the spectrally much broader soliton, which was very strongly shifted to lower optical frequencies. Raising the power significantly above P_c created an initial temporal narrowing and spectral broadening of the soliton, which in turn greatly increased its spectral shift to lower frequencies. The overall frequency shifts were huge; for example, for $\tau = 560$ fs and 260 fs at fiber input and output, respectively, the soliton was down-shifted by 8 THz at the fiber's output [145].

We quickly came to realize that the only plausible explanation was a self-Raman effect, whereby the higher frequency components of the pulse provide Raman gain for the lower frequency ones, with the result that the pulse's energy is continuously transferred to ever lower frequencies [145].[2] Note from Fig. 3.4 that the Raman gain coefficient in optical fibers is roughly in direct proportion to the frequency difference between pump and signal, and essentially extends right down to zero frequency difference. In that approximation, the effective differential gain across the pulse's spectrum should increase as the square of its spectral width, and hence as τ^{-2}. Since the soliton's peak power also scales as τ^{-2}, the net rate of soliton self-frequency shift should scale as τ^{-4}. Shortly after he had been shown the experimental results, Jim Gordon produced a theoretical model [147] that predicted the following rate of soliton self-frequency shift:

$$d\nu_0/dz = 0.0436h(\tau)/\tau^4. \tag{B.10}$$

Here $d\nu_0/dz$ is in THz/km, the soliton pulse width τ is in ps, and $h(\tau)$ is a function, derived from the actual Raman gain curve, that oscillates between limits of about 0.4 and 1 as the pulse width ranges from 20 fs to 10 ps (see Fig. B.7). [$h(\tau)$ is set = 1 in the approximation that the Raman gain coefficient is directly proportional to the difference between pump and signal frequencies.] Equation (B.10)

[2] After publication of our paper [145], we learned that E. M. Dianov *et al.* from the GPI in Moscow had made similar observations of the self-Raman shift of solitons about a year earlier [146].

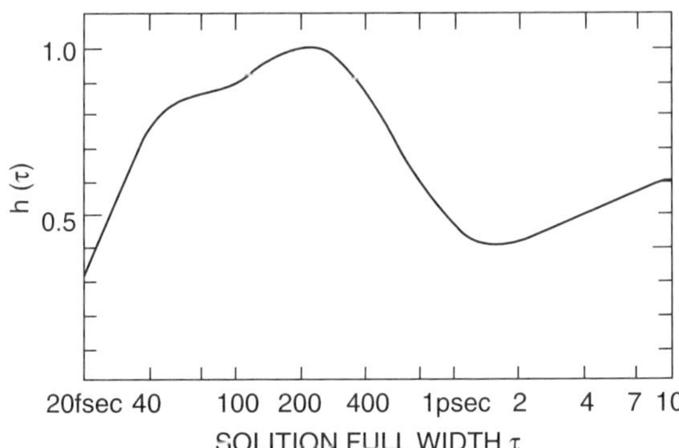

FIGURE B.7 The function $h(\tau)$ for silica fibers, numerically computed from the Raman gain curve.

provided values that were in agreement with the experimental data to within a factor of 2 or better.

It should be noted that as large as the soliton self-frequency shift is for pulses in the sub-picosecond regime, because of its scaling as τ^{-4}, it tends to be negligibly small for the sort of pulse widths normally used in telecommunications. For example, according to Eq. (B.10), an ordinary soliton with $\tau = 25$ ps would shift frequency at rate of $\approx 1 \times 10^{-5}$ GHz/km, so the net shift in 10,000 km would be all of about 0.1 GHz, and the frequency shifts of a dispersion-managed soliton of the same unchirped pulse width would be still smaller. But even if the frequency-shifting rate were to become several orders of magnitude higher (as it might for pulse widths suitable for bit rates of 40 Gbit/s or higher), the frequency shifts could be easily countered by the use of frequency-guiding filters.

B.3.7. *First Demonstration of Ultra-long-haul Soliton Transmission*

Theoretical predictions that solitons could be transmitted over thousands of kilometers of fiber with periodic (Raman) amplification [148–150] had existed for a number of years before the first experimental demonstration. Part of the reason for this delay rested in the fact that the theoretical ideas were far ahead of hardware development. In the engineering world, however, predictions tend to grow stale and become discredited without the hard proof of an experimental test. With that thought in mind, in early 1988, postdoc Kevin Smith and I set about to make the first test [116], despite the impending difficulties.

The experimental setup was deceptively simple. It consisted of a 42-km length of standard SMF, closed on itself with 95% efficiency at the signal wavelength of

1600 nm, and brought to unity gain by Raman pumping at 1500 nm by 300 mW of cw power from a color center laser. A second color center laser provided $\tau = 55$ ps pulses at a 100-MHz repetition rate. The pulse trains emerging from this primitive recirculating loop were detected and their quality determined from their microwave spectra, as detailed in Section 8.5 of Chapter 8. Acousto-optic modulators turned the pump and signal sources on and off at appropriate times to allow for repeated transmissions of N round-trips each. The microwave spectrum analyzer was triggered electronically to begin measurements for a fixed time interval at the end of each transmission.

Since a WDM coupler suitable for closing the loop on itself was not available, we had to fashion our own in the form of an all-fiber, Mach–Zehnder interferometer from two 3-dB splitters. The interferometer had to be carefully temperature tuned to enable nearly 100% efficient coupling of the pump light into the loop, while simultaneously closing the loop on itself with nearly 95% efficiency at the signal wavelength. To avoid Brillouin backscattering of the pump light, we had to broaden the spectrum of the pump laser with a homemade lithium niobate frequency modulator driven at a frequency of several hundreds of megahertz. Coupling of the free-space laser beams into the fibers was accomplished by dipping the fiber ends into a well of index-matching halocarbon oil (for zero back reflection and zero absorption at 1500 nm) whose bottom was formed by one surface of the microscope objective used for mode-matching into the fiber. Those touchy and sensitive couplings had to be constantly maintained at the highest possible efficiency, since every milliwatt of pump power was precious. Finally, at least to an outsider, the sight of a very large optical table loaded with bulky free-space lasers (which were constantly breaking down) must have seemed more than a bit incongruous for the task at hand. In our minds, however, those lasers were just temporary substitutes for the vastly more efficient and smaller semiconductor lasers we were sure would become available in due course. Perhaps it should also be noted that success of the experiment was due in no small way to the persistence and skill of Kevin Smith.

Despite the fact that the fiber's dispersion ($\approx$18 ps/nm-km at 1600 nm) was more than 30$\times$ too large (thereby making both the signal energies and the Gordon–Haus jitter much greater than necessary), the experiment was able to demonstrate faithful maintenance of the solitons over more than 4000 km (see Fig. B.8). At the time, this was considered a most satisfactory result.

A short while later, with the same apparatus, Smith and I were able to demonstrate adiabatic compression and expansion of the soliton pulses [117] in a convincing demonstration of just how very robust the solitons were. The demonstration in turn was a most useful foil to some of the misconceptions held by some about solitons. Also at that time, we observed the "long-range interaction" [118] (the acoustic effect) already discussed in Chapter 3, Section 3.4.5.

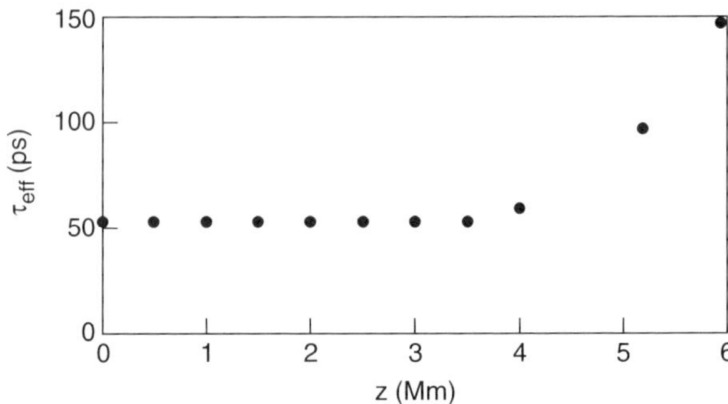

FIGURE B.8 Plot of the effective transmitted pulse width, τ_{eff}, as a function of distance, in the first demonstration of long-haul soliton transmission.

B.3.8. A Public Relations Coup

Those who are required to present ideas on the grand scale tend to prefer them in capsule form. Thus, it should have come as no surprise that one day, the request came down to me through the chain of command that the then director of Bell Labs needed a single-view graph that "would enable him to explain solitons to bankers." Nevertheless, I was at a total loss to know how to respond. How could the NLS equation, the canceling nonlinear and dispersive phase shifts, and all the intervening development (essentially, the core of Chapter 1 of this book) be crammed into a single picture? Shortly after hearing of the challenge, however, postdoc Stephen Evangelides said that he had an inspiration. His idea was based on the well known view that the soliton is a self-trapped pulse. That is, the little blip of index change it produces through the nonlinear term forms a potential well to trap the pulse. Evangelides' simple but brilliantly conceived analog was that of a group of runners on a soft mattress, as shown in Fig. B.9. Perhaps needless to say, the runners viewgraph became an instant hit, and was passed down rapidly through the various levels of management. Since the analog works so very well, it was a great help, not just for explaining what the soliton is, but also for getting across the idea that the fundamental soliton is a truly stable and robust entity.

B.4. The Soliton Legacy

The soliton community has influenced fiber optic transmission far beyond the creation of any one particular system, such as Lucent's *LambdaXtreme*. Many

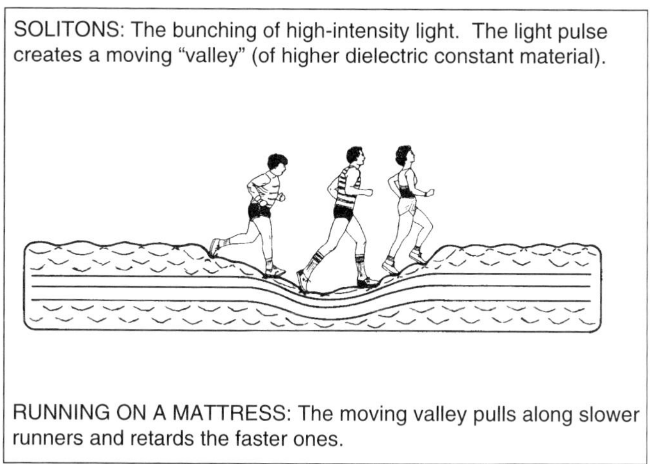

FIGURE B.9 A group of runners on a soft mattress as an analog of the soliton as a self-trapped pulse. The heavy-set fellow at the rear, who represents the low frequency components of the pulse, and would normally fall behind, can't, because he is running downhill, while the woman in front, who represents the high-frequency components, and would normally go racing ahead, can't, because she is always running uphill!

ideas and concepts, now commonplace, originated with the soliton community. This section represents an attempt to recognize at least some of those contributions.

B.4.1. Introduction of the NLS Equation

Today, prepackaged programs for solving the NLS equation numerically are readily available commercially, and just about everyone in the business of engineering fiber optic transmission systems makes use of them. It has not always been that way, however. For a long time after Zakharov introduced the NLS equation for the study of optical pulse propagation in nonlinear dielectric media, and Hasegawa's first application of it to fiber optics, the NLS equation tended to remain the province of the soliton fraternity. In those days, numerical simulation using the NLS equation was a "do it yourself" activity, i.e., everyone had to write his own program. It was not until the early 1990s that outsiders (primarily those designing undersea cable systems) began to use numerical simulation of the NLS equation as an inexpensive way to check performance of an already designed system. Even there, however, the NLS equation was not the fountainhead of ideas that it had been for such a long time in the soliton world. In support of this view, a glance at the index of the two-volume set "Optical Fiber Telecommunications III" (Kaminov and Koch, eds.) [151], published in 1997, and something of a bible in the industry, is most revealing: The only reference to the NLS equation is in Chapter 12 (on solitons, written by ourselves), while Chapter 8, "Fiber Nonlinearities and Their Impact on Transmission Systems" never mentions the NLS equation, not even once!

So what changed attitudes, and so quickly? Almost certainly, it was the confluence in the late 1990s of ultra-long-haul transmission and dense WDM, especially in the world of terrestrial transmission. The manifold nonlinear effects encountered there forced use of the NLS equation as a *sine qua non* of system design. The rising importance and success of dispersion management probably played a major role as well, since it was surely recognized that pulse behavior in that context was not fully predictable without recourse to the NLS equation or its ODE equivalent.

B.4.2. Understanding of the Nonlinear Interaction of Signals and Noise

In a strictly linear transmission line, noise and signal fields can be treated completely independently of each other, and need to be added only at the detector. As the seminal paper [55] of Gordon and Haus showed, however, in real fibers, one does so at his peril: As already discussed in Chapter 3, Section 3.4.3, the often severely limiting Gordon–Haus jitter arises from more or less continual nonlinear interaction between the signals and the noise all along the transmission path. We have also seen how this same dynamic is echoed in the Gordon–Mollenauer effect (Section 3.4.6). Unfortunately, when the Gordon and Haus paper [55] first appeared, at least some got the misguided idea that the effect applies to solitons only. In fact, it applies, and equally so, to *all* modes of transmission. The Gordon–Haus effect was calculated in terms of solitons only because solitons represent the one transmission mode for which the calculation is tractable analytically. Thus, nonlinear simulations, whether they involve solitons or some other transmission mode, are just plain wrong and can generate seriously misleading results if the ASE noise is not added as it occurs all along the transmission line.

B.4.3. Transparency and the Dream of the All-optical Network

It will probably be conceded, without much argument, that the dream of a truly all-optical, ultra-long-haul network arose from the soliton fraternity. It is only natural that the idea should have arisen there. After all, the limitations of ASE noise growth aside, solitons are the only pulses expected to be able to go on more or less indefinitely without significant change or the need for electronic regeneration. They are also the only pulses truly amenable to purely optical regeneration, as with the use of sliding-frequency guiding filters, for example. Thus, it should come as no surprise that the first commercial, truly ultra-long-haul transmission system

(Lucent's *LambdaXtreme*), designed to be the backbone of an all-optical network, is based on (dispersion-managed) solitons. Perhaps it should be noted that the paper by Mollenauer *et al.* [150], written in 1986, for all its naiveté, was nevertheless remarkably prescient. That is, in its most general features and performance, the envisioned system was remarkably similar to *LambdaXtreme*: It was to be a dense WDM system, using solitons and all-Raman amplification, with a projected reach of many thousands of kilometers and a capacity of at least 100 Gbit/s. (In 1986, 100 Gbit/s was considered to be many times greater than would ever be needed.)

Perhaps it should also be pointed out that throughout the intervening years, there were some individuals who thought they had air-tight arguments as to why such transparency either would never be achieved or should not be the goal. A combination of technological advances on the one hand, and the economic pressures against extensive electronic regeneration in ultra-long-haul dense WDM on the other, finally put all of those arguments to rest.

References

[1] V. E. Zakharov and A. B. Shabat, "Exact theory of two-dimensional self-focusing and one-dimensional self-modulation of waves in nonlinear media," *Zh. Eksp. Teor. Fiz.*, vol. 61, no. 1, pp. 118–134, 1971.

[2] V. E. Zakharov and A. B. Shabat, "Exact theory of two-dimensional self-focusing and one-dimensional self-modulation of waves in nonlinear media," *Sov. Phys. JETP*, vol. 34, no. 1, pp. 62–69, 1972.

[3] J. C. Knight, J. Broeng, T. A. Birks, and P. S. J. Russell, "Photonic band gap guidance in optical fibers," *Science*, vol. 282, pp. 1476–1478, Nov. 20, 1998.

[4] R. F. Cregan, B. J. Mangan, J. C. Knight, T. A. Birks, P. S. J. Russell, P. J. Roberts, and D. C. Allan, "Single-mode photonic band gap guidance of light in air," *Science*, vol. 285, pp. 1537–1539, Sept. 3, 1999.

[5] P. S. J. Russell, "Photonic crystal fibers," *Science*, vol. 299, pp. 358–362, Jan. 17, 2003.

[6] B. J. Eggelton, C. Kerbage, P. S. Westbrook, R. S. Windeler, and A. Hale, "Microstructured optical fiber devices," *Opt. Express*, vol. 9, pp. 698–713, Dec. 17, 2001.

[7] J. C. Knight, J. Arriaga, T. A. Birks, A. Ortigosa-Blanch, W. J. Wadsworth, and P. S. J. Russell, "Anomalous dispersion in photonic crystal fiber," *IEEE Photon. Technol. Lett.*, vol. 12, pp. 807–809, July 2000.

[8] W. H. Reeves, J. C. Knight, P. S. J. Russell, and P. J. Roberts, "Demonstration of ultra-flattened dispersion in photonic crystal fibers," *Opt. Express*, vol. 10, pp. 609–613, July 15, 2002.

[9] L. F. Mollenauer, J. P. Gordon, and M. N. Islam, "Soliton propagation in long fibers with periodically compensated loss," *IEEE J. Quantum Electron.*, vol. QE-22, pp. 157–173, Jan. 1986.

[10] L. F. Mollenauer, M. J. Neubelt, S. G. Evangelides, J. P. Gordon, J. R. Simpson, and L. G. Cohen, "Experimental study of soliton transmission over more than 10,000 km in dispersion shifted fiber," *Opt. Lett.*, vol. 15, pp. 1203–2105, Nov. 1990.

[11] L. F. Mollenauer, S. G. Evangelides, and H. A. Haus, "Long-distance soliton propagation using lumped amplifiers and dispersion shifted fiber," *J. Lightwave Technol.*, vol. 9, pp. 194–197, Feb. 1991.

[12] K. J. Blow and N. J. Doran, "Average soliton dynamics and the operation of soliton systems with lumped amplifiers," *Photon. Technol. Lett.*, vol. 3, pp. 369–371, 1991.

[13] A. Hasegawa and Y. Kodama, "Guiding center soliton in optical fibers," *Opt. Lett.*, vol. 15, pp. 1443–1445, 1990.

[14] M. Suzuki, I. Morita, N. Edagawa, S. Yamamoto, H. Taga, and S. Akiba, "Reduction of Gordon–Haus timing jitter by periodic dispersion compensation in soliton transmission," *Electron. Lett.*, vol. 31, no. 23, pp. 2027–2028, 1995.

[15] N. J. Smith, F. M. Knox, N. J. Doran, K. J. Blow, and I. Bernnion, "Enhanced power solitons in optical fibres with periodic dispersion mangement," *Electron. Lett.*, vol. 32, pp. 54–55, Jan. 1996.

[16] N. J. Smith, N. J. Doran, F. M. Knox, and W. Forysiak, "Energy-scaling characteristics of solitons in strongly dispersion-managed fibers," *Opt. Lett.*, vol. 21, pp. 1981–1983, Dec. 1996.

[17] V. S. Grigoryan, T. Yu, E. A. Golovchenko, C. R. Menyuk, and A. N. Pilipetskii, "Dispersion-managed soliton dynamics," *Opt. Lett.*, vol. 22, pp. 1609–1611, Nov. 1997.

[18] T.-S. Yang and W. L. Kath, "Analysis of enhanced-power solitons in dispersion-managed optical fibers," *Opt. Lett.*, vol. 22, pp. 985–987, July 1997.

[19] I. R. Gabitov and S. K. Turitsyn, "Averaged pulse dynamics in a cascaded transmission system with passive dispersion compensation," *Opt. Lett.*, vol. 21, pp. 327–329, March 1996.

[20] N. J. Smith, W. Forysiak, and N. J. Doran, "Reduced Gordon–Haus jitter due to enhanced power solitons in strongly dispersion managed systems," *Electron. Lett.*, vol. 32, pp. 2085–2086, Oct. 1996.

[21] I. R. Gabitov, E. G. Shapiro, and S. K. Turitsyn, "Optical pulse dynamics in fiber links with dispersion compensation," *Opt. Commun.*, vol. 134, pp. 317–329, Jan. 1997.

[22] E. A. Golovchenko, J. M. Jacob, A. N. Pilipetskii, C. R. Menyuk, and G. M. Carter, "Dispersion-managed solitons in a fiber loop with in-line filtering," *Opt. Lett.*, vol. 22, pp. 289–291, March 1997.

[23] T. Yu, E. A. Golovchenko, A. N. Pilipetskii, and C. R. Menyuk, "Dispersion-managed soliton interactions in optical fibers," *Opt. Lett.*, vol. 22, pp. 793–795, June 1997.

[24] S. K. Turitsyn, N. F. Smyth, and E. G. Turitsyna, "Solitary waves in nonlinear dispersive systems with zero average dispersion," *Phys. Rev. E*, vol. 58, pp. R44–R47, July 1998.

[25] J. M. Jacob and G. M. Carter, "Error-free transmission of dispersion-managed solitons at 10 Gbit/s over 24500 km without sliding-frequency filters," *Electron. Lett.*, vol. 33, pp. 1128–1129, June 1997.

[26] G. M. Carter and J. M. Jacob, "Dynamics of solitons in filtered dispersion-managed systems," *IEEE Photon. Technol. Lett.*, vol. 10, pp. 546–548, April 1998.

[27] I. Morita, K. Tanaka, N. Edagawa, S. Yamamoto, and M. Suzuki, "40 Gbit/s single-channel soliton transmission over 8600 km using periodic dispersion compensation," *Technical Digest of the Third Optoelectronics and Communications Conf. (OECC'98)* pp. PD1–1, July 1998.

[28] G. M. Carter, R. M. Mu, V. S. Grigoryan, C. R. Menyuk, P. Sinha, T. F. Caruthers, M. L. Dennis, and I. N. Duling III, "Transmission of dispersion-managed solitons at 20 Gbit/s over 20,000 km," *Electron. Lett.*, vol. 35, pp. 233–234, Feb. 1999.

[29] I. S. Penketh, P. Harper, S. B. Alleston, A. M. Niculae, I. Bennington, and N. J. Doran, "10-Gbit/s dispersion-managed soliton transmission over 16,500 km in standard fiber by reduction of soliton interactions," *Opt. Lett.*, vol. 24, pp. 802–805, June 1999.

[30] R. M. Mu, V. S. Grigoryan, C. R. Menyuk, G. M. Carter, and J. M. Jacob, "Comparison of theory and experiment for dispersion-managed solitons in a recirculating fiber loop," *IEEE J. Select. Top. Quantum Electron.*, vol. 5, no. 6, 1999.

[31] W. Forysiak, J. F. L. Devaney, N. J. Smith, and N. J. Doran, "Dispersion management for wavelength-division-multiplexed soliton transmission," *Opt. Lett.*, vol. 22, pp. 600–602, May 1997.

[32] E. A. Golovchenko, A. N. Pilipetskii, and C. R. Menyuk, "Periodic dispersion management in soliton wavelength-division multiplexing transmission with sliding filters," *Opt. Lett.*, vol. 22, pp. 1156–1159, Aug. 1997.

[33] J. F. L. Devaney, W. Forysiak, A. M. Niculae, and N. J. Doran, "Soliton collisions in dispersion-managed wavelength-division-multiplexed systems," *Opt. Lett.*, vol. 22, pp. 1695–1697, Nov. 1997.

[34] T.-S. Yang, W. L. Kath, and S. K. Turitsyn, "Optimal dispersion maps for wavelength-division-multiplexed soliton transmission," *Opt. Lett.*, vol. 23, pp. 597–599, April 1998.

[35] P. V. Mamyshev and L. F. Mollenauer, "Soliton collisions in wavelength-division-multiplexed dispersion-managed systems," *Opt. Lett.*, vol. 24, pp. 448–450, April 1999.

[36] A. R. Chraplyvy, A. H. Gnauck, T. W. Tkach, and R. M. Derosier, "8 $\times$ 10 Gb/s transmission through 280 km of dispersion-managed fiber," *IEEE Photon. Technol. Lett.*, vol. 5, pp. 1233–1235, Oct. 1993.

[37] A. R. Chraplyvy, A. H. Gnauck, R. W. Tkach, and R. M. Derosier, "One-third terabit/s transmission through 150 km of dispersion-managed fiber," *IEEE Photon. Technol. Lett.*, vol. 7, pp. 98–100, Jan. 1995.

[38] N. S. Bergano and C. R. Davidson, "Wavelength division multiplexing in long-haul transmission systems," *J. Lightwave Technol.*, vol. 14, pp. 1299–1308, June 1996.

[39] P. V. Mamyshev and N. A. Mamysheva, "Pulse-overlapped dispersion-managed data transmission and intrachannel four-wave mixing," *Opt. Lett.*, vol. 24, pp. 1454–1456, Nov. 1999.

[40] D. Anderson, "Variational approach to nonlinear pulse propagation in optical fibers," *Phys. Rev. A*, vol. 27, p. 3135, 1983.

[41] S. K. Turitsyn and E. G. Shapiro, "Variational approach to the design of optical communication systems with dispersion management," *Opt. Fiber Technol.*, vol. 4, pp. 151–188, 1998.

[42] J. N. Kutz, P. Holmes, S. G. Evangelides, and J. P. Gordon, "Hamiltonian dynamics of dispersion-managed breathers," *J. Opt. Soc. Am. B*, vol. 15(1) p. 87, Jan. 1998.

[43] J. P. Gordon and L. F. Mollenauer, "Scheme for the characterization of dispersion-managed solitons," *Opt. Lett.*, vol. 24, pp. 323–325, Feb. 1999.

[44] L. F. Mollenauer, A. Grant, X. Liu, X. Wei, C. Xie, and I. Kang, "Experimental test of dense wavelength-division multiplexing using novel, periodic-group-delay-complemented dispersion compensation and dispersion-managed solitons," *Opt. Lett.*, vol. 28, pp. 2043–2045, Nov. 2003.

[45] L. F. Mollenauer, P. V. Mamyshev, and J. P. Gordon, "Effect of guiding filters on the behavior of dispersion-managed solitons," *Opt. Lett.*, vol. 24, pp. 220–222, Feb. 1999.

[46] D. J. Dougherty, F. X. Kärtner, H. A. Haus, and E. P. Ippen, "Measurement of the Raman gain spectrum of optical fibers," *Opt. Lett.*, vol. 20, pp. 31–33, Jan. 1995.

[47] R. J. Mears, L. Reekie, I. M. Jauncie, and D. N. Payne, "Low-noise, erbium-doped, fibre amplifier operating at 1.54 μm," *Electron. Lett.*, vol. 23, pp. 1026–1027, Sept. 1987.

[48] E. Desurvire, J. R. Simpson, and P. C. Becker, "High-gain Erbium-doped, travelling-wave fiber amplifier," *Opt. Lett.*, vol. 12, pp. 888–890, Nov. 1987.

[49] E. Desurvire, *Erbium-Doped Fiber Amplifiers*. New York: John Wiley & Sons, Inc., 1994.

[50] P. C. Becker, N. A. Olsson, and J. R. Simpson, *Erbium-Doped Fiber Amplifiers*. San Diego: Academic Press, 1999.

[51] A. K. Srivastava and Y. Sun, "Advances in erbium-doped fiber amplifiers," in *Optical Fiber Telecommunications IVA* (I. P. Kaminow and T. L. Koch, eds.), vol. IVA, ch. 4, pp. 174–212. San Diego: Academic Press, 2002.

[52] K. H. Ylä-Jarkko, C. Codemard, J. Singleton, P. W. Turner, I. Godfrey, S. U. Alam, J. Nilsson, J. K. Sahu, and A. B. Grudinin, "Low-noise intelligent cladding-pumped L-band EDFA," *IEEE Photonics Tech. Lett.*, vol. 15, pp. 909–911, July 2003.

[53] L. F. Mollenauer, R. Bonney, J. P. Gordon, and P. V. Mamyshev, "Dispersion-managed solitons for terrestrial transmission," *Opt. Lett.*, vol. 24, pp. 285–287, March 1999.

[54] J. P. Gordon and L. F. Mollenauer, "Effects of fiber nonlinearities and amplifier spacing on ultra-long distance transmission," *J. Lightwave Technol.*, vol. 9, pp. 170–173, Feb. 1991.

[55] J. P. Gordon and H. A. Haus, "Random walk of coherently amplified solitons in optical fiber transmission," *Opt. Lett.*, vol. 11, pp. 665–667, Oct. 1986.

[56] K. Smith and L. F. Mollenauer, "Experimental observation of soliton interaction over long fiber paths: Discovery of a long-range interaction," *Opt. Lett.*, vol. 14, pp. 1284–1286, Nov. 1989.

[57] E. M. Dianov, A. V. Luchnikov, A. N. Pilipetskii, and A. N. Starodumov, "Electrostriction mechanism of soliton interaction in optical fibers," *Opt. Lett.*, vol. 15, pp. 314–316, 1990.

[58] E. M. Dianov, A. V. Luchnikov, A. N. Pilipetskii, and A. M. Prokhorov, "Long-range interaction of solitons in ultra-long communication systems," *Sov. Lightwave Commun.*, vol. 1, pp. 235–246, 1991.

[59] E. M. Dianov, A. V. Luchnikov, A. N. Pilipetskii, and A. M. Prokhorov, "Long-range interaction of picosecond solitons through excitation of acoustic waves in optical fibers," *Appl. Phys. B*, vol. 54, pp. 175 –180, 1992.

[60] C. Xu, X. Liu, and X. Wei, "Differential phase-shift keying for high spectral efficiency optical transmissions," *IEEE J. Select. Top. Quantum Electron.*, vol. 10, pp. 281–293, Mar./April 2004.

[61] C. Xu, X. Liu, and L. F. Mollenauer, "Comparison of return-to-zero differential phase-shift keying and on–off keying in long-haul dispersion management," *IEEE Photon. Technol. Lett.*, vol. 15, pp. 617–619, April 2003.

[62] X. Wei, X. Liu, and C. Xu, "Numerical simulation of the SPM penalty in a 10-Gbit/s RZ-DPSK system," *IEEE Photon. Technol. Lett.*, vol. 15, pp. 1636–1638, Nov. 2003.

[63] J. P. Gordon and L. F. Mollenauer, "Phase noise in photonic communications systems using linear amplifiers," *Opt. Lett.*, vol. 15, pp. 1351–1353, Dec. 1990.

[64] L. F. Mollenauer, E. Lichtman, G. T. Harvey, M. J. Neubelt, and B. M. Nyman, "Demonstration of error-free soliton transmission over more than 15,000 km at 5 Gbit/s, single-channel, and over 11,000 km at 10 Gbit/s in a two-channel wdm," *Electron. Lett.*, vol. 28, pp. 792–794, 1992.

[65] A. Mecozzi, J. D. Moores, H. A. Haus, and Y. Lai, "Soliton transmission control," *Opt. Lett.*, vol. 16, pp. 1841–1843, 1991.

[66] Y. Kodama and A. Hasegawa, "Generation of asymptotically stable optical solitons and suppression of the Gordon–Haus effect," *Opt. Lett.*, vol. 17, pp. 31–33, 1992.

[67] L. F. Mollenauer, J. P. Gordon, and S. G. Evangelides, "The sliding-frequency guiding filter: An improved form of soliton jitter control," *Opt. Lett.*, vol. 17, pp. 1575–1577, Nov. 1992.

[68] E. A. Golovchenko, A. N. Pilipetskii, C. R. Menyuk, J. P. Gordon, and L. F. Mollenauer, "Soliton propagation with up- and down-sliding-frequency guiding filters," *Opt. Lett.*, vol. 20, pp. 539–541, March 1995.

[69] L. F. Mollenauer, P. V. Mamyshev, and M. J. Neubelt, "Measurement of timing jitter in soliton transmission at 10 Gbit/s and achievement of 375 Gbit/s-mm, error-free, at 12.4 and 15 Gbit/s," *Opt. Lett.*, vol. 19, pp. 704–705, May 1994.

[70] A. Mecozzi, "Soliton transmission control by butterworth filters," *Opt. Lett.*, vol. 20, pp. 1859–1861, 1995.

[71] D. LeGuen, F. Fave, R. Boittin, J. Debeau, F. Devaux, M. Henry, C. Thebault, and T. Georges, "Demonstration of sliding-filter-controlled soliton transmission at 20 Gbit/s over 14 mm," *Electron. Lett.*, vol. 31, pp. 301–302, 1995.

[72] P. V. Mamyshev and L. F. Mollenauer, "Stability of soliton propagation with sliding-frequency guiding filters," *Opt. Lett.*, vol. 19, pp. 2083–2085, Dec. 1994.

[73] M. Nakazawa, E. Yamada, H. Kubotra, and K. Suzuki, "10 Gbit/s soliton transmission over one million kilometers," *Electron. Lett.*, vol. 27, pp. 1270–1271, 1991.

[74] T. Widdowson and A. D. Ellis, "20 Gbit/s soliton transmission over 125 mm," *Electron. Lett.*, vol. 30, pp. 1866–1867, 1992.

[75] J. P. Gordon, "Dispersive perturbations of solitons of the nonlinear Schrödinger equation," *J. Opt. Soc. Am. B*, vol. 9, pp. 91–97, Jan. 1992.

[76] J. P. Gordon, "Interaction forces among solitons in optical fibers," *Optics Lett.*, vol. 8, pp. 596–598, Nov. 1983.

[77] L. F. Mollenauer, S. G. Evangelides, and J. P. Gordon, "Wavelength division multiplexing with solitons in ultra long distance transmission using lumped amplifiers," *J. Lightwave Technol.*, vol. 9, pp. 362–367, March 1991.

[78] P. V. Mamyshev and L. F. Mollenauer, "Pseudo-phase-matched four-wave mixing in soliton WDM transmission," *Opt. Lett.*, vol. 21, pp. 396–398, March 1996.

[79] J. S. G. Evangelides and J. P. Gordon, "Energy transfers and frequency shifts from three soliton collisions in a multiplexed transmission line with periodic amplification," *J. Lightwave Technol.*, vol. 14, pp. 1639–1643, July 1996.

[80] A. Mecozzi and H. A. Haus, "Effect of filters on soliton interactions in wavelength-division-multiplexing systems," *Opt. Lett.*, vol. 17, pp. 988–990, 1992.

[81] P. V. Mamyshev and L. F. Mollenauer, "WDM channel energy self-equalization in a soliton transmission line by guiding filters," *Opt. Lett.*, vol. 21, pp. 1658–1660, Oct. 1996.

[82] L. F. Mollenauer, P. V. Mamyshev, and M. J. Neubelt, "Demonstration of soliton WDM transmission at 6 and 7 $\times$ 10 Gbit/s, error-free over transoceanic distances," *Electron. Lett.*, vol. 32, pp. 471–473, 1996.

[83] L. F. Mollenauer, P. V. Mamyshev, and M. J. Neubelt, "Demonstration of soliton WDM transmission at 8 $\times$ 10 Gbit/s, error-free over transoceanic distances," in *Technical Digest of the 1996 Optical Fiber Conference, San Jose, CA*, pp. PD22-2–PD22-5, Optical Society of America and IEEE, Feb./March 1996.

[84] J. J. Veselka and S. K. Korotky, "Optical soliton generator based on a single mach-zehnder generator," *Technical Digest of the 1994 OSA & IEEE Integrated Photonics Research Topical Meeting, San Francisco, CA, Optical Society of America and IEEE*, pp. 190–192, Feb. 17–19, 1994.

[85] C. J. McKinstrie, "Frequency shifts caused by collisions between pulses in dispersion-managed systems," *Opt. Communications*, vol. 205, pp. 123–137, 2002.

[86] X. Liu, X. Wei, L. F. Mollenauer, C. J. McKinstrie, and C. Xie, "Collision-induced time shift of a dispersion-managed soliton and its minimization in wavelength-division-multiplexed transmission," *Opt. Lett.*, vol. 28, pp. 1412–1415, Aug. 2003.

[87] X. Wei, X. Liu, C. Xie, and L. F. Mollenauer, "Reduction of collision-induced timing jitter in dense WDM by the use of periodic group delay dispersion compensators," *Opt. Lett.*, vol. 28, pp. 983–985, June 2003.

[88] M. Shirasaki, "Chromatic-dispersion compensator using virtually imaged phased array," *IEEE Photon. Technol. Lett.*, vol. 9, pp. 1598–1600, 1997.

[89] C. R. Doerr, L. W. Stulz, S. Chandrasekhar, and L. Buhl, "Multichannel integrated tunable dispersion compensator employing a thermooptic lens," in *Technical Digest of the Optical Fiber Communication Conference OFC 2002*, pp. PD FA6-2, Optical Society of America and IEEE, March 17–22, 2002.

[90] C. K. Madsen and G. Lenz, "Optical all-pass filters for phase response design with applications for dispersion compensation," *IEEE Photon. Technol. Lett.*, vol. 10, pp. 994–996, 1998.

[91] D. J. Moss, S. McLaughlin, G. Randall, M. Lamont, M. Ardekani, P. Colbourne, S. Kiran, and C. A. Hulse, "Multichannel tunable dispersion compensation using all-pass multicavity etalons," in *Technical Digest of the Optical Fiber Communication Conference OFC 2002*, pp. 132–133, Optical Society of America and IEEE, March 17–22, 2002.

[92] L. F. Mollenauer, A. R. Grant, X. Liu, X. Wei, C. Xie, and I. Kang, "Experimental test of dense wavelength-division-multiplexing using novel, periodic-group-delay-complemented dispersion compensation and dispersion-managed solitons," *Opt. Lett.*, vol. 28, pp. 2043–2045, Nov. 2003.

[93] L. F. Mollenauer, A. R. Grant, X. Liu, X. Wei, C. Xie, and I. Kang, "Novel, periodic-group-delay-complemented dispersion compensation and dispersion-managed solitons enable dense WDM over 20,000 km," in *Proceedings of the 29th European Conference on Optical Communication, ECOC-IOOC 2003, Rimini, Italy*, vol. 2, AEI (Associazione Elettrotecnica ed Elettronica Italiana), Milan, Sept. 21–25, 2003.

[94] L. F. Mollenauer, A. R. Grant, X. Liu, X. Wei, C. Xie, and I. Kang, "Demonstration of 109 × 10G dense WDM over more than 18,000 km using novel, periodic-group-delay-complemented dispersion compensation and dispersion-managed solitons," in *Proceedings of the 29th European Conference on Optical Communication, ECOC-IOOC 2003, Rimini, Italy*, vol. 6 (Post Deadline), AEI (Associazione Elettrotecnica ed Elettronica Italiana), Milan, Sept. 21–25, 2003.

[95] M. Chertov, I. Gabitov, and J. Moeser, "Pulse confinement in optical fibers with random dispersion," *Proc. Nat. Acad. Sci. U.S.A.*, vol. 98, pp. 14208–14211, Dec. 2001.

[96] M. Chertov, I. Gabitov, P. M. Lushnikov, J. Moeser, and Z. Toroczkai, "Pinning method of pulse confinement in optical fiber with random dispersion," *J. Opt. Soc. Am. B*, vol. 19, pp. 2538–2550, Nov. 2002.

[97] J. P. Gordon and H. Kogelnik, "PMD fundamentals: Polarization mode dispersion in optical fibers," *Proc. Nat. Acad. Sci. U.S.A.*, vol. 97, pp. 4541–4550, April 2000.

[98] L. F. Mollenauer and J. P. Gordon, "Birefringence-mediated timing jitter in soliton transmission," *Opt. Lett.*, vol. 19, pp. 375–377, March 1994.

[99] C. D. Poole, "Statistical treatment of polarization dispersion in single-mode fiber," *Opt. Lett.*, vol. 13, pp. 687–689, 1988.

[100] G. J. Forschini and C. D. Poole, "Statistical theory of polarization dispersion in single-mode fibers," *J. Lightwave Technol.*, vol. 9, pp. 1439–1456, 1991.

[101] C. D. Poole, "Measurement of polarization-mode dispersion in single-mode fibers with random mode coupling," *Opt. Lett.*, vol. 14, pp. 523–525, 1989.

[102] L. F. Mollenauer, K. Smith, J. P. Gordon, and C. R. Menyuk, "Resistance of solitons to the effects of polarization dispersion in optical fibers," *Opt. Lett.*, vol. 14, pp. 1219–1221, Nov. 1989.

[103] P. K. A. Wai, C. R. Menyuk, and H. H. Chen, "Stability of solitons in randomly varying birefringent fibers," *Opt. Lett.*, vol. 16, pp. 1231–1233, 1991.

[104] S. V. Manakov, "On the theory of two-dimensional stationary self-focusing of electromagnetic waves," *Zh. Eksp. Teor. Fiz.*, vol. 65, p. 505, 1973.

[105] S. V. Manakov, "On the theory of two-dimensional stationary self-focusing of electromagnetic waves," *Sov. Phys. JETP*, vol. 38, pp. 248–263, 1974.

[106] L. F. Mollenauer, J. P. Gordon, and F. Heismann, "Polarization scattering by soliton–soliton collisions," *Opt. Lett.*, vol. 20, pp. 2060–2062, Oct. 1995.

[107] S. G. Evangelides, L. F. Mollenauer, J. P. Gordon, and N. S. Bergano, "Polarization multiplexing with solitons," *J. Lightwave Technol.*, vol. 10, pp. 28–35, Jan. 1992.

[108] L. F. Mollenauer and C. Xu, "Time-lens timing-jitter compensator in ultra-long haul DWDM dispersion-managed soliton transmission," in *Post Deadline Papers of the Conference on Lasers and Electro-Optics*, pp. CPDB1–1, Opt. Soc. Am., May 19–24, 2002.

[109] L. F. Mollenauer, C. Xu, C. Xie, and I. Kang, "The temporal lens as jitter-killer," in *Technical Digest of the Nonlinear Optics Conference*, pp. 66–68, Opt. Soc. Am., July 29–Aug 2, 2002.

[110] L. A. Jiang, M. E. Grein, B. S. Robinson, E. P. Ippen, and H. A. Haus, "Experimental demonstration of a timing jitter eater," in *Technical Digest of the Conference on Lasers and Electro-Optics*, p. CTuF7, Opt. Soc. Am., May 19–24, 2002.

[111] M. Romagnoli, P. Franco, R. Corsini, A. Schiffini, and M. Midrio, "Time-domain Fourier optics for polarization-mode dispersion compensation," *Opt. Lett.*, vol. 24, pp. 1197–1199, Sept. 1999.

[112] S. Nishi and M. Saruwatari, "Technique for measuring the distributed zero-dispersion wavelength of optical fibers using pulse amplification caused by modulational instability," *Electron. Lett.*, vol. 31, pp. 225–226, Feb. 1995.

[113] R. M. Jopson, M. Eiselt, R. H. Stolen, R. M. Derosier, A. M. Vengsarkar, and U. Koren, "Non-destructive dispersion-zero measurements along an optical fiber," *Electron. Lett.*, vol. 31, pp. 2115–2116, Nov. 1995.

[114] L. F. Mollenauer, P. V. Mamyshev, and M. J. Neubelt, "Method for facile and accurate measurement of optical fiber dispersion maps," *Opt. Lett.*, vol. 21, pp. 1724–1726, Nov. 1996.

[115] J. Gripp and L. F. Mollenauer, "Enhanced range for OTDR-like dispersion map measurements," *Opt. Lett.*, vol. 23, pp. 1603–1605, Oct. 1998.

[116] L. F. Mollenauer and K. Smith, "Demonstration of soliton transmission over more than 4000 km in fiber with loss periodically compensated by Raman gain," *Opt. Lett.*, vol. 13, pp. 675–677, Aug. 1988.

[117] K. Smith and L. F. Mollenauer, "Experimental observation of adiabatic compression and expansion of soliton pulses over long fiber paths," *Opt. Lett.*, vol. 14, pp. 751–753, July 1989.

[118] K. Smith and L. F. Mollenauer, "Experimental observation of soliton interaction over long fiber paths: Discovery of a long-range interaction," *Opt. Lett.*, vol. 14, pp. 1284–1286, Nov. 1989.

[119] L. F. Mollenauer, M. J. Neubelt, S. G. Evangelides, J. P. Gordon, J. R. Simpson, and L. G. Cohen, "Experimental study of soliton transmission over more than 10,000 km in dispersion-shifted fiber," *Opt. Lett.*, vol. 15, pp. 1203–1205, Nov. 1990.

[120] A. R. Grant, "Calculating the Raman pump distribution to achieve minimum gain ripple," *IEEE J. Quantum Electron.*, vol. 38, pp. 1503–1509, Nov. 2002.

[121] L. F. Mollenauer, A. R. Grant, and P. V. Mamyshev, "Time-division multiplexing of pump wavelengths to achieve ultra-broad band, flat, backward-pumped Raman gain," *Opt. Lett.*, vol. 27, pp. 592–594, April 2002.

[122] J. S. Russell, "Report on waves," *Report of the fourteenth meeting of the British Association for the Advancement of Science*, pp. 311–391, Sept. 1844.

[123] D. J. Korteweg and G. de Vries, "On the change of form of long waves advancing in a rectangular canal and on a new type of long stationary waves," *Philosoph. Magazine*, no. 36, pp. 422–443, 1895.

[124] N. J. Zabusky and M. D. Kruskal, "Interaction of 'solitons' in a collisionless plasma and the recurrence of initial states," *Phys. Rev. Lett.*, vol. 15, pp. 240–243, 1965.

[125] V. L. Ginsburg and L. D. Landau, "On the theory of superconductivity," *Zh. Eksp. Teor. Fiz. (Sov. JETP)*, vol. 20, pp. 1064–1082, 1950.

[126] R. Y. Chao, E. Garmire, and C. H. Townes, "Self trapping of optical beams," *Phys. Rev. Lett.*, vol. 13, pp. 479–481, 1964.

[127] P. L. Kelly, "Self focusing of optical beams," *Phys. Rev. Lett.*, vol. 15, pp. 1005–1007, 1965.

[128] V. I. Talanov, "Self focusing of wave beams in nonlinear media," *Sov. Phys. JETP Lett.*, vol. 2, pp. 138–140, 1965.

[129] E. P. Gross, "Structure of a quantized vortex in Boson systems," *Nuovo Cimento*, vol. 20(3), pp. 454–477, 1961.

[130] L. P. Pitaevskii, "Vortex lines in an imperfect Bose gas," *Soviet Physics JETP-USSR*, vol. 13(2), pp. 451–454, 1961.

[131] E. P. Gross, "Hydrodynamics of a superfluid condensate," *J. Math Phys.*, vol. 4, pp. 195–207, Feb. 1963.

[132] V. E. Zakharov, "Instability of the light self-focusing," *Zh. Eksp. Teor. Fiz.*, vol. 53, pp. 1745–1743, 1967.

[133] D. J. Benney and A. C. Newell, "The propagation of nonlinear wave envelopes," *J. Math. Phys. (Stud. Appl. Math.)*, vol. 46, pp. 133–139, 1967.

[134] A. Hasegawa and F. Tappert, "Transmission of stationary nonlinear optical pulses in dispersive dielectric fibers. I. Anomalous dispersion," *Appl. Phys. Lett.*, vol. 23, pp. 142–144, Aug. 1973.

[135] A. Hasegawa and F. Tappert, "Transmission of stationary nonlinear optical pulses in dispersive dielectric fibers. II. Normal dispersion," *Appl. Opt.*, vol. 23, pp. 171–172, Sept. 1973.

[136] J. Satsuma and N. Yajima, "Initial value problems of one-dimensional self-modulation of nonlinear waves in dispersive media," *Suppl. Prog. Theor. Phys.*, no. 55, pp. 284–306, 1974.

[137] L. F. Mollenauer, "Color center lasers," in *Tunable Lasers* (L. F. Mollenauer and J. C. White, eds.), vol. 59 (Topics in Applied Physics), ch. 6, pp. 225–277, Berlin: Springer-Verlag, 1987.

[138] L. F. Mollenauer, R. H. Stolen, and J. P. Gordon, "Experimental observation of picosecond pulse narrowing and solitons in optical fibers," *Opt. Lett.*, vol. 45, pp. 1095–1097, Sept. 1980.

[139] L. F. Mollenauer, "Solitons in optical fibres and the soliton laser," *Philosoph. Trans. Roy. Soc. Lond.*, vol. A 315, pp. 437–450, 1985.

[140] L. F. Mollenauer, R. H. Stolen, J. P. Gordon, and W. J. Tomlinson, "Exteme picosecond pulse narrowing by means of soliton effect in single-mode optical fibers," *Opt. Lett.*, vol. 8, pp. 289–291, May 1983.

[141] L. F. Mollenauer and R. H. Stolen, "The soliton laser," *Opt. Lett.*, vol. 9, pp. 13–15, Jan. 1984.

[142] F. M. Mitschke and L. F. Mollenauer, "Stabilizing the soliton laser," *IEEE J. Quantum Electron.*, vol. QE-22, pp. 2242–2250, Dec. 1986.

[143] J. Mark, L. Y. Liu, K. L. Hall, H. A. Haus, and E. P. Ippen, "Femtosecond pulse generation in a laser with a nonlinear external resonator," *Opt. Lett.*, vol. 14, pp. 48–50, Jan. 1989.

[144] E. P. Ippen, H. A. Haus, and L. Y. Liu, "Additive pulse mode locking," *J. Opt. Soc. Am. B*, vol. 6, pp. 1736–1745, Sept. 1989.

[145] F. M. Mitschke and L. F. Mollenauer, "Discovery of the soliton self-frequency shift," *Optics Lett.*, vol. 11, pp. 659–661, Oct. 1986.

[146] E. M. Dianov, A. V. Karasik, P. V. Mamyshev, A. M. Prokhorov, V. N. Serkin, M. F. Stel'makh, and A. A. Fomichev, "Stimulated-Raman conversion of multisoliton pulses in quartz optical fibers," *JETP Lett.*, vol. 4, pp. 294–297, March 1985.

[147] J. P. Gordon, "Theory of the soliton self-frequency shift," *Opt. Lett.*, vol. 11, pp. 662–664, Oct. 1986.

[148] A. Hasegawa, "Amplification and reshaping of optical solitons in a glass fiber—IV: Use of the stimulated Raman process," *Opt. Lett.*, vol. 8, pp. 650–652, Dec. 1983.

[149] A. Hasegawa, "Numerical study of optical soliton transmission amplified periodically by the stimulated Raman process," *Appl. Opt.*, vol. 23, pp. 3302–3309, Oct. 1984.

[150] L. F. Mollenauer, J. P. Gordon, and M. N. Islam, "Soliton propagation in long fibers with periodically compensated loss," *IEEE J. Quantum Electron.*, vol. QE-22, pp. 157–173, Jan. 1986.

[151] I. P. Kaminow and T. L. Koch, eds., *Optical Fiber Telecommunications*, vols. IIIA, IIIB. San Diego: Academic Press, 1997.

Index

A

B

C

D

E

F

G

H

I

J

K

L

M

N

O

P

Q

R

S

T

U

V

W

X